INTERNATIONAL BACCALAUREATE

BIOLOGY

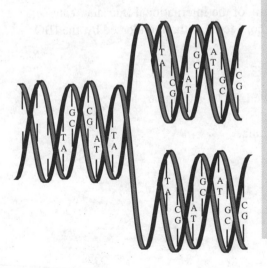

Minka Peeters Weem

ACKNOWLEDGMENTS

Copyright ©IBID Press, Victoria. First published in 1999 by IBID Press, Victoria,
Published by IBID Press, Victoria.
Library Catalogue:

Peeters Weem, M.
1. Biology, 2. International Baccalaureate. Series Title: International Baccalaureate in Detail
ISBN: 0 9585686 8 5

All rights reserved except under the conditions described in the Copyright Act 1968 of Australia and subsequent amendments. No part of this publication may be reproduced, stored in a retrieval system, or transmitted in any form or by any means, without the prior permission of the publishers.

While every care has been taken to trace and acknowledge copyright, the publishers tender their apologies for any accidental infringement where copyright has proved untraceable. They would be pleased to come to a suitable arrangement with the rightful owner in each case. Syllabus statements are reproduced with the permission of the IBO.

This book has been developed independently of the International Baccalaureate Organisation (IBO). The text is in no way connected with, or endorsed by, the IBO.

Cover design by Adcore.

Published by IBID Press, Box 9, Camberwell, 3124, Australia.

Printed by Shannon Books, Victoria, Australia.

PREFACE

This book has been prepared to help students in their studies of Biology. It has been designed around the IB syllabus and the section numbers correspond to the syllabus numbers. In most cases the book follows the order of the syllabus except in cases where a slight re-ordering has seemed to help in the logical development of ideas.

The tags in the margins identify the main sections of the syllabus.

SSC   Subject specific core.

AHL   Additional higher level material.

Option Optional material. Students should seek the guidance of their teachers as to which of these they should study.

This book is intended as a framework for studying this course. However, you should realise that if you are to get the best out of your studies you will need to consult as wide a range of resources as possible.

In particular, your own practical work will assist you in developing understanding as well as an appreciation of the depth and richness of this fascinating subject. Chapter 6 will give you a flavour of this, though a much wider range of activities are covered in the resources mentioned in that Chapter. In addition, some other information can be found on our website: ibid.com.au

In addition, you are encouraged to make effective use of school and public libraries to widen the basis of your knowledge. The rapidly developing internet is also a valuable source, if you have access to it. There are some addresses given in the text.

This is particularly important when it comes to your preparations for the examinations. The questions in the text have been designed to help you develop key ideas and recall important facts. The are not necessarily of the same level of difficulty as examination questions.

The main author of this text is Minka Peeters Weem. Chapter 6 is a short extract from *Biology: A Portfolio of Investigations* by John Gibson and David Greig (IBID Press, ISBN: 0 9585686 3 4. The multiple choice questions and chapters 22 and 25 have been contributed by Barbara Free.

CONTENTS

CELLS
1.1	Cells	2
1.2	Prokaryotic cell structure	6
1.3	Eukaryotic cell structure	8
1.4	Membranes	12
1.5	Cell division - Mitosis	16

THE CHEMISTRY OF LIFE
2.1	Elements of life	26
2.2	Carbohydrates, lipids and proteins	29
2.3	Enzymes	35
2.4	DNA structure	37
2.5	DNA replication	40
2.6	Transcription and translation	42
2.7	Genetic engineering, DNA fingerprinting, gene therapy	45

GENETICS
3.1	Chromosomes, genes and alleles	58
3.2	Gene mutation	60
3.3	Meiosis	62
3.4	Theoretical genetics	65
3.5	Applied genetics	70

ECOLOGY
4.1	Communities and ecosystems	78
4.2	Photosynthesis, respiration and energy relationships	81
4.3	Populations, natural selection and evolution.	90
4.4	Human impact	94
4.5	Ecological techniques	99

HUMAN HEALTH AND PHYSIOLOGY
5.1	Digestion and nutrition	112
5.2	The transport system	116
5.3	Defense against infectious disease	119
5.4	Gas exchange	122
5.5	Homeostasis	127
5.6	Reproduction	131

PRACTICALS
6.1	Practical work	144

CELLS

7.1	Membranes	148
7.2	Cell division - Mitosis	149
7.3	Differentiation and functional specialisation of cells	151

NUCLEIC ACIDS AND PROTEINS

8.1	DNA structure	158
8.2	DNA replication	160
8.3	Transcription	163
8.4	Translation	167
8.5	Proteins	170
8.6	Enzymes	171

CELL RESPIRATION AND PHOTOSYNTHESIS

9.1	Cell respiration	180
9.2	Photosynthesis	186

GENETICS

10.1	Meiosis	200
10.2	Dihybrid crosses	203
10.3	Autosomal gene linkage and gene mapping	204
10.4	Statistical analysis	207
10.5	Polygenic inheritance	210
10.6	Applications of genetics to agriculture and horticulture	212

HUMAN REPRODUCTION

11.1	Production of gametes	220
11.2	Fertilisation and pregnancy	224

DEFENSE AGAINST INFECTIOUS DISEASE

12.1	Agents that cause infectious disease	230
12.2	Types of defense	232

CLASSIFICATION AND DIVERSITY

13.1	Classification	242
13.2	Diversity	245

NERVES, MUSCLES AND MOVEMENT

14.1	Nerves	252
14.2	Muscles and movement	257

EXCRETION

15.1	Excretion	266
15.2	The human kidney	267

PLANT SCIENCE

16.1	Dicotyledenous plant structure	276
16.2	Transport in Angiosperms	279
16.3	Germination	285
16.4	Plants and people	287

OPTION A: DIET AND HUMAN NUTRITION

A.1	Diet	294
A.2	Biochemistry of nutrition	300
A.3	Diet and health	304

OPTION B: PHYSIOLOGY OF EXERCISE

B.1	The skeleton, joints and muscles	314
B.2	Coordination of muscle activity	319
B.3	Muscles and energy	322
B.4.	Fitness and training	324
B.5	Injuries	326

OPTION C: CELLS AND ENERGY

C.1	Membranes	330
C.2	Proteins	331
C.3	Enzymes	333
C.4	Photosynthesis	336
C.5	Cell respiration	344

OPTION D: EVOLUTION

D.1	Origin of life on Earth	354
D.2	The origin of species	358
D.3	Evidence for evolution	361
D.4	Human evolution	368
D.5	Neo-Darwinism	375
D.6	The Hardy-Weinberg Principle	379

OPTION E: NEUROBIOLOGY AND BEHAVIOUR

E.1	Introduction and examples of behaviour	386
E.2.	Perception of stimuli	389
E.3	Innate behaviour	395
E.4	Learned behaviour	400
E.5	Social behaviour	402
E.6	The autonomic nervous system	404
E.7	Neurotransmitters and synapses	406

OPTION F: APPLIED PLANT AND ANIMAL SCIENCE

F.1	Plant and animal science and the world food problem	412
F.2	Applied plant science	415
F.3	Applied animal science	420
F.4	Science applied to horticulture	423
F.5	Modern methods and techniques of plant and animal science	426
F.6	Wider biological and ethical issues	427

OPTION G: ECOLOGY AND CONSERVATION

G 1.1	The ecology of species	432
G.2	The ecology of communities	440
G.3	The ecology of ecosystems	443
G.4	Biodiversity and conservation	445
G.5	Microbial ecology	453
G.6	Reducing harmful impacts of humans on ecosystems	458

OPTION H: FURTHER HUMAN PHYSIOLOGY

H.1	Homeostasis	466
H.2	Digestion	469
H.3	Absorption of digested food	473
H.4	The functions of the liver	475
H.5	Transport	481
H.6	Gas exchange	486

THEORY OF KNOWLEDGE

The scientific method	496

CELLS

1

Chapter contents

- Cell Theory
- Prokaryotic Cell Structure
- Eukaryotic Cell Structure
- Membranes
- Cell Division - Mitosis

CELLS

1.1 CELL THEORY
1.1.1 A HISTORICAL PERSPECTIVE

State one contribution made by each of the following: Robert Hooke, Anton van Leeuwenhoek, Matthias Schleiden, Theodor Schwann and Rudolph Virchow.

© IBO 1996

A series of steps led to the discovery of cells, most of them being related to the advances in technology.

1590: Jansen invents the compound microscope. A compound microscope has 2 lenses which makes it more powerful.

1665: Robert Hooke studies cork with a compound microscope and names the structures "cells".

1675: Anton(ie) van Leeuwenhoek discovers unicellular organisms.

1838: Schleiden suggests that all plants were made of cells.

1839: Schwan suggests that all animals were also made of cells.

1840: Purkinje names the cell content "protoplasm"

1855: Virchow suggests that "all cells come from cells".

Virchow's statement completed the cell theory which is usually considered to have three aspects:
- all organisms are made of one or more cells
- cells are the units of life
- cells only arise from pre-existing cells

1.1.2 & 1.1.3

Describe three advantages of using light microscopes.
Describe two advantages of using electron microscopes.
In comparing electron and light microscopes, the terms 'resolution' and 'magnification' need to be explained. Details of differences between the types of electron microscopes or the principles of how they work is not required.

© IBO 1996

The resolving power of the unaided eye is approximately 0.1 mm. This means that 2 dots or lines which are more than 0.1 mm apart can be seen as separate. If they are closer together, they will be perceived as a single line or dot. So to see cells, we need to magnify them. Magnification alone is not the answer. A grainy newspaper picture can be blown up to view a specific detail but will still be as grainy, just the dots become larger. To see more detail, a finer grain film should have been used.

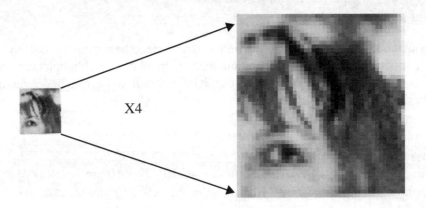

Since eukaryotic cells (cells with nuclei) measure approximately 20 μm, you need a microscope to see them. Although many different microscopes exist, they can be put into two main categories:

	LIGHT microscope	**ELECTRON microscope**
to create image	uses light (λ=0.4-0.7μm)	uses electrons (λ=0.005nm)
to focus beams	lenses	electromagnets
resolving power	200 nm	0.2 nm
object	can be living	dead (in vacuum)
colour	staining not required	staining (metal dyes) required
	real colours visible	black and white only
thickness	thin - fixation not required	very thin - fixation required
expenses	within range of high school	universities and research labs

Overall the compound light microscope is an extremely useful tool for studying cells. It is relatively simple and cheap. The electron microscope, developed in the 1930's, requires a larger budget and skilled people to process the material to be viewed. The result is a black and white picture of extremely high resolution.

1.1.4
Define **organelle**.

© IBO 1996

When cells are studied in more detail, they are seen to contain small structures. These structures seem to be to the cell what our organs are to us. They are called **cell organelles**. Cell organelles are structures within the cell that take part in carrying out the cell's life functions.

Cells

1.1.5
Compare the relative sizes of molecules, cell membrane thickness, viruses, bacteria, organelles and cells, using appropriate SI units.

© IBO 1996

Some organelles can be seen with the light microscope, others cannot. To study their internal structure, we need an electron microscope in all cases. Although even the largest cell is too small to see with the unaided eye, it is important to have an understanding of the relative sizes of cells and organelles.

Eukaryotic cell	: 10 - 100 µm	=	$10 - 100 \times 10^{-6}$ m
Prokaryotic cell	: 1 - 5 µm	=	$1 - 5 \times 10^{-6}$ m
Nucleus	: 10 - 20 µm	=	$10 - 20 \times 10^{-6}$ m
Chloroplast	: 2 - 10 µm	=	$2 - 10 \times 10^{-6}$ m
Mitochondrion	: 0.5 - 5 µm	=	$0.5 - 5 \times 10^{-6}$ m
Large virus (HIV)	: 100 nm	=	100×10^{-9} m
Ribosome	: 25 nm	=	25×10^{-9} m
Membrane	: 7.5 nm thick	=	7.5×10^{-9} m
DNA double helix:	2 nm	=	2×10^{-9} m
H atom	: 0.1 nm	=	0.1×10^{-9} m

Some eukaryotic cells are larger than is indicated above. Animal cells are often smaller than plant cells. The yolk of an egg is one cell. Each sap filled vesicle of an orange is one cell.

1.1.6
Explain the importance of the surface area to volume ratio as a factor limiting cell size.

© IBO 1996

The size of a cell is limited by its need to exchange materials with its environment. If a cell becomes too large, its diffusion distance become to long to be efficient and its surface to volume ration becomes too small to allow the necessary exchange.

A cube with sides of 1 cm has a surface area of $6 \times 1 \times 1 = 6$ cm² and a volume of $1 \times 1 \times 1 = 1$ cm³ which means a surface area: volume ratio of 6:1. This means that every 1 cm³ of volume has 6 cm² of surface area.

A cube with sides of 10 cm has a surface area of $6 \times 10 \times 10 = 600$ cm² and a volume of $10 \times 10 \times 10 = 1000$ cm³ which gives a surface area: volume ratio of 600:1000 = 0.6/1.

This means that every 1 cm³ of volume has 0.6 cm² of surface area, which is less than the smaller cube.

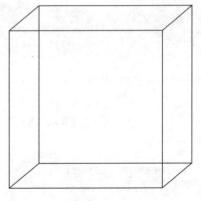

1 cm cube
Surface area = 6cm²
Volume = 1cm³

1 cm cube
Surface area = 600cm²
Volume = 1000cm³

As you can see the volume increases more rapidly than the surface area which eventually creates a problem for the cell.

Ways of dealing with this are to increase the surface area by protruding extensions or by flattening the cell. Multicellular organisms face the same problem. This is why, for example, we have lungs (structures in lungs increase the surface area available for gaseous exchange) and a circulatory system (blood carries materials round the body, reducing the diffusion distance).

The 1930 essay *'On Being the Right Size'* by JBS Haldane explores some of these issues.

The Pelican

There are few birds larger than the pelican that are able to fly.

Cells

1.2 PROKARYOTIC CELL STRUCTURE

1.2.1
Draw a generalised prokaryotic cell as seen in electronmicrographs.

© IBO 1996

Prokaryotic organisms are commonly called bacteria.

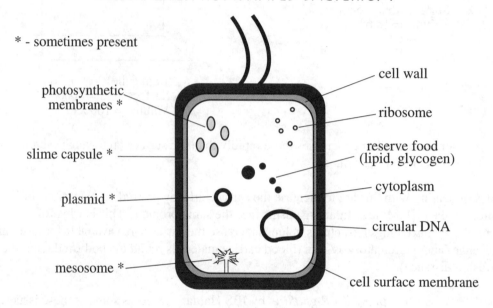

1.2.2
State one function for each of the following: ribosomes, mesosome, slime capsule, cell wall, flagellum, cell surface membrane, plasmid and naked nucleic acid.

© IBO 1996

The following cell organelles can be found in prokaryotic cells. Some are also found in eukaryotic cells (see Section 1.3.3).

Ribosomes: synthesise proteins using information from mRNA.

Mesosome: pocket of cell surface membrane, possible function in cellular respiration.

Slime capsule: capsule outside cell wall, made of polysaccharide or polypeptide, enables prokaryote to stick to surfaces such as soil particles, rocks in streams, cells of host animals, teeth.

Cell wall:	made of amino sugars, protects cell (from harm and bursting in hypotonic media) and gives it shape.
Flagellum:	long, thin helix shaped "thread", rotated to propel cell.
Cell surface membrane:	separates cell from the environment, maintains homeostasis.
Plasmid:	small circular section of DNA found in bacteria, allows exchange of genetic material between individuals.
Naked nucleic acid:	carries genetic information and control centre of the cell.
Cytoplasm:	contains few organelles, contains enzymes
Photosynthetic membranes:	created by invagination of the cell surface membrane, possess pigments.

A CELL AT APPROXIMATELY X15 000 MAGNIFICATION

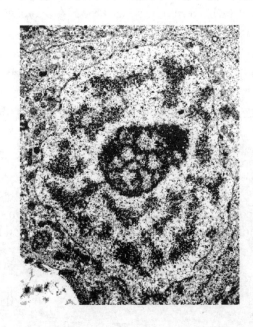

1.3 EUKARYOTIC CELL STRUCTURE

1.3.1
Discuss the possible origin of eukaryotic cells, referring to the theory of endosymbiosis.
© IBO 1996

The oldest fossil records of prokaryotic cells are approximately 3.6 billion years old; the oldest fossil records of eukaryotic cells are approximately 1.5 billion years old.

The **endosymbiont theory** suggests that mitochondria and chloroplasts originally were independent prokaryotes which developed an extremely close symbiotic relationship with another prokaryote (probably one without a cell wall) until they became part of the cell.

This theory is supported by the following observations:

- *Cyanophora* is known to have a symbiotic relationship with a cyanobacterium with lives in its cytoplasm; the bacterium produces food by photosynthesis, the protist (unicellular eukaryote) moves them into the light with its flagellum.
- many similarities between prokaryote cells and chloroplasts and mitochondria, while they are different from eukaryote cells. This is shown in the table below:

Comparison of prokaryotes, chloroplasts, mitochondria and eukaryotes

	prokaryotes, chloroplasts and mitochondria	eukaryotes
DNA	circular not contained in chromosomes not contained in nucleus	linear contained in chromosomes contained in nucleus
Ribosomes	70S	80S
Size	0.5 - 5 µm	20 µm

The endosymbiont theory is not accepted by all but offers some interesting points.

1.3.2
Draw a diagram to show the ultrastructure of a generalised animal cell as seen in electron micrographs.
© IBO 1996

Biology

Below is a diagram of a generalised animal cell as seen with the electron microscope.

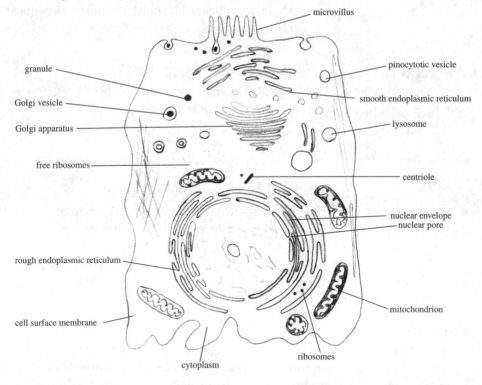

1.3.3
State one function of each of these organelles: ribosome, rough endoplasmic reticulum (rER), lysosome, Golgi apparatus mitochondrion, nucleus and chloroplast.

© IBO 1996

Below is a list of cell organelles with their function, commonly found in eukaryotic cells.

nucleus: largest cell organelle, contains chromatin. Controls all activities of the cell.

ribosome: small cell organelle made of 2 subunits, made from protein and rRNA attached to endoplasmic reticulum or free in cytoplasm.

rough endoplasmic reticulum: system of membranes forming sheets continuous with the nuclear envelope. A site for protein synthesis.

lysosome: small vesicle containing digestive enzymes. Intracellular digestion after fusion with food vacuole, 'suicide bag'.

Golgi apparatus: system of membranes for intracellular transport. Vesicle formation for exocytosis.

mitochondrion: double membrane around matrix. The site for the Krebs cycle and electron transport chain and oxidative phosphorylation (cristae).

Cells

chloroplast: double membrane around stroma and thylakoids. The site for light dependent (thlakoids) and light independent (stroma) reactions.

1.3.4 & 1.3.5
State two similarities between prokaryotic and eukaryotic cells. State two differences between the eukaryotic nucleus and prokaryotic nuclear material.

© IBO 1996

The table below summarises some of the similarities and differences between prokaryotic and eukaryotic cells.

feature	similarities	differences	
		prokaryotes	eukaryotes
form	cells	unicellular	uni/multicellular
genetic material	DNA	circular	linear
		no chromosomes	chromosomes
		no nucleus	nucleus
protein synthesis	ribosomes	70S	80S
flagellum	possible	simple	complex
cell wall	possible	polysaccharides	cellulose
respiration	process	mesosomes	mitochondria
photosynthesis	process	membranes	chloroplasts

1.3.6
Describe three differences between plant and animal cells.

© IBO 1996

The table below summarises some of the differences between plant and animal cells.

feature	plant cell	animal cell
cell wall	present	absent
chloroplasts	present	absent
vacuole	large, permanent	small temporary
reserve food	starch	glycogen

1.3.7
State the composition and function of the plant cell wall.

© IBO 1996

As indicated in the table above, plant cells have a cell wall. The main function of the cell wall is to provide mechanical support for the cell. In addition it plays a very important role in cell **turgor**. Plant cells need to be firm because of the absence of any (endo- or exo-) skeleton which supports animals. The cell wall itself is not sufficient. Plant cells are hypertonic to their environment and will take up water by osmosis. The cell will swell up and push against the cell wall. An animal cell would burst in this situation but the cell wall of a plant cell will push back until the amount of water taken in by osmosis is equal to the amount pushed out by the pressure of the cell wall. The cell is then said to be **turgid**. The cell wall also allows the movement of water and mineral salts across and laterally.

The cell wall is made of cellulose microfibrils. These polysaccharide molecules are polymers of glucose which form long straight chains with few covalent bonds from one chain to the next. This helps create their strength. Humans also find this useful. Paper and cotton are mainly cellulose.

One of the reasons that these cut flowers 'stand up' is turgor in the cells of their stems.

The cells are fed water through the cuts in their stems.

These are planted in a moist medium at the base of the display.

Cells

1.4 MEMBRANES

1.4.1
Draw a diagram showing the fluid mosaic model of a cell membrane including the phospholipid bilayer, cholesterol, glycoproteins and intrinsic and extrinsic proteins.

© IBO 1996

In 1972, Singer and Nicolson proposed a model for membrane structure in which a mosaic of protein molecules float in a fluid lipid layer.

Diagram of the fluid mosaic model of the cell surface membrane, using EM and biochemical data.

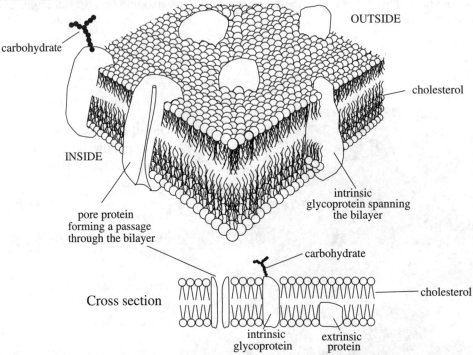

As you can see in the diagram above of the fluid mosaic model of the cell surface membrane, it consists of a phospholipid bilayer with a protein molecules found anywhere between the lipid molecules. In between the lipid molecules, you find the (hydrophobic) cholesterol molecules. If a membrane contains a lot of cholesterol, the membrane will be less fluid (because of the large inflexible ring of cholesterol). The exact composition of the cell membrane differs according to the type of membrane (cell surface, chloroplast, etc.) and the function of the cell.

1.4.2
Explain how the hydrophobic and hydrophilic properties of phospholipids help to maintain the structure of cell membranes.

© IBO 1996

The phospholipid molecules have a hydrophobic and a hydrophilic section. They are therefore considered to be **amphipatic**. The phosphate group is polar and therefore hydrophilic. This part of the molecule readily interacts with water and is found on the surface of the membrane, outside and inside. The non polar hydrophobic lipid "tails" gather together on the inside of the molecule to avoid contact with water molecules. Some lipids have carbohydrate groups attached on the outside. They are then called glycolipids. Many proteins also have these carbohydrate groups attached to the outside. They are then called **glycoproteins**. The carbohydrates often function as receptors for e.g. hormones.

Integral (or intrinsic) proteins are found embedded in the membrane, partially or spanning the membrane; peripheral **(or extrinsic) proteins** are found just inside the membrane. The integral proteins are usually hydrophilic on either side, with a hydrophobic section in the middle, near the lipid tails. Peripheral proteins are hydrophilic and do not interact with the hydrophobic core of the phospholipid bilayer.

The function of the proteins can be structural and/or functional. They can "anchor" a cell organelle to a certain position. Some proteins act as carriers, transporting certain substances through the membrane. Some proteins are hollow, forming hydrophilic channels or pores, also facilitating transport. Several proteins can function as electron carriers as we will see in section 4.2, C 4/9.2, and C 5/9.1.

1.4.3
Define diffusion.

© IBO 1996

Diffusion is the movement of particles from an area of higher concentration to an area of lower concentration down a concentration gradient.

1.4.4
State that osmosis is the passive movement of water molecules, across a partially-permeable membrane from a region of lower solute concentration to a region of higher solute concentration.

© IBO 1996

Osmosis is the (passive) movement of water molecules from a region of lower solute concentration to a region of higher solute concentration across a partially permeable membrane.

Structures can be virtually impermeable (e.g. plastic foil), partially permeable (also called semi-permeable or selectively permeable) (e.g. membranes, coffee filters) or totally permeable (e.g. cell wall). Partially permeable structures will have a certain structure to allow some molecules to pass through while others cannot.

1.4.6
Describe passive transport across membranes including the roles of protein carriers, ATP and a concentration gradient.

© IBO 1996

Cells

The role of the cell surface membrane is to keep the cell content together and to maintain homeostasis by controlling what enters and leaves the cell. By the nature of the cell surface membrane, substances can easily pass through it only when they are small and carry little or no electric charge. This rules out many substances.

Small uncharged molecules can **diffuse** through the cell membrane. This is **passive transport**, i.e. it does not require energy but the molecules only can move from a higher to a lower concentration. Ions and hydrophilic molecules cannot pass through the lipid barrier of the cell surface membrane even to go from a higher to a lower concentration. Special transport proteins will assist in moving some of these across the membrane, often via hydrophilic channels. This process is called **facilitated diffusion**. It does not require energy and is driven by diffusion forces. Water can move into or out of the cell by **osmosis**.

1.4.6 & 1.4.7
Describe active transport across membranes including the roles of protein carriers, ATP and a concentration gradient.
Compare endocytosis (phagocytosis/pinocytosis) and exocytosis.
Phagocytosis could be likened to the cell 'eating' and pinocytosis to the cell 'drinking'.

© IBO 1996

As a result of the above, cells need to have mechanisms to allow substances which cannot diffuse through the membrane to enter or leave the cell. In addition, they might need to absorb materials which are present already in a higher concentration than in the surrounding medium. The various ways in which this can be achieved are collectively called **active transport**. Active transport requires energy.

Several different mechanisms are recognised:
- **Carrier proteins:** (sometimes called membrane pumps) e.g. the sodium-potassium pump. A protein in the cell surface membrane can move sodium out of the cell and potassium into it even though the concentration of sodium outside is higher than inside and vice versa for potassium. This is important in the functioning of nerve cells. The proteins involved in this process are selective.
- **Endocytosis:** the process by which the cell takes up a substance by surrounding it with membrane.

  Two different types of endocytosis are recognised: **pinocytosis** (when the substance is fluid), which is sometimes called "cell drinking" and **phagocytosis** (when the substance is solid), which is sometimes called "cell eating".

- **Exocytosis** is the reverse of endocytosis.

	ATP required	Concentration gradient
Diffusion	no	with
Facilitated diffusion	no	with
Osmosis	no	with (water)
Carrier proteins	yes	against possible
Endocytosis	yes	against possible

SCHEMATIC DIAGRAM OF ENDOCYTOSIS

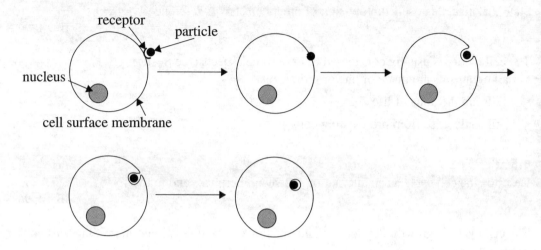

Cells

1.5 CELL DIVISION - MITOSIS

1.5.1
State that all cells arise from division of other cells (cross reference 1.1.1).

© IBO 1996

The cell theory is usually considered to have three aspects (see Section 1.1.1):
- all organisms are made of one or more cells.
- cells are the units of life.
- cells only arise from pre-existing cells.

1.5.2
Describe the cell cycle as an alternation between interphase and mitosis.

© IBO 1996

The **cell cycle** of growth, duplication and division can be divided into many stages as you can see below.

SCHEMATIC DIAGRAM OF THE CELL CYCLE

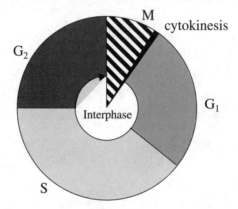

In the diagram above the letters represent the following stages:
M: mitosis = nuclear division.
cytokinesis: cell division (often (incorrectly) included in mitosis).
G$_1$: stage of cell growth and cell organelles increase in number.
S: stage of synthesis of DNA = DNA replication.
G$_2$: stage of preparation for mitosis.

1.5.3
State that interphase is an active period in the life of a cell, where many biochemical reactions, DNA transcription and DNA replication occur.

© IBO 1996

G$_1$, S and G$_2$ together are called **interphase**. Interphase is a very active period in the life of a cell, where many biochemical reactions, DNA transcription and DNA replication occur.
The duration of the cell cycle varies greatly between different cells. The cell cycle of

bacteria is 20 min, beans take 19 hours and mouse fibroblasts take 22 hours. Also the different stages in interphase can take different amounts of time. However, usually interphase lasts longer than mitosis.

1.5.4
Outline how replicated DNA molecules (chromosomes) are moved to opposite ends of the cell by microtubules.

© IBO 1996

For nuclear division to occur, the chromosomes need to be moved to the centre of the cell. Before this can happen, the genetic material which is in the form of chromatin needs to be replicated and condensed into chromosomes. The histones (proteins) associated with the DNA, make sure that the condensation process takes place properly and that the chromosomes can unwind again to form chromatin.

The chromosomes are moved through the cytoplasm by **microtubules**. (Microtubules, centrioles and microfilaments together make the cytoskeleton). The microtubules contribute to the shape of the non-dividing cell and form the spindle fibres during mitosis. (See section 7.2) The spindle fibres originate at the poles. Some span the entire cell - pole to pole, while others run from the pole to a chromosome. The spindle fibres contract, separating the identical copies of the chromosome and pulling them to opposite poles.

NB: Not on the syllabus

The following terms may be confusing:

- **centriole**: one of a pair of small organelles lying at right angles to each other just outside the nucleus of animal cells (and lower plants). Centrioles are self-replicating during the S phase. Each pair of centrioles then moves to a pole and forms the aster.
- **centromere**: a structure on the chromosome containing complex fibres (kinetochore). It becomes duplicated when chromosomes divide into chromatids and attaches spindle fibres.
- **centrosome**: an area of cell cytoplasm near the nucleus involved in nuclear division. It plays a role in assembling microtubules. At the start of nuclear division the centrosome divides into two, each moving to a pole (with the centriole, if present). It is an old term for "centriole + surrounding area".

MITOSIS (ANIMAL CELL)

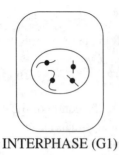
INTERPHASE (G1)

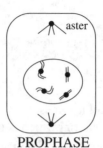
PROPHASE

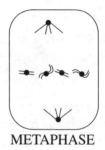
METAPHASE

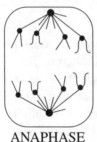
ANAPHASE

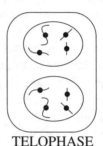
TELOPHASE

The genetic material in Interphase is in reality NOT in the form of chromosomes as is suggested here. The formation of chromosomes from chromatin happens during Prophase. The reason why they are drawn as chromosomes is to illustrate the changes in the amount of DNA and chromosome numbers during the process of mitosis.

1.5.5
State that the products of mitosis are two genetically identical nuclei.

© IBO 1996

The result of the above is two nuclei. Since identical chromosomes were separated and divided, the two nuclei are identical to each other and to the original nucleus (before DNA replication).

1.5.6
State that tumours (cancers) are the result of uncontrolled cell division and that these can occur in any organ.

© IBO 1996

Mitosis and cell division are usually under strict control, only producing new cells needed for growth or repair. Tumour repressor genes produce proteins which inhibit cell division. Proto-oncogenes produce proteins which stimulate growth and cell division. If mutations occur in these genes, cell division can become uncontrolled resulting in a **tumour**.

Radiation and certain chemicals are known to be **carcinogenic**: mostly, they increase the chances of mutation although some seem to increase the effect of mutations already present. Viruses inserting their genetic material into the chromosomes of the host, may also contribute to the formation of tumour cells. A series of changes in a cell is needed before it becomes a tumour cell. However, the changes accumulate over the years making cancer a very common cause of death.

Some tumours are harmless, e.g. warts. Others may become malignant and are then called **cancer**. Cancer can be treated by surgery: remove the tumour cells before they can grow further and spread; by radiation therapy: use a strong radiation beam which (directed to a precise point) will "burn" all cells in the area; by chemotherapy: destroying all rapidly dividing cells by medication (including cells which grow the hair, cells which form the lining of the gut and sperm producing cells).

EXERCISE

1. Unicellular organisms were discovered by:

    A   Robert Hooke
    (B) Anton van Leeuwenhoek
    C   Matthias Schleiden
    D   Theodor Schwann

2. The resolving power of a light microscope is closest to:

    A   20nm
    (B) 200nm
    C   2000nm
    D   2mm

3. If it was desired to observe the true colour of a coral polyp of diameter 0.02mm it would be best to use

    A   the unaided eye
    B   a light microscope with staining
    (C) a light microscope without staining
    D   an electron microscope

4. The structures within a cell are known as:

    A   cytoplasm
    B   nuclei
    C   granules
    (D) organelles

Cells

5. The size of the nucleus of a cell is about:

- (A) 10 - 20 μm
- B 50 - 70 μm
- C 100 - 200 μm
- D 10 - 20 mm

6. Two cubes have side lengths in the ratio 1:3. Their surface areas are in the ratio:

- A 1:3
- B 1:6
- (C) 1:9
- D 1:27

7. Large fish have to have gills because:

- A Their bodies have a large surface area through which they can exchange gases with the surrounding water.
- (B) Their bodies have a small surface area compared with their volume. Gills provide additional surface are through which they can exchange gases with the surrounding water.
- C Their bodies have a small surface area through which they can exchange gases with the surrounding water.
- D They use them to adjust their buoyancy

8. A long, thin thread used to propel a cell is known as:

- A a plasmid
- B a mesosome
- C a tail
- (D) a flagellum

9. The DNA of eukaryotic cells is:

- A linear but not contained in the nucleus
- (B) linear and contained in the nucleus
- C circular but not contained in the nucleus
- D circular and contained in the nucleus

10. The largest cell organelle is generally the

- A Golgi apparatus
- B rough endoplasmic reticulum
- (C) nucleus
- D ribosome

11. The Krebs cycle occurs in the:

    A   rough endoplasmic reticulum
    B   Golgi apparatus
    C   chloroplast
    (D)   mitochondrion

12. The main reserve food for animal cells is:

    (A)   glycogen
    B   DNA
    C   starch
    D   RNA

13. The Singer and Nicolson model for membrane structure says that a membrane

    A   consists of a phospholipid layer
    (B)   consists of a phospholipid bilayer
    C   consists of a protein bilayer
    D   consists of a protein layer

14. The movement of a substance from an area of high concentration to an area of low concentration is known as:

    A   equalisation
    B   osmosis
    (C)   diffusion
    D   expansion

15. Endocytosis is:

    (A)   the process by which the cell takes up a substance by surrounding it with membrane
    B   the process by which the cell takes up a substance by osmosis
    C   the process by which the cell takes up a substance by diffusion
    D   a disease of single celled animals

16. In the cell cycle, the interphase is:

    A   a passive phase between two active phases
    (B)   an active phase during which DNA replication and many other processes occur
    C   a period when mitosis occurs
    D   a period when cytokinesis occurs

Cells

17. Replicated DNA molecules (chromosomes) are moved to opposite ends of the cell by:

A diffusion
B osmosis
C the nucleus
D spindle fibres

18. a. Outline the cell theory

b. Explain how the advances of technology were related to the development of the cell theory. Include the contributions of Hooke, van Leeuwenhoek, Schleiden, Schwann and Virchow in the correct chronological order.

19. Arrange the following by size. Start with the largest.

atom, cell organelle, DNA double helix, eukaryotic cell, prokaryotic cell, thickness of membranes.

20. A chihuahua (small dog) was wrapped in a tightly fitting paper cylinder. Her surface area was estimated to be approximately 0.13 m^2. The volume was estimated (using a clay model) at 2 dm^3.

The same method was used to estimate the surface area and volume of a child. The child was estimated to have a surface area of 0.9 m^2 and a volume of 24 dm^3.

a. Determine the surface area over volume ratio for the dog and the child.
b. Which has the largest surface area?
c. Who would need the most food per kg bodyweight? Explain.

21. a. Draw a diagram of a prokaryotic cell as seen with the electron microscope.
b. Draw a diagram of a eukaryotic cell as seen with the electron microscope.
c. List the cell organelles found in eukaryotic cells but not in prokaryotic cells and state their structure and functions.
d. Explain how two of these cell organelles relate to the endosymbiont theory.

22. a. Describe the structure of a membrane according to the fluid mosaic model. Use a diagram to illustrate your answer.
b. Explain why intrinsic proteins in the membrane are arranged so that their non polar amino acids are in the middle of the protein molecule.

23. During which stage of the cell cycle (Interphase or Mitosis) does each of the following occur?

creation of two genetically identical nuclei

biochemical reactions

separation of sister chromatids

DNA replication
chromosomes moving to the equator
protein synthesis

Cells

Biology

THE CHEMISTRY OF LIFE

2

Chapter contents

- Elements of life
- Carbohydrates, lipids and proteins
- Enzymes
- DNA structure
- DNA replication
- Transcription and translation
- Genetic engineering
- Genetic fingerprinting
- Gene therapy

The Chemistry of Life

2.1 ELEMENTS OF LIFE

Some materials are made of more than one substance. Others are made of only one substance and therefore cannot be decomposed into different components by chemical means. These substances are **elements** and their particles are **atoms**. They are listed in the periodic table.

Two or more different atoms can combine to form **molecules**. Molecules make up compounds. In biological systems, atoms often combine by forming covalent bonds. In covalent bonds atoms share electrons between them. Atoms can also combine to form ionic bonds. In ionic bonds an electron is transferred from one atom to another. The result is one positively charged particle and one negatively charged particle. These particles are called **ions**.

Methane. A molecule with covalent bonding.

Na^+Cl^- Sodium Chloride. A molecule with ionic bonding.

Na^+ A sodium ion

2.1.5 ORGANIC MOLECULES
Define organic.

© IBO 1996

Organic molecules are all molecules containing carbon, except a few such as carbon dioxide (CO_2) which are inorganic. Inorganic molecules are all other molecules. Organic molecules often have covalent bonds; inorganic molecules can also have other bonds, e.g. ionic bonds.

2.1.1 COMMON ELEMENTS
State that the three commonest elements of life are carbon, hydrogen and oxygen.

© IBO 1996

The three most common element in living organisms are:
- Hydrogen   H: forms 1 covalent bond with other atoms
- Oxygen   O: forms 2 covalent bonds with other atoms
- Carbon   C: forms 4 covalent bonds with other atoms and is capable of forming long chains with other carbon atoms

2.1.2 & 2.1.3 OTHER IMPORTANT ELEMENTS
State that a variety of other elements are needed by living organisms including nitrogen, sulfur, phosphorus, iron and potassium.
State one role for each of the elements mentioned in 2.1.2.

© IBO 1996

Other elements are required to make up compounds.
Proteins need nitrogen (N) and often contain sulfur (S).
Nucleic acids need phosphorus (P).

Haemoglobin needs iron (Fe).

Sodium (Na) and potassium (K) are needed for transmitting nervous impulses and maintaining the electrical potential across membranes.

2.1.6 WATER

Outline the significance of water in biology including transparency, cohesion, surface tension, solvent properties and thermal properties, referring to the polarity of water molecules and hydrogen bonding where relevant.

© IBO 1996

Water is made up hydrogen and oxygen. Hydrogen is a flammable gas and oxygen is a gas which makes combustion possible. Together they form a fluid which can be use to put out a fire!

The chemical formula of water is H_2O. It looks like this:

As you can see, water has a slightly negative side and a slightly positive side. This is caused by oxygen attracting the shared electrons more strongly than does hydrogen.

Oxygen also has two pairs of electrons not used in a covalent bond. They contribute to this distribution of charge. As a result, water is a **polar** molecule. This means that water molecules attract each other: the δ^+ side of one molecule is attracted to the δ^- side of another.

Attractions between slightly negative atoms (nitrogen or oxygen) and slightly positive hydrogen atoms are called **hydrogen bonds**.

N.B.: *These are not real bonds*. There is no sharing or transferring of electrons. Instead it is an attraction force between a slightly positive atom and a slightly negative one. Hydrogen bonds are much weaker than covalent or ionic bonds.

The very many hydrogen bonds in water are the most important factor in determining its properties.

Some properties of water that have a great affect on biological systems are listed below.

1. Water is transparent. As a result, light penetrates water. This means that plants can live in water and still receive light for photosynthesis (water is a habitat). It also means that we can see through the film of water covering our eyes.

2. Water has strong cohesion forces. Cohesion describes the attraction forces between molecules of the same kind. Water has strong cohesion forces because the molecules are polar. This results in a high surface tension. Surface tension is the force between the molecules of water at the surface. It is sufficiently strong for small insects to be able to walk on water.

3. Water is a good solvent for polar molecules. The positive side of any polar molecule will attract the negative side of the water molecule and vice versa. For example salt

(Na^+Cl^-) dissolves well in water while butter (large organic non-polar molecules) does not. Many organic molecules are somewhat polar and therefore soluble in water. This makes water a good transport medium. Blood is approximately 50% water. Plants transport minerals dissolved in water from the soil to the leaves and dissacharides dissolved in water from leaves to other areas.

4. Water has a high specific heat and a high heat of melting and evaporation. The specific heat is the amount of energy needed to raise the temperature of a standard amount of substance by 1°C. Raising the temperature means that the molecules move faster. This means that hydrogen bonds need to be broken and reformed more frequently. Breaking the hydrogen bonds at a higher rate requires a lot of energy. The heat of melting is the amount of energy required to change solid water (ice) at 0°C into liquid water at 0°C. The heat of evaporation is the amount of energy required to change liquid water at 100°C into gaseous water (water vapour or steam) at 100°C. Both of these changes of state reduce the interactions between the molecules and therefore must break hydrogen bonds. This requires a great deal of energy because there are so many hydrogen bonds.

5. A consequence of the above is that lakes and seas have to absorb a large amount of energy (heat) to change temperature or state. Therefore, large bodies of water act as 'buffers' in temperature changes. When you live near the coast, your summers will be cooler and your winters warmer compared to someone living inland.

6. Another aspect of the same property is the use of water as a coolant. When we are hot, we sweat. Our skin secretes water which is led to the surface. There, it uses heat from the skin surface to evaporate. Since the heat of evaporation for the water in the sweat is large, a lot of heat is removed from the body in evaporating a small amount of sweat. Therefore water is an effective coolant.

7. Ice floats. This obvious statement has far reaching effects on life. Life may not have been possible otherwise. If a lake freezes, the ice remains on top, protecting the water below (and the organisms living in it) from freezing. In spring the sun will heat up the ice and melt it. If ice sank, the entire lake would freeze in winter (killing most organisms) and it would require a lot of heat to melt it again in spring.

2.2 CARBOHYDRATES, LIPIDS AND PROTEINS

Carbohydrates are made of carbon (C), hydrogen (H) and oxygen (O) in the proportions 1:2:1 which makes the generalized empirical formula for carbohydrates: $(CH_2O)_n$

Monosaccharides are the simplest carbohydrates. An example of a monosaccharide is glucose.

Long molecules of repeating subunits are generally called polymers. The subunits are called monomers.

The empirical formula of glucose is $C_6H_{12}O_6$

2.2.2 STRUCTURE
Draw the ring structure of α-D-glucose.

© IBO 1996

Structural formula of α-D-glucose:

2.2.8 EXAMPLES
List two examples each of monosaccharides, disaccharides and polysaccharides.

© IBO 1996

Other monosaccharides with the same empirical but different structural formulae are fructose and galactose. Dissacharides include sucrose and maltose. Polysaccharides (long chains of monosaccharide molecules) include the energy storing glycogen and the important structural compound, cellulose.

2.2.9 FUNCTIONS
State one function for a monosaccharide and one for a polysaccharide.

© IBO 1996

Glucose is the principal product of photosynthesis. It also gives a sweet flavour to some fruits. Cellular respiration utilises glucose to produce carbon dioxide and water and release energy. Maltose occurs in germinating seeds and is used in making beer.

2.2.6 SOLUBILITY
Explain the relative solubility of carbohydrates, lipids and protein in water.

© IBO 1996

The Chemistry of Life

Monosaccharides dissolve very well in water. This is caused by the hydrogen bonds which their hydroxyl groups can form with the water molecules. Larger molecules such as polysaccharides are much less water soluble (e.g. cellulose).

2.2.4 CONDENSATION AND HYDROLYSIS
Outline the role of condensation and hydrolysis in the relationships between monosaccharides and disaccharides; fatty acids, glycerol and triglycerides; amino acids, dipeptides and polypeptides.

© IBO 1996

Two monosaccharides can combine to form a disaccharide plus water.
glucose + glucose -> maltose + water

This is a **condensation** (or **dehydration**) reaction.

In a condensation reaction the bond between an oxygen and a hydrogen is broken. The hydrogen will move to a hydroxyl group (OH- group) and attach to the oxygen. This causes the bond between this oxygen and the carbon to be broken, and the water molecule thus formed moves away. This leaves one molecule with an oxygen with one lone electron and another molecule with a carbon with one lone electron. The two electrons join and form a new covalent bond, linking the two molecules.

The reverse reaction is also possible:
disaccharide + water -> 2 monosaccharides
This is called a **hydrolysis reaction**.

2.2.8 EXAMPLES
List two examples each of monosaccharides, disaccharides and polysaccharides.

© IBO 1996

Examples of **disaccharides** are maltose (glucose +glucose), sucrose (glucose + fructose) and lactose (glucose + galactose). Generally disaccharides are used by plants for transporting carbohydrates. We use sucrose for its sweet flavour and lactose is a sugar found in milk.

2.2.6 SOLUBILITY
Explain the relative solubility of carbohydrates, lipids and protein in water.

© IBO 1996

Disaccharides dissolve well in water. Approximately 5 kg sugar (sucrose) can be dissolved in 1 dm^3 of boiling water. As for monosaccharides, the hydroxyl groups are responsible for the solubility.

2.2.4 CONDENSATION AND HYDROLYSIS
Outline the role of condensation and hydrolysis in the relationships between monosaccharides and disaccharides; fatty acids, glycerol and triglycerides; amino acids, dipeptides and polypeptides.

© IBO 1996

Polysaccharides are long chains of monosaccharides. They are made by condensation reactions and they can be broken apart in hydrolysis reactions. For each bond that is to be broken between the monomers, one molecule of water is produced.

2.2.8 EXAMPLES
List two examples each of monosaccharides, disaccharides and polysaccharides.

© IBO 1996

Monosaccharides can be linked in different ways creating different polysaccharide structures. Examples of **polysaccharides** are starch, glycogen and cellulose.

2.2.9 FUNCTIONS
State one function for a monosaccharide and one for a polysaccharide.

© IBO 1996

Starch and **glycogen** are used to store carbohydrates. Plants generally store starch in underground tissues like bulbs, or as potato tubers. Animals store glycogen in their liver and muscles.

Cellulose is a structural part of cell walls in plants. The cellulose lends strength to the cell wall.

2.2.6 SOLUBILITY
Explain the relative solubility of carbohydrates, lipids and protein in water.

© IBO 1996

Although the smaller carbohydrates dissolve well in water, the solubility of large polysaccharides such as starch is relatively low due to the large size of the molecules. This is an advantage in that the cell wall of a plant does not dissolve and that stored starch does

The Chemistry of Life

not contribute significantly to the osmotic value of the cell. **Lipids** are commonly known as fats and oils.

2.2.10 LIPIDS
State three functions of lipids.

© IBO 1996

Lipids have various functions in the body. In mammals and birds, a subcutaneous fat layer (layer of fat under the skin) helps to reduce heat loss. Phospholipids are an important component of cell membranes and steroid hormones (e.g. sex hormones such as testosterone) cannot be formed without lipids. Lipids are also a way to store energy. (see section 2.2.7)

2.2.3 STRUCTURE
Draw the basic structure of glycerol and a generalised fatty acid.

© IBO 1996

Lipids are made from glycerol and fatty acids by way of condensation reactions. Although lipids can be very large molecules, they are not organised in the same way as carbohydrates and proteins. Lipids are not long chains or repeating units but are made from 1 molecule of glycerol and 1, 2 or 3 molecules of fatty acids as indicated below.

$$\begin{array}{c}\text{H} \\ \text{H}-\text{C}-\text{OH} \\ | \\ \text{H}-\text{C}-\text{OH} \\ | \\ \text{H}-\text{C}-\text{OH} \\ | \\ \text{H}\end{array} \quad \begin{array}{c}\text{O} \\ \| \\ \text{HO}-\text{C}-\text{R}_1 \\ \text{O} \\ \| \\ \text{HO}-\text{C}-\text{R}_2 \\ \text{HO}-\text{C}-\text{R}_3 \\ \| \\ \text{O}\end{array} \quad \longrightarrow \quad \begin{array}{c}\text{H} \quad \text{O} \\ | \quad \| \\ \text{H}-\text{C}-\text{O}-\text{C}-\text{R}_1 \\ | \quad \text{O} \\ \quad \| \\ \text{H}-\text{C}-\text{O}-\text{C}-\text{R}_2 \\ | \quad \| \\ \text{H}-\text{C}-\text{O}-\text{C}-\text{R}_3 \\ | \quad \| \\ \text{H} \quad \text{O}\end{array} + 3\,\text{H}_2\text{O}$$

glycerol   fatty acids   Lipid
(1,2,3-propanol or   (alkanoic acid)   (or triglyceride)
propan-1,2,3-triol)

[R1, R2, R3 are hydrocarbon chains or rings. When the R group is a chain, the fatty acid is called **aliphatic**, when the R group is a ring, it is called **aromatic**). In any lipid, the R groups may be the same, or they may be different.]

As you can see, the above is a condensation reaction. Breaking a lipid into glycerol and fatty acids requires water and is a hydrolysis reaction.

2.2.6 SOLUBILITY
Explain the relative solubility of carbohydrates, lipids and protein in water.

© IBO 1996

Lipids are not very soluble in water. They are non-polar molecules and as such do not dissolve well in a polar solvent. The long hydrocarbon chains of the fatty acids are hydrophobic. Phospholipids have a phosphate in the place of one fatty acid and as a result have a polar section to the lipid molecule. Phospolipids will float on water with the polar phosphate 'head' immersed in the water and the non-polar lipid 'tail' sticking out.

2.2.7 ENERGY
Compare the energy content of carbohydrates, lipids and proteins.

© IBO 1996

Although both carbohydrates and proteins can be used as a source of energy, the most economic way of storing (or obtaining) energy is as lipids. Lipids contain almost twice as much energy per gram as carbohydrates or proteins.

Proteins are also polymers, in some ways similar to carbohydrates. The building blocks, or monomers, of proteins are called amino acids. Unlike carbohydrates, the monomers of proteins are different from each other. Approximately 20 different amino acids are commonly used to create a protein. A protein can be made of one or more polypeptide chains.

Proteins are sensitive to heat and pH. Too high a temperature will alter the structure of the protein and the drastic change will be irreversible. A cold boiled egg is not the same as a raw egg. A change in pH is also likely to change the structure of a protein. This change may be less drastic than boiling. In both cases, the protein has been **denatured** and will not be able to function properly.

2.2.1 STRUCTURE
Draw the ring structure of α-D-glucose.

© IBO 1996

The generalised structure of an amino acid is given below.

$$\begin{array}{c} O=C-OH \\ | \\ R-C-H \\ | \\ H-N-H \end{array} \quad \text{or} \quad \begin{array}{c} COOH \\ | \\ R-C-H \\ | \\ NH_2 \end{array}$$

An amino acid has an amine group (NH_2) and a carboxylic acid group (COOH).

R stands for 'Restgroup' and can be many things: e.g. '-H', which gives glycine; or tryptophan for which the following Restgroup is needed:

The Chemistry of Life

2.2.4 & 2.2.5 CONDENSATION AND HYDROLYSIS

Outline the role of condensation and hydrolysis in the relationships between monosaccharides and disaccharides; fatty acids, glycerol and triglycerides; amino acids, dipeptides and polypeptides.
Draw the structure of a generalised dipeptide, showing the peptide linkage.

© IBO 1996

A dipeptide is made by combining two amino acids in a condensation reaction.

$$H_2N-\underset{\underset{R}{|}}{\overset{\overset{H}{|}}{C}}-C\overset{\overset{O}{\parallel}}{\underset{OH}{\diagdown}} + H_2N-\underset{\underset{R}{|}}{\overset{\overset{H}{|}}{C}}-C\overset{\overset{O}{\parallel}}{\underset{OH}{\diagdown}} \rightarrow H_2N-\underset{\underset{R}{|}}{\overset{\overset{H}{|}}{C}}-\overset{\overset{O}{\parallel}}{C}-\underset{\underset{H}{|}}{\overset{\overset{}{|}}{N}}-\underset{\underset{R}{|}}{\overset{\overset{H}{|}}{C}}-C\overset{\overset{O}{\parallel}}{\underset{H}{\diagdown}} + H_2O$$

amino acid 1 + amino acid 2 → dipeptide + water

Again this reaction can be reversed and a dipeptide can be split into two amino acids in a **hydrolysis reaction**.

The bond between the C(O) and NH is called the peptide bond. In a polypeptide, you will find x amino acids and $x-1$ peptide bonds.

2.2.6 SOLUBILITY

Explain the relative solubility of carbohydrates, lipids and protein in water.

© IBO 1996

Proteins are usually moderately soluble in water. Although their size would suggest limited solubility, the fact that some of the R groups are often polar improves solubility.

2.3 ENZYMES

2.3.1 DEFINITION
Define enzyme

© IBO 1996

Enzyme: a globular protein molecule that accelerates a specific chemical reaction. Enzymes are biological catalysts. A catalyst speeds up a reaction without changing it in any other way. Adding an enzyme to a reaction does not create different products and does not alter the reaction's equilibrium. It only assists in reaching this equilibrium faster. Enzymes are not used up in the reactions they catalyse. They can be used again and again.

2.3.2 ACTIVE SITE
Define active site

© IBO 1996

Active site: the region of an enzyme surface that binds the substrate (reacting substance) during the reaction catalysed by the enzyme.

Enzymes make it easier for a reaction to take place. The substrate will bind to a special area of the enzyme called the active site.

2.3.3 LOCK AND KEY MODEL
Describe the 'lock and key model'

© IBO 1996

Enzymes are specific to a certain reaction. According to the **lock and key model**, this is caused by the three-dimensional structure of the active site which is complementary to its substrate like a key to a lock.

The shape of the active site is caused by the tertiary and quaternary level of organisation of the protein (see Section C.2.1/8.5.1) According to this model, the shape active site is so specific that it can only catalyse one reaction.

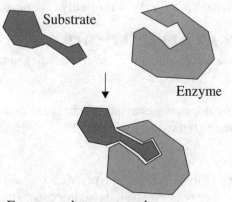

Enzyme-substrate complex

2.3.4 & 2.3.5 ENZYME ACTIVITY
List three factors that affect enzyme activity. Outline the effects of temperature and substrate concentration on enzyme activity.

© IBO 1996

The Chemistry of Life

The speed of a reaction can be measured in two ways: how fast does the substrate disappear and how fast is the product formed? Since both of these should give information about the speed of reaction, it is a matter of convenience which is chosen.

Generally, biological reactions are extremely slow at room temperature without the help of enzymes. Therefore the speed of reaction is a measurement of the enzyme activity.

A number of factors affect enzyme activity.
- **Concentration of substrate**. Using a given amount of enzyme, the reaction will go faster if the substrate is more concentrated.
- **Concentration of enzyme**. Using a given amount of substrate, the reaction will go faster if the enzyme is more concentrated.
- **Temperature**. Every enzyme has an optimum temperature where its activity is highest. The optimum temperature of our enzymes is mostly around 37°C. Almost every enzyme will be denatured above 60°C although some bacteria live in hot springs at 80°C and have enzymes suitable for this environment.
- **pH**. Every enzyme works best at a certain pH. Pepsin works best at approximately pH 2 but trypsin works best at approximately pH 8.

2.3.6 DENATURATION
Define denaturation.

© IBO 1996

Denaturation: the loss of the native configuration of a macromolecule resulting, for instance, from heat treatment, extreme pH changes, chemical treatment, or other denaturing agents. It is usually accompanied by a loss of biological activity.

2.3.7 APPLICATIONS
Explain two applications of enzymes in biotechnology.

© IBO 1996

We use enzymes as any other organism does. However, we also have biotechnological uses for enzymes.

For example:
- Biological washing powders contain enzymes as well as the usual detergents. The enzymes assist in breaking down stains (mainly proteins and fats) into smaller components which dissolve more easily in the water.
- Fruit juices are extracted using the enzyme pectinase. Pectin keeps plant cells together. Treating the fruit with pectinase makes the cells separate and it becomes much easier to squeeze the juice out. It also makes the juice clear rather than cloudy.
- Certain proteases are used in the preparation of baby food. When the large proteins are broken into shorter chains and amino acids they become much easier for babies to absorb.

2.4 DNA STRUCTURE

2.4.1 DNA
Outline DNA nucleotide structure in terms of sugar (deoxyribose), base and phosphate.
© IBO 1996

Genetic information is stored by nucleic acids. There are two kinds of nucleic acids: deoxyribonucleic acid (DNA) and ribonucleic acid (RNA). For the majority of species genetic information is stored in DNA in the nucleus and RNA is found in the cytoplasm. (Some viruses and prokaryotes store genetic information in RNA.)

Nucleic acids are long chain molecules (like proteins, but longer) and their building blocks are called nucleotides (like amino acids but again more complicated).

Nucleotides themselves are complex molecules consisting of three molecules linked together:
- a pentose sugar
- a phosphate
- an organic base

The sugar: two possibilities: ribose gives RNA and deoxyribose gives DNA.
Name: deoxyribose   ribose
Empirical formula: $C_5H_{10}O_4$ $C_5H_{10}O_5$

The phoshate:
Empirical formula: H_3PO_4 (P forms 5 bonds with other atoms.)

2.4.2 BASES
State the names of the four bases in DNA.
© IBO 1996

The **organic base**: (also known as the **nitrogenous base**)

There are 5 different bases: adenine, cytosine, guanine, thymine and uracil. A, T, C, G are found in DNA; A, U, C, G are found in RNA.

The structural formulae of the organic/nitrogenous bases are shown below. You do not need to memorise these.

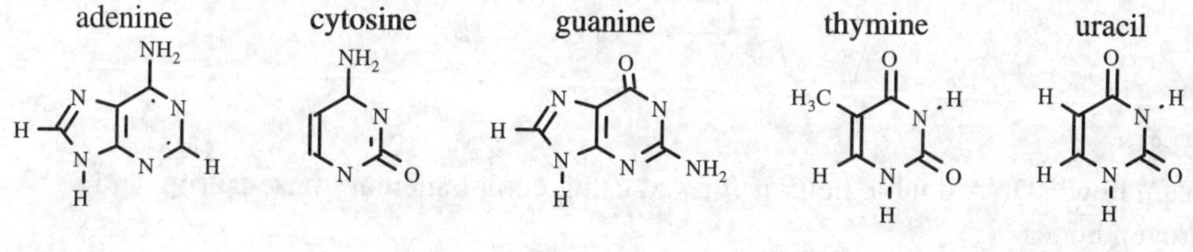

The Chemistry of Life

2.4.1. DNA
Outline DNA nucleotide structure in terms of sugar (deoxyribose), base and phosphate.
© IBO 1996

Building a nucleotide:

The nucleotide is usually schematically represented as follows:

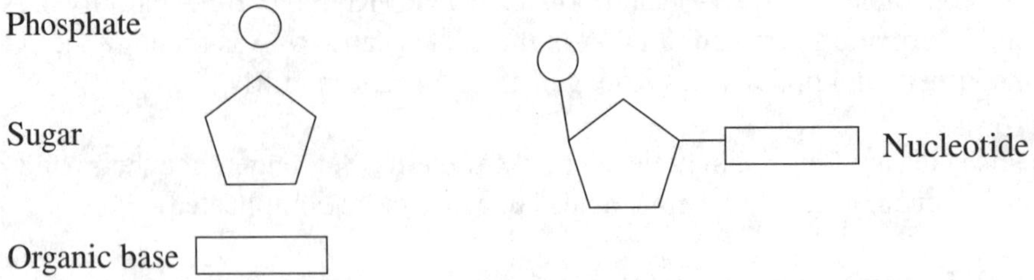

Phosphate

Sugar

Nucleotide

Organic base

Since the reactions involved are (again) condensation reactions, the equation becomes:

phosphate + sugar + organic base ⇒ nucleotide + 2 water

2.4.3 BUILDING A NUCLEIC ACID
Outline how the DNA nucleotides are linked together by covalent bonds into a single strand.
© IBO 1996

The nucleotides can be linked together by a condensation reaction between the phosphate of one nucleotide and the sugar of another. This way the sugar and phosphate form the 'backbone' of the nucleic acid with the organic bases sticking out. Covalent bonds are formed between the phosphate and sugar and a single strand is made with a 'backbone' of phosphate and (deoxy)ribose and an organic base attached to every ribose.

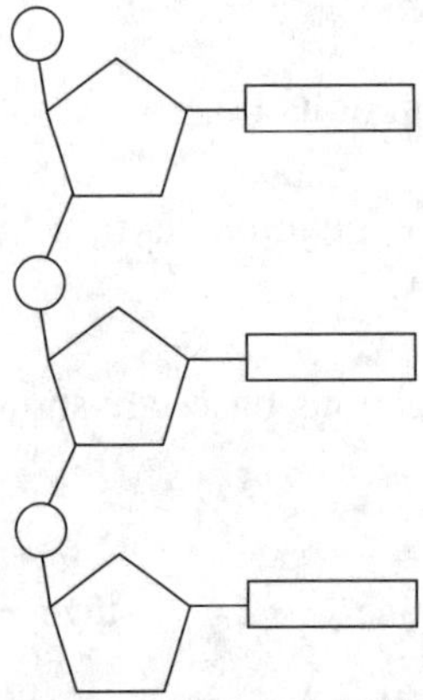

2.4.4
Explain how a DNA double helix is formed using complementary base pairing and hydrogen bonds
© IBO 1996

Knowing that the structure of DNA is a 'double helix' (a twisted ladder) we need to fit two DNA molecules together. Since some of the organic bases can form bonds with certain others, we can combine the DNA strands this way. The sugar and phosphate 'backbones' run antiparallel forming the sides of the ladder with the organic base pairs as rungs in between.

2.4.5
Draw a simple diagram of the molecular structure of DNA.

© IBO 1996

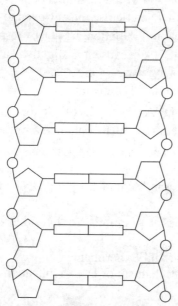

DNA - double helix

The diagram above is DNA as it would look if we 'untwisted' it. The DNA molecules coils in such a way that approximately 10 nucleotides complete one turn of the helix. Adenine, thymine and uracil are capable of forming 2 Hydrogen bonds, cytosine and guanine can form 3 Hydrogen bonds. So adenine (A) can form a basepair with thymine (T) (or uracil (U) in RNA) and cytosine (C) can form a basepair with guanine (G).

All organic bases have a complicated molecular structure involving ring compounds (see section 2.4.2). Adenine and guanine are purines, 'big' 2-ring structures; cytosine, thymine and uracil are pyrimidines, 'small' 1-ring structures.

In a DNA molecule, the backbone is made of alternating phosphate - sugar groups. The 'rungs' of the twisted ladder which DNA resembles are the organic base pairs. The available space for these rungs is the equivalent of three of the rings described in the previous paragraph.

This means that to fill the space, an organic base pair must have a total of 3 rings. To allow them to have hydrogen bonding, they must be capable of forming the same number of bonds. Which, in DNA, makes A and T one of the possible base pairs and C and G the other. When the base pairing involves RNA, the pairs are A and U, C and G.

The Chemistry of Life

2.5 DNA REPLICATION

2.5.1
State that DNA replication is semi-conservative.

© IBO 1996

Before mitosis occurs, DNA replication occurs during interphase. The nature of the process of DNA replication was shown by Meselson and Stahl (1957, California Institute of Technology) in a very elegant experiment. It is possible to grow bacteria in a medium containing the 'heavy' isotope of nitrogen: ^{15}N. If the medium contains no other source of nitrogen, it means that the growing E. coli bacteria use this nitrogen to incorporate it in their DNA. Although the isotope does not emit radiation, ^{15}N-DNA responds slightly different in a centrifugation test than ^{14}N-DNA because it is slightly heavier. Meselson and Stahl utilized heavy nitrogen in the following way.

- They grew cultures of E. coli bacteria in a ^{15}N medium.

    Results: - all cells contain only ^{15}N-DNA ('heavy DNA') in the centrifuge test.

- They moved the cultures to a new medium which contained only ^{14}N.

    Results: - First generation of *E. coli* bacteria: all DNA strands contain ^{15}N-^{14}N-DNA ('hybrid DNA').
    - Second generation: half the DNA strands contain 'hybrid DNA'; the other half contain 'light DNA' (^{14}N-DNA).
    - Third generation: one quarter of the DNA strands are 'hybrid DNA'; the other three quarters are 'light DNA'.

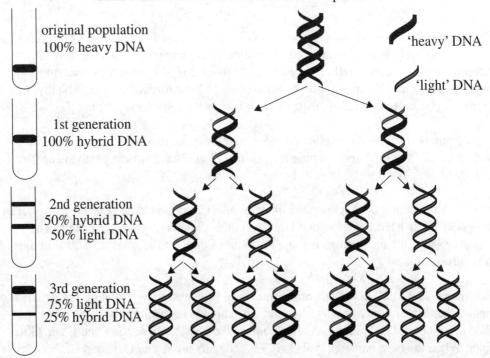

Meselson & Stahl Experiment
semi-conservative mechanism of DNA replication

The above results of the experiment supports one explanation: the replication of DNA is a semiconservative process. This means that during interphase (before mitosis starts), when DNA replication occurs, the DNA unwinds and then 'unzips' (which means that the hydrogen bonds between the base-pairs are broken and the two strands separate). The enzyme **helicase** is responsible for this.

Free nucleotides which are 'floating around' in the nucleus form complementary bonds with the nucleotides of the DNA strands (this happens to both strands) and then join to form a new strand of DNA. The enzyme **DNA polymerase** is responsible for this. All this results in two identical DNA double helices, each consisting of one 'old' and one 'new' strand.

If DNA replication had followed a 'conservative process', i.e. keeping the original strand intact and forming a complete new one, what would the results of the above experiment have been after 1 generation?

2.5.2 2.5.3 DNA REPLICATION

Outline DNA replication in terms of unwinding the double helix and separation of the strands by helicase followed by formation of the new complementary strands by DNA polymerase.
Explain the significance of complementary base pairing in the conservation of the base sequence of DNA.

© IBO 1996

Since every organic base will only 'fit' with one other one (complementary base-pairing), the 'new' strand of DNA will be identical to one from which the 'old' strand just separated. So (theoretically) this process can continue for ever without any change to the genetic material.

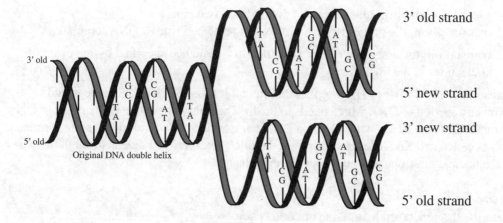

The Chemistry of Life

2.6 TRANSCRIPTION AND TRANSLATION

2.6.1 RNA AND DNA.
Compare the structure of RNA and DNA.

© IBO 1996

As was mentioned in section 2.4, there are two different kinds of nucleic acids: DNA and RNA.

	Deoxyribonucleic acid	*Ribonucleic acid*
	double helix (2 strands)	single strand
sugar:	deoxyribose	ribose
bases:	A, T, C and G	A, U, C and G

2.6.2 FUNCTIONS
State one function of messenger RNA and one function of transfer RNA.

© IBO 1996

Three kinds of RNA are commonly found in cells:
rRNA: **ribosomal RNA**: major component of ribosomes
tRNA: **transfer RNA**: folded upon itself, carries amino acids to mRNA
mRNA: **messenger RNA**: sequence of nucleotides determines the primary sequence of a polypeptide.

2.6.3
Outline DNA transcription in terms of the formation of a RNA strand complementary to the DNA strands by RNA polymerase.

© IBO 1996

The process of protein synthesis can be divided into two parts:
- **transcription**: the process by which RNA is produced from a DNA template.
- **translation**: the assembly of a polypeptide in a sequence specified by the order of nucleotides in the mRNA.

Transcription is similar to DNA replication in that it takes place in the nucleus and involves a section of DNA which needs to unzip. Then only one of the two strands of DNA is transcribed (the 'sense' strand) and a RNA complementary strand to this strand is made. This is called **mRNA**. After transcription, the mRNA leaves the nucleus through the pores in the nuclear envelope and goes into the cytoplasm.

2.6.4
Describe the genetic code in terms of codons composed of triplets of bases.

© IBO 1996

The genetic code is based on sets of 3 nucleotides called codons. Since 20 amino acids are commonly found in cells, it is not possible to have 4 different nucleotide codes for all of them. The possible permutations are $4^1 = 4$. If we tried using a set of 2 nucleotides, we could create $4^2 = 16$ different combinations, which is insufficient for the 20 different amino acids. So, therefore, we need to use a set of 3 nucleotides which will create $4^3 = 64$ possibilities. This is more than is required for the 20 different amino acids. As a result, the code is said to be **degenerate**, meaning that more than one codon can code for a single amino acid. The sequence of the codons in the mRNA determines the sequence of the amino acids in the polypeptide.

2.6.5
Describe translation including the roles of mRNA codons, tRNA anticodons and ribosomes leading to peptide linkage formation.

© IBO 1996

After the mRNA has reached the cytoplasm, translation takes place. In the cytoplasm, ribosomes attach to the mRNA. The ribosome covers an area of 2 codons on the mRNA. tRNA carrying an amino acid will come in and the anti-codon exposed on the tRNA will have complementary binding with the codon of the mRNA. The second tRNA with its own anti-codon, and consequently carrying a specific amino acid, will complementary bind to the second codon on the mRNA, filling the other site in the ribosome. The second amino acid will attach to the first and the first amino acid will be released from the tRNA.

The ribosome and mRNA will move relative to each other and the tRNA will separate from the mRNA. The ribosome now covers the second and third codon. A new tRNA with an amino acid comes in, the second tRNA is released after its amino acid has attached to the third one. This process continues until a STOP codon is reached. The STOP codon does not code for an amino acid but terminates translation.

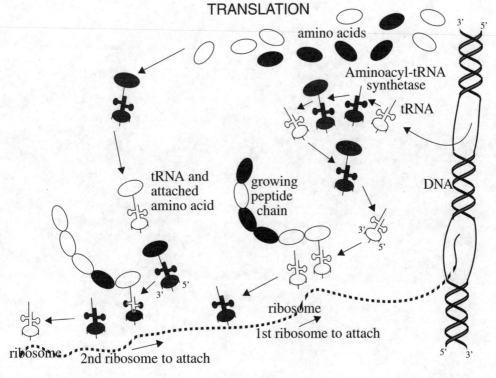

The Chemistry of Life

2.6.6
Define the terms degeneracy and universal as they relate to the genetic code.

© IBO 1996

The triplet code is said to be **degenerate** since more than one codon codes for a single amino acid. The code is also said to be **universal** since, for every organism, the same codon codes for the same amino acid. E.g. UUU and UUC on the mRNA both code for the amino acid phenylalanine, CUU, CUC, CUA and CUG all code for leucine, regardless of the species.

2.6.7
Explain the relationship between one gene and one polypeptide and its significance.

© IBO 1996

Since all reactions in biological systems require enzymes to proceed at a reasonable speed, controlling the enzymes essentially means controlling the rate of reaction. It has become clear that every polypeptide chain in an enzyme is coded for by a specific gene. By transcribing and translating the genes, we can control the rate of enzyme production and hence the rate of any reaction.

2.7 GENETIC ENGINEERING, DNA FINGERPRINTING, GENE THERAPY

2.7.1
State that genetic material can be transferred between species because the genetic code is universal (cross reference 2.6.6).

© IBO 1996

Genetic engineering: the deliberate manipulation of genetic material.

Since the genetic code is universal (see section 2.6.6), it is possible to transfer genetic material from one species to another. It is possible to introduce a human gene for making insulin into a bacterium and have it produce the protein hormone which is responsible for making cells change glucose into glycogen (See section 8.3.9).

Genetic engineering creates new combinations in DNA. In meiosis, crossing over can cause **recombinant DNA** and create new combinations of genes. In genetic engineering, scientists change DNA using genetic materials from different species. Mutation and recombination occur randomly in nature and hence most new (combinations of) traits are unfavourable to the individual and will not remain in the population. Genetic engineering will direct the process of recombination or mutation and the chances of a favourable new (combinations of) trait(s) are increased.

2.7.2
Outline a basic technique used for gene transfer involving plasmids, a host cell (bacterium, yeast or other cell), restriction enzymes (endonuclease) and DNA ligase.

© IBO 1996

Gene transfer involves the following elements: a vector, a host cell, restriction enzymes and DNA ligase.

The **vector** is what is needed to carry the gene into the host cell. Two vectors are commonly used: phages (**bacteriophage**: a virus which is a parasite on certain bacteria) and plasmids.

Bacteria carry all the required genetic information on one large circular DNA. However, most bacteria also possess extra DNA in the form of plasmids. Plasmids are circular bits of genetic material carrying 2-30 genes. Plasmids may replicate at the time the chromosome replicates or at other times. It is therefore possible for a cell to possess several identical plasmids. Alternatively a cell can only have one at a time of cytokinesis which will result in one of the daughter cells not receiving a plasmid. Many different plasmids have been found in *E. coli* alone. The best known are the sex factor (the F plasmid) and the drug resistance, or R, plasmid. The F plasmid consists of a double strand of DNA. The strands can separate and one goes to another cell which does not yet posses the F plasmid. Both cells will form the complementary strand and then possess a complete F plasmid. This is called conjugation and is a natural way for transferring DNA from one cell to another.

The Chemistry of Life

Plasmids are also used to clone a desired gene. You splice the desired gene into a plasmid and transfer it into a bacterial cell. Culture these bacteria and many of them will have a plasmid with the desired gene. Use a section of nucleotides complementary to the desired gene but also attached to a (e.g. radioactive) label to find which plasmids have the gene and which do not. Use restriction enzymes to cut the desired gene out of the plasmids. Purify using gel electrophoresis (see section 2.7.6)

The **host cell** is the cell which is to receive the genetic material.

A bacterium may be the host cell and produce a protein desired by humans. Plant cells or animal cells can also be host cells as can be seen in section 2.7.3. Human cells can also be the host cell as we will see in section 2.7.9

Restriction enzymes are used to cut the desired section of the DNA. Bacteria produce these enzymes naturally as a defense against invading viruses. One commonly used restriction enzyme recognises the sequence $\begin{smallmatrix}GAATTC\\CTTAAG\end{smallmatrix}$ and cuts both strands of the DNA between G and A.

The result is an uneven cut in the DNA. If the same sequence is found on another section of DNA and is cut in the same way, the two strands can be combined as is shown below.

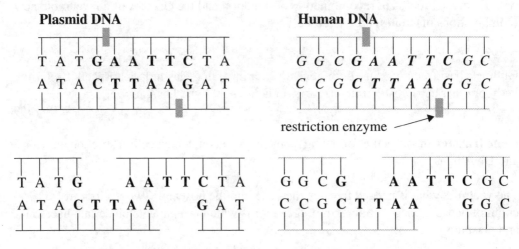

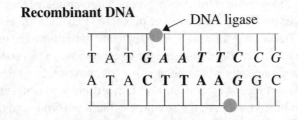

To attach the two cut sections of DNA, the enzyme DNA ligase is used to create the required covalent bonds.

2.7.3
State two examples of the current uses of genetic engineering in agriculture and/or pharmacy.

© IBO 1996

Genetic engineering is becoming more common in many areas of our life.

Medical use: In section 2.7.9 the example is given of a girl who cannot live a normal life without genetically modified white blood cells.

Pharmaceutical use: Higher level students can see how insulin is made by bacteria (section 8.3.9). By splicing a section of human DNA (coding for the production of insulin) into an *E. coli* cell, the pharmaceutical industry now produces 'human' insulin (made by bacteria). This is preferable to cow insulin extracted from slaughtered cattle.
The next step is to splice a human gene into the DNA of a cow or goat and create transgenic animals who will produce a desired enzyme or hormone in their milk. The yield of this is high and continuous. However, the costs of setting this up are staggering and some people object to the practice (see section 2.7.4). An enzyme to dissolve blood clots is already being produced by transgenic goats and is extracted from their milk.

Pioneering work of a similar nature is now being done on plants because many people are concerned about transmitting diseases like BSE (mad cow disease) from animals to people.

Agricultural use: Bacteria have also been modified to produce the growth hormone of cows. When this is injected into cows, they produce 10-20% more milk. Plants can be made more resistant to diseases thus increasing their yield. This has been done with, for example, potatoes and tomatoes.

In agriculture, people have used selective breeding for as long as we can remember. They chose to eat the calf from the cow that did not produce much milk and to breed the daughter of a good milker with a bull whose mother was a good milker. In addition, people would always be looking for spontaneous mutations which we might want. Think of all the different breeds of dogs and horses we bred without the use of genetic engineering.
The effect of genetic engineering is that we can now direct the mutations. Instead of having to wait for them to happen and being unsure of how they would turn out, we can now control them. This saves a lot of time.

Looking back at history, it was only just over 40 years ago that Watson and Crick suggested the structure of DNA (1953). Only 25 years ago, the first restriction enzymes were discovered, and in 1994 the first genetically manipulated organisms became available for food.

The Chemistry of Life

2.7.4
Explain one potential harmful result of genetic engineering.

© IBO 1996

In the above example of the cows producing more milk when injected with growth hormone, it has been shown that these cows need larger amounts of antibiotics to keep them healthy. Some of these antibiotics, or even some growth hormone, may end up in the milk which we drink. Also, since many industrialised countries already produce a surplus of milk, it seems pointless to try and improve the yield.

Hospitals all over the world fear 'plasmids'. These segments of DNA so useful for cloning genes can cause hospitals the worst problems they can imagine. Plasmids can carry genes which cause the bacterium to be resistant to antibiotics. Plasmids reproduce rapidly and can be passed from one species to another. In hospitals, this may mean that many diseases can quickly become resistant to (almost) all known antibiotics.

As a result, many people fear that genetically altered bacteria will 'escape' from research facilities and have unknown effects on us, e.g. cause incurable diseases. Some 'animal rights' groups also object to using cattle or goats to produce medicine and fear that soon we will have blond cattle with blue eyes. Ultimately some people fear that anyone who is less than perfect will not be acceptable anymore if all disease and genetical disorders can be diagnosed and repaired before a child is born. This uncertainty also applies to genetically altered organisms. What will they do to the existing ecosystem?

When 'new' organisms are introduced into an ecosystem, we can never fully predict the effect of them. This is the case for existing and for genetically modified organisms. Many examples can be given e.g. rabbits (and cats) in Australia which have no natural predators. Uncontrolled breeding has made them become a pest. Kudzu plants deliberately introduced in sections of the US to control soil erosion, grew so well that they destroyed large parts of the natural forest. This uncertainty also applies to genetically modified organisms. What effect will they have on existing ecosystems?

2.7.5
State that PCR (polymerase chain reaction) copies and amplifies minute quantities of nucleic acid.

© IBO 1996

PCR stands for **polymerase chain reaction**. When researchers want to study a particular sequence of DNA, they need many (identical) copies of it. The traditional method of cloning (using plasmids) takes a lot of time and work.

Instead, a 'photocopier for DNA' is used, which is essentially what the PCR does. It works on the following principle:

> The desired DNA is heated which breaks the hydrogen bonds between the strands of the double helix so that they separate. Primers are added to start the process of DNA replication (see section 8.2). As the mixture is cooled, the primers bond to the original

but now single single strand DNA molecules (hydrogen bonding, complementary base pairing). Nucleotides and DNA polymerase are added. The nucleotides will bond with the 'exposed' organic bases of the single strand DNA (hydrogen bonding, complementary base pairing) and the DNA polymerase will then join them into a DNA strand. This way, each of the original strand has formed a new complementary strand. These strands can be heated and separated and will function as a template for more DNA strands to be formed.

2.7.6
State that gel electrophoresis involves the separation of fragmented pieces of DNA according to their charge and size.

© IBO 1996

Gel electrophoresis is a process commonly used in biochemistry. Electrophoresis is a technique used to separate molecules based on their different rates of movement in an electric field caused by a combination of their charge and their size.

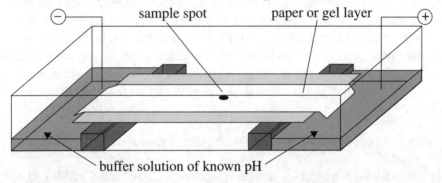

2.7.7
State that gel electrophoresis of DNA is used in DNA profiling.

© IBO 1996

In gel electrophoresis, the mixture of DNA cut into segments by restriction enzymes is placed on a special gel and a current is run through it. The DNA mixture will separate into bands. By running a control of known DNA segments under identical conditions, you know which segments end up where and can compare this with the results of the unknown mixture.

2.7.8
Describe two applications of DNA profiling.

© IBO 1996

DNA profiling or DNA fingerprinting is used in criminal investigations. A small sample of blood or semen from the criminal might be collected from the victim or the area. Gel electrophoresis can be used to compare the DNA found, with that of a suspect. DNA is as individual as a fingerprint so a match in the 2 samples can be used as evidence.

The Chemistry of Life

This technique has also been used to try to determine the identity of the remains of dead people. Many people claimed that the Tsar of Russia and his family were shot during the Russian revolution and bodies were shown to prove this. The identity of these bodies could not be proven until DNA fingerprinting was brought in. By taking blood samples of (distant) relatives of the Romanov's, DNA patterns could be established. Samples from the bodies showed similar DNA patterns and the conclusion was that the bodies were likely to be the Romanov family.

2.7.9
Outline the process of gene therapy using a named example.

© IBO 1996

Gene therapy had already been used in experimental medicine in 1990. The principle of gene therapy is as follows:

A girl has a genetic disease which makes her unable to produce a certain enzyme. This means that her leucocytes (white blood cells) could not function properly. Some blood was taken from her and a copy of the proper gene was inserted in her leucocytes. These leucocytes were returned to her and it quickly became evident that her immune system now worked properly. The girl needs to have this treatment repeated regularly because leucocytes only live for a few months. It would be better to change the genes of the stem cells which make the leucocytes because then she would be permanently cured. However, this is not yet possible.

Other examples of genetic diseases treated with gene therapy are:

- SCID stands for severe combine immune deficiency. Often both T and B cell production is affected. This usually results in one or more serious infections during the first few months of life. These infections (pneumonia, meningitis or bloodstream infections are common examples) can be fatal.

The purpose of the replaced gene is to allow for the production of the enzyme **ADA** (adenosine deaminase).

- Thalassaemia is a genetic blood disorder. People suffering from thalassaemia are anaemic and need regular blood transfusions because their genes for making haemoglobin (A or B) are dysfunctional. Gene therapy for thalassaemia is likely to be successful in the future.

EXERCISE

1. An organic molecule is one that contains the element:

    A   hydrogen.

    B   oxygen.

    (C)   carbon.

    D   iron.

2. The three most common elements in the 'molecules of life' are:

    A   carbon, hydrogen & sodium.

    B   iron, hydrogen & oxygen.

    C   carbon, nitrogen & oxygen.

    (D)   carbon, hydrogen & oxygen.

3. In addition to carbon, hydrogen & oxygen, nucleic acids contain the element:

    A   sodium.

    B   potassium.

    (C)   phosphorus.

    D   iron.

4. The comparatively high melting and boiling points of water are due to:

    (A)   the polarity of the water molecule.

    B   the covalent bonds of the water molecule.

    C   the fact that water is a compound of oxygen and hydrogen.

    D   the fact that water is inorganic.

5. When we cool ourselves by sweating we are using:

    A   the high surface tension of water.

    B   the high boiling point of water.

    C   the high melting point of water.

    (D)   the high heat of evaporation of water.

6. Sucrose and maltose are examples of

    A   monosaccharides.

    (B)   disaccharides.

    C   polysaccharides.

    D   inorganic molecules.

The Chemistry of Life

7. Generally, if we are considering solubility in water:

A Monosaccharides are more soluble than disaccarides which are more soluble than polysaccharides.

B Monosaccharides are less soluble than disaccarides which are more soluble than polysaccharides.

C Monosaccharides are more soluble than disaccarides which are less soluble than polysaccharides.

D Monosaccharides are less soluble than disaccarides which are less soluble than polysaccharides.

8. Which one of the following is not a major function of lipids in the human body?

A as a short term enegy source.

B synthesis of hormones.

C long term energy storage.

D insulation.

9. An example of a structural protein in the human body is:

A haemoglobin.

B amylase.

C cellulose.

D collagen.

10. In the lock and key model for enzyme action, the enzyme:

A fits very closely onto a site on the substrate.

B has a hydrophobic head which dissolve in the substrate.

C has a hydrophilic head which dissolve in the substrate.

D attaches itself to the substrate by a chemical bond.

11. Which one of the following statements is not true?

A An enzyme catalysed reaction will go faster if the concentration of the enzyme is increased.

B An enzyme catalysed reaction will go faster if the concentration of the substrate is increased.

C An enzyme catalysed reaction will go faster if the temperature is increased from 10°C to 20°C.

D The rate of an enzyme catalysed reaction generally depends on pH.

12. The cooking of a 'hard boiled egg' is an example of:

  A   an enzyme reaction.
  (B) denaturation.
  C   hydrolysis.
  D   condensation.

13. Which one of the following is not a component of a nucleotide molecule?

  A   a pentose sugar.
  B   a phosphate.
  C   an organic base.
  (D) an inorganic base.

14. The process by which RNA is produced from a DNA template is known as:

  (A) transcription.
  B   translation.
  C   division.
  D   coding.

15. The number of possible sequences of three nucleotides is:

  A   3
  B   4
  C   16
  D   64

16. Which one of the following graphs best represents the reaction rate of an enzyme catalysed reaction.

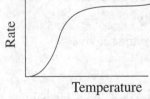 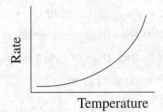

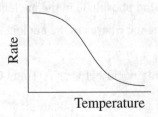 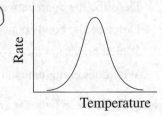

The Chemistry of Life

17. Which one of the following graphs represents the reaction rate of an enzyme catalysed reaction.

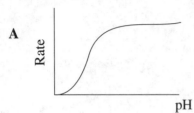

A

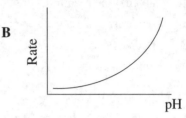

B

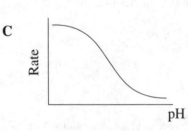

C

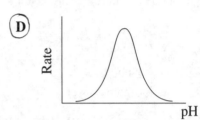

(D)

18. Explain the properties of water in terms of the structure of its molecules.

19. a. Which elements are found in

carbohydrates.

lipids.

proteins.

b. Name and draw the molecular structure of the building blocks of

carbohydrates.

lipids.

proteins.

c. How do the condensation reactions in forming disaccharides and lipids differ from those forming dipeptides?

20. Describe in terms of the lock and key model of enzyme action :

a. the effect of temperature on enzyme activity.

b. the effect of substrate concentration on enzyme activity.

c. denaturation.

21. a. Describe the bond between the sugar and the base in the nucleotide.

b. Describe the bond between the sugar and the phosphate in the nucleotide.

c. Describe the bonds between the complementary bases in the nucleic acid.

d. Which bonds are the strongest and why?

e. Why does complementary base pairing only occur between A-T and C-G?

22. Distinguish between helicase and DNA polymerase in DNA replication.

23. Haemoglobin consists of 4 polypeptide chains: 2 a chains and 2 b chains.

a. How many genes need to be transcribed to make the mRNA needed for one molecule of haemoglobin?

b. How many times does each mRNA chain need to be translated to make one molecule of haemoglobin?

GENETICS

3

Chapter contents

- Chromosomes, genes and alleles
- Gene mutation
- Meiosis
- Theoretical genetics
- Applied genetics

Genetics

3.1 CHROMOSOMES, GENES AND ALLELES

3.1.1
State that an eukaryote chromosome is made of DNA and protein

© IBO 1996

Eukaryotic chromosomes are made of DNA and protein.

3.1.2
State that chromosomes can be stained to show banding.

© IBO 1996

Chromosomes can be stained to show banding. This is used in the process of karyotyping.

3.1.3
State that the chromosome structure and banding can be used to arrange the chromosomes in their pairs.

© IBO 1996

Chromosome structure and banding can be used to arrange the chromosomes in their pairs.

3.1.4
Describe one application of karyotyping (cross reference 3.3.4).

© IBO 1996

One application of **karyotyping** can be found in an **amniocentesis**. The risk of having a child with Down's syndrome increases with the mother's age. Non-disjunction (which can cause Down's syndrome) also seems to have a genetic component so that it can be said that people with relatives with Down's syndrome have an increased chance of having a baby with this genetic disorder.

In either situation, it is possible to do an amniocentesis around the 16th week of the pregnancy. A sample of the amniotic fluid (containing foetal cells) is taken and a culture is made. When sufficient cells have been obtained, a karyotype can be done to detect chromosome abnormalities. The dividing cells are photographed and, using these pictures, the chromosomes are arranged in homologous pairs. Some genetic disorders can be detected this way (e.g. Down's Syndrome). This process takes approximately 3 weeks.

3.1.5
Define gene.

© IBO 1996

Gene: a heritable factor that controls a specific characteristic (such as eye colour), consisting of a length of DNA occupying a position on a chromosome known as a locus.

3.1.6
Define allele.

© IBO 1996

Allele: one specific form of a gene, differing from other alleles by one or a few bases only and occupying the same gene locus as other alleles of the gene.

3.1.7
Define genome.

© IBO 1996

Genome: the total genetic material of an organelle, cell or organism.

Genetics

A cat is a cat

Cats have a comparatively large 'gene pool'. There are large variations in markings etc. of domestic cats.

and not a cheetah

Cheetahs have a very small 'gene pool'. There is very little difference between individuals. This makes this animal particularly vulnerable to extinction.

because of its genes........................

Genetics

3.2 GENE MUTATION

3.2.1
Define gene mutation.

© IBO 1996

Generally, a broad distinction between two kinds of mutations is made: chromosome mutations and **gene mutations**. Chromosome mutations involve (a large section of) a chromosome. Examples are Down's syndrome (extra copy of chromosome no. 21) and Turner's syndrome (sex chromosomes XO instead of XX or XY). A gene mutation is a much smaller change involving the genetic material of only one gene.
Gene mutation: a change in the base sequence of a gene.

3.2.2
Outline the difference between an insertion and a deletion.

© IBO 1996

Gene mutations can be point mutations involving a change in only one organic base. This might have no effect for several reasons:
- it involves part of the non-sense DNA.
- it involves part of the DNA which that particular cell does not use.
- it changes the third (or second) organic base of a codon, and since the genetic code is degenerate, the same amino acid is still coded for.

Generally more serious are the changes known as insertion and deletion. As the names already suggest, in these cases an organic base is added or removed from the DNA. The result of this is that all the codons following this change will have been altered.

E.g. original sequence on DNA: A T G T C G A A G C C C

This is translated into mRNA: U A C A G C U U C G G G

This codes for the following amino acid sequence:
                    thy - ser - phe - gly

Adding an organic base after the first T will give me the sequence:
                    A T A G T C G A A G C C C

This is translated into mRNA: U A U C A G C U U C G G

This codes for the following amino acid sequence:
                    thy - glu - leu - arg

You can see how drastic the results of this kind of mutation can be. A deletion is the opposite process with equally dramatic consequences.

3.2.3.
Explain the consequence of a base substitution mutation in relation to the process of transcription and translation, using the example of sickle cell anaemia.

© IBO 1996

Substitution seems to be less far reaching. Yet even the replacement of one amino acid with another can have severe consequences.

Haemoglobin is made of 4 polypeptide chains: 2 α chains and 2 β chains. When a base substitution is made in the gene coding for the 6th amino acid in the β chain, the codon then codes for glutamic acid instead of the original valine. A substitution may have taken place, for example changing GAA into GUA.

The resulting polypeptide is different and the haemoglobin formed is commonly known as HbS.

The result of this is a slightly different structure of the haemoglobin molecule which makes it crystallise at low oxygen levels (e.g. in the capillaries). The erythrocyte in which the haemoglobin can be found will then change from a biconcave shape into a sickle cell shape and can block the small capillaries, and is less efficient in transporting oxygen. The symptoms are acute anaemia which causes physical weakness. The lack of oxygen may be severe enough to cause damage to heart and kidneys or even death (in homozygous individuals). Sickle cell anaemia is codominant with the 'normal' allele although the latter is expressed more strongly in the heterozygous individual. Heterozygous individuals (carriers) have some HbS but more normal haemoglobin. They may suffer from mild anaemia. The selective advantage of being a carrier is found in malaria infested areas. Plasmodium (the protist causing malaria) cannot live in erythrocytes with HbS. This means that individuals heterozygous for the sickle cell trait have a reduced chance of contracting malaria. Natural selection has ensured that the sickle cell trait is more common among people living in malaria infested areas such as West-Africa. as the african-american population largely originates from this area, the trait is also found in frequencies higher than usual in this group.

Genetics

3.3 MEIOSIS

3.3.1 DEFINITION
State that meiosis is a reduction division in terms of diploid and haploid numbers of chromosomes.

© IBO 1996

Meiosis is a reduction division in terms of diploid and haploid numbers of chromosomes. The purpose of meiosis is to produce gametes. Meiosis only occurs in diploid (or tetraploid) cells and reduces the number of chromosomes per cell. Gametes contain half the number of chromosomes (one chromosome from each homologous pair). When two gametes fuse, in the process of fertilisation, the original number of chromosomes is restored.

3.3.2 OUTLINE OF THE PROCESS OF MEIOSIS.
Outline the process of meiosis including pairing of chromosomes followed by two divisions which result in four haploid cells.

© IBO 1996

A key issue of meiosis is the pairing of the homologous chromosomes followed by two divisions which result in four haploid cells. See the diagrams on the next page.

3.3.3 GENETIC VARIETY
Explain how the movement of chromosomes during meiosis can give rise to genetic variety in the resulting haploid cells.

© IBO 1996

Since the chromosomes pairing up are homologous and not identical, different combinations are possible.

In a diploid individual, one of the two chromosomes of a homologous pair comes from the mother, the other chromosome comes from the father. This is the case for all chromosomes, i.e. every homologous pair.

In theory it is possible that all paternal chromosomes end up at one pole and all maternal chromosomes at the other. In practice, this does not happen. Since the number of permutations possible is $2n$ when n is the haploid number, the chance of this one combination occurring is remote. For a human gamete, this would happen once in every $246 = 7 \times 10^{13}$ cases for all paternal chromosomes and the same for all maternal chromosomes.

Copy the diagrams below and colour of the straight chromosomes red and the other blue. Colour one of the wavy chromosomes green and the other yellow. Can you see that the chances for a straight blue chromosome to end up in a gamete with a wavy green chromosome are the same as for it joining a wavy yellow chromosome?

INTERPHASE

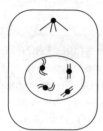
PROPHASE I

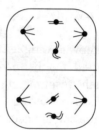
PROPHASE II

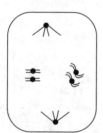
METAPHASE I

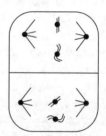
METAPHASE II

MEIOSIS I

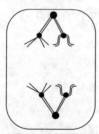
ANAPHASE I

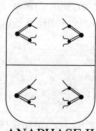
ANAPHASE II

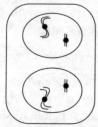
TELOPHASE I

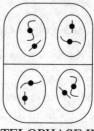
TELOPHASE II

MEIOSIS II

Genetics

3.3.4
Explain that non-disjunction can lead to changes in chromosome number, illustrated by reference to Down's Syndrome (trisomy 21).

© IBO 1996

The pairing up of the homologous chromosomes during Prophase I is called **synapsis**. The resulting pair of homologous chromosomes (each chromosome consisting of two identical chromatids) is called **tetrad** or **bivalent**. The separation of homologous chromosomes during Anaphase I is called **disjunction**. Failure of chromosomes to separate can lead to **aneuploidy** (one chromosome missing or one extra chromosome) or in the case of total non-disjunction, **polyploidy**.

One of the best known examples of aneuploidy is Down's Syndrome. When non-disjunction occurred in either parent and one of the two gametes carried two copies of chromosome 21, the resulting zygote will contain three copies of chromosome 21 and will be aneuploid. It is also referred to as trisomy 21. The resulting symptoms are called **Down's Syndrome** and are accompanied by (varying degrees of) retardation.

Unlike the production of sperm cells in the male (which begins at puberty), the female gametes develop before birth. At birth, the future female gametes are present as primary oocytes in prophase I of meiosis. They remain in this stage until ovulation. This means that for a twenty year old woman, her egg cells have collected twenty years of damage (chemicals, radiation, etc.). But for a forty year old woman, the cells have collected forty years of damage and the chance of non-disjunction resulting in a baby with Down's Syndrome is increased.

Since the sperm cells of the male are produced constantly, the age of the father seems to have less effect than the age of the mother. Genetic factors can play a role in either parent: generally if a person has a relative with Down's Syndrome, this person may have an increased chance of having a baby with Down's Syndrome.

3.3.5
State Mendel's Law of Segregation.

© IBO 1996

Mendel's Law of Segregation (first law): The separation of the pair of parental factors, so that one factor is present in each gamete.

3.3.6
Explain the relationship between Mendel's Law of Segregation and meiosis.

© IBO 1996

The factors which Mendel mentioned are now known to be genes, located on chromosomes. Each gene is present twice (diploid) on pairs of homologous chromosomes. Since the homologous chromosomes pair up (synapsis) and then separate (disjunction), one chromosome of each homologous pair ends up in a gamete.

3.4 THEORETICAL GENETICS

3.4.1 - 3.4.10 DEFINITIONS
Define phenotype, dominant allele, recessive allele, codominant allele, locus, homozygotis, heterozygous, carrier, test cross.

© IBO 1996

Genotype: the alleles possessed by an organism.

Phenotype: all the characteristics of an organism.

Dominant allele: an allele which has the same effect on the phenotype whether it is present in the homozygous or heterozygous state.

Recessive allele: an allele which only has an effect on the phenotype when present in the homozygous state.

Codominant alleles: alleles which have a partial effect on the phenotype when present in heterozygotes but a greater effect in homozygous individuals.

(the terms 'incomplete' and 'partial' dominance are no longer used)

Locus: the particular position of a gene on homologous chromosomes.

Homozygous: having the two identical alleles of a gene.

Heterozygous: having two different alleles of a gene.

Carrier: an individual that has a recessive allele of a gene that does not have an effect on the phenotype.

Test cross: testing a suspected heterozygote by crossing with a known homozygous recessive (the term 'backcross' is no longer used.)

3.4.11
Draw a Punnett Grid.

© IBO 1996

A **Punnett grid** is a way of finding the expected ratio of the offspring, given certain parental phenotypes.

We can study one of the characteristics Mendel used in his experiment. He studied the size of pea plants and found that 'tall' is dominant over 'short'. If we start with 2 pure breeding (homozygous) plants of contrasting traits (tall and short), we will obtain an F1 which has the dominant phenotype (tall) but is heterozygous. When self-fertilising the F1, we will obtain an F2 which will appear 3/4 dominant (tall) and 1/4 recessive (short).

Possible phenotypes	Corresponding Genotypes
Tall	TT or Tt
Short	tt

Genetics

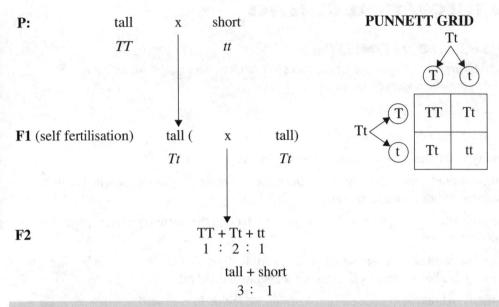

3.4.12
Draw a pedigree chart.

© IBO 1996

A **Pedigree chart** is used to show the inheritance of certain traits over several generations of humans. A classic example of a pedigree chart is the one starting with Queen Victoria and concerning the disease haemophilia (recessive, sex-linked).

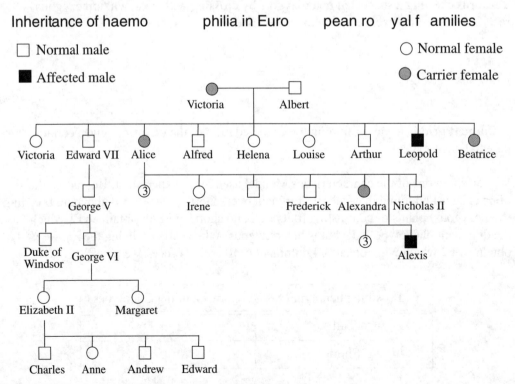

Males are either 'normal' or 'affected' while females are 'normal' or 'carrier'. This is because the haemophilia allele is homozygous lethal (as well as sex linked). Since the allele is carried on the x chromosome, males can be $x^h y$ ('normal') or $x^h y$ ('affected'). females can be $x^h x^h$ ('normal') or $x^h x^h$ ('carrier'). See section 3.4.19.

3.4.13
State that some genes have more than two alleles (multiple alleles).

© IBO 1996

Some genes have more than two alleles. **Multiple alleles** can be found e.g. in blood groups (see 3.4.14).

3.4.14
Describe ABO blood groups as an example of codominance and multiple alleles.

© IBO 1996

The ABO blood group system is based on 4 different phenotypes (group A, B, AB and O) caused by different combinations of 3 different alleles (I^A, I^B and i).
We have the following possibilities:

Phenotypes	Genotypes
A	$I^A I^A$ or $I^A i$
B	$I^B I^B$ or $I^B i$
AB	$I^A I^B$
O	ii

Using a Punnett Grid, you can work out how a female with blood group A and a male with blood group B can have four children, each with a different blood group.

You can also explain why a female with blood group O and a male with blood group AB cannot have children with either of the parents' blood group.

3.4.15
Outline how the sex chromosomes determine gender, by referring to the inheritance of X and Y chromosomes of humans.

© IBO 1996

Gender determination in humans is decided by the sex chromosomes. Females have two X chromosomes (XX), males have one X and one Y chromosome (XY). A Punnett Grid can be used to predict the chances of the gender of a child.

So the gender determining factors (the X and Y chromosomes) can be treated like any other trait and predictions can be made accordingly.
See also Section 3.4.12 on haemophilia, a recessive and sex linked trait.

PUNNETT GRID

	X	X
X	XX	XX
Y	XY	XY

Genetics

3.4.16
State that some genes are present on the X chromosome and absent from the shorter Y chromosome in humans.

© IBO 1996

The X chromosome is relatively large, the Y chromosome is much smaller. Several genes are located on the X chromosome e.g. colourblindness and haemophilia but will be absent from the Y chromosome. Only few genes are located on the Y chromosome (hairy ears) and you will normally not come across them.

3.4.17
Define sex linkage.

© IBO 1996

Sex linkage: genes carried on the sex chromosomes, most often on the X chromosome.

3.4.18
State two examples of sex linkage.

© IBO 1996

Examples of sex linked genes are colourblindness and haemophilia (see Section 3.4.12). Both of these traits are recessive and found on the X chromosome.

3.4.19
State that a human female can be homozygous or heterozygous, with respect to sex-linked genes.

© IBO 1996

A human female has one of three possible genotypes for a sex linked trait: homozygous dominant, heterozygous, or homozygous recessive. Human males cannot be heterozygous since they will only have one copy of one allele.

E.g. 1 colourblindness:
existing alleles: X^b for colourblindness, X^B for normal vision.
A female can be: $X^B X^B$ or $X^B X^b$ or $X^b X^b$
A male can be: $X^B Y$ or $X^b Y$

E.g. 2 haemophilia
existing alleles: X^h for haemophilia, X^H for normal bloodclotting.
A female can be: $X^H X^H$ or $X^H X^h$ or $X^h X^h$ (this does not exist)
A male can be: $X^H Y$ or $X^h Y$

The haemophilia allele is homozygous lethal, meaning that having this allele present twice in a cell means that the individual will not exist. Sometimes there does not seem to be a reason for this (some fur colour in mice is homozygous lethal) but in this case there is. A haemophiliac female would bleed to death at her first menstrual period. See 3.4.12.

3.4.20
Explain that female carriers are heterozygous for X-linked alleles.

© IBO 1996

Heterozygous females for sex linked traits are said to be carriers since they will not show the disease but can pass it on to the next generation. Again, see Section 3.4.12.

3.4.21
Calculate and predict the genotypic and phenotypic ratios of offspring of monohybrid crosses involving any of the above patterns of inheritance.

© IBO 1996

The accepted notation is the following:
Sickle cell: Hb^A = normal; Hb^S = sickle cell
Colourblindness: X^b = colourblindness; X^B = normal vision.
Haemophilia: X^h = haemophilia; X^H = normal bloodclotting

blood groups:

Phenotypes	Genotypes
A	$I^A I^A$ or $I^A i$
B	$I^B I^B$ or $I^B i$
AB	$I^A I^B$
O	ii
Rhesus positive	Rh^+Rh^+ or Rh^+Rh^-
Rhesus negative	Rh^-Rh^-

Codominance: main letter should relate to the gene, suffix to allele. both should be capital letters

e.g. C^R = red flowers C^W = white flower
Drosophila a.o.   a+ = dominant allele   a = recessive allele
e.g. vg^+ = normal wing   vg = vestigial wing

Remember these two laws:

Mendel's first law = Law of segregation
'Parental factors (genes) are in pairs and split so that one factor is present in each gamete.'

Mendel's second law = Principle of independent assortment
'Any of one pair of characteristics may combine with either one of another pair.' (dihybrid inheritance)

3.4.22
Deduce the genotypes or phenotypes of individuals in pedigree charts.

© IBO 1996

Go back to the pedigree chart of 3.4.12 and write in all the possible genotypes using the correct notation.

3.5 APPLIED GENETICS

3.5.1
Define genetic screening.

© IBO 1996

Genetic screening: testing a population for the presence or absence of a gene.

3.5.2
Discuss three advantages and/or disadvantages of genetic screening.

© IBO 1996

Advantages of genetic screening include possible pre-natal diagnosis of genetic disease (followed by treatment or abortion), determining if people possess certain alleles which can cause genetic disease. This might stop people from unwittingly passing on genetic diseases. Examples of this could be Huntington's (dominant) which only becomes apparent around the age of 40 and will lead to death around 50 years of age. Another example could be sickle cell anaemia (recessive) for which one could be a carrier without being aware of it.

3.5.3
State that the Human Genome Project is an international cooperative venture to sequence the complete human genome.

© IBO 1996

Genome: the total genetic material of an organelle, cell or organism. The Human Genome Project is a commitment undertaken by the American scientific community to determine the location and structure of all genes in the human chromosomes. It is part of the international Human Genome Organisation.

It was first suggested in 1985 and the expected financial and time investment were staggering. Some new developments have made the process faster and cheaper than originally expected but the combined effort required is still staggering. The total expected cost will be over $3 billion and scientists hope to have a complete map of all human genes by the year 2005.

3.5.4
Describe two possible advantageous outcomes of this project.

© IBO 1996

Advantages of having a map of all human genes will include an understanding of many genetic diseases and hopefully to the production of pharmaceuticals (based on DNA sequences) to cure them. It should then also be possible to determine fully which genetic diseases any individual is prone to (genetic screening). This information could be valuable, but it could also be abused (e.g. by insurance companies or prospective employers) and society faces the challenge of coming to terms with this issue.

Biology

3.5.5
Define clone.

© IBO 1996

Clone: a group of organisms of identical genotype. OR
Clone: a group of cells descended from a single parent cell.

3.5.6
Outline a technique used in the cloning of farm animals.

© IBO 1996

A technique used in the cloning of farm animals uses the cluster of 8 cells produced by cleavage divisions of the zygote. These 8 cells are totipotent (they are totally undifferentiated and could give rise to any kind of cell) and can be separated. If each cell is then placed in the uterus of a suitable female, 8 genetically identical individuals can result.

This way, a genetically manipulated zygote can give rise to 8 individuals. Even without genetic manipulation, it is an advantage to be able to increase the number of offspring from one genetically desirable individual. It could mean half a dozen calves a year from one good cow, rather than just one.

3.5.7
Discuss the ethical issues of cloning human embryos.

© IBO 1996

Monozygotic twins (identical twins) are genetically identical and therefore can be considered to be 'spontaneous clones'. The technique described above for farm animals will also work for humans. Is this acceptable or does it lead to a selection of those fit to be cloned vs. those who are not?

A couple which will have children suffering from Huntington's, cystic fibrosis, Alzheimer's or some other genetic disorder could produce a genetically manipulated zygote. If they want more than one child, if would be relatively simple to separate the cells resulting from the first cleavage divisions. These cells could be stored (frozen) and used for later pregnancies. Ethical or not?

Clone embryos could be frozen and the genetic material could later be used to grow a new individual who could donate a kidney to the earlier born version of the clone. Or, it might be possible to grow only a specific organ which could then be used for transplantation.

3.5.8
Discuss the ethical issues of cloning human embryos.

© IBO 1996

Selective breeding has produced organisms which cannot reproduce without our help.

Genetics

Wild beans will disperse their seeds when they are ripe by having the pod burst open. This minor explosion will move the seeds away from the parent plant. Crop beans are much easier to harvest since they stay in the pod. They have been selected for generations for this.

A breed of French beef cattle produces such large calves that most cannot be born the normal way and need the vet to perform a caesarian.

For centuries, cows have been selected to produce milk with a high fat content and pigs to produce fatty meat. Due to changes in society, people now want milk with less fat (but high protein) and lean bacon. There has been a resulting shift in the direction of the selection process.

EXERCISE

1. Eukaryotic chromosomes are made of:

    A   DNA and lipids.
    B   DNA and protein.
    C   DNA and polysaccharides.
    D   lipids and protein.

2. A heritable factor that controls a specific characteristic, consisting of a length of DNA occupying a position on a chromosome known as a locus is a:

    A   chromosome
    B   gene
    C   allele
    D   genome

3. The total genetic material of an organelle, cell or organism is known as a:

    A   chromosome
    B   gene
    C   allele
    D   genome

4. A change in the base sequence of a gene is known as a:

    A   replication
    B   translation
    C   genome
    D   genetic mutation

5. The pairing up of the homologous chromosomes during Prophase I is known as:

A   meiosis
B   synapsis
C   interphase
D   prophase

6. Mendel's Law of Segregation is:

A   The separation of the pair of parental factors, so that one factor is present in each gamete.
B   The separation of the pair of parental factors, so that two factors are present in each gamete.
C   Offsping resemble their parents.
D   Offsping resemble the female parent.

7. The alleles possessed by an organism are known as the:

A   Phenotype
B   Dominant allele
C   Genotype
D   Recessive allele

8. An allele which has the same effect on the phenotype whether it is present in the homozygous or heterozygous state is known as a:

A   Phenotype
B   Dominant allele
C   Genotype
D   Recessive allele

9. If a tall garden pea is crossed with a dwarf garden pea, the F1 are all tall. Predict the result of self fertilisation of the F1, using a Punnett square.

10. In Drosophila, black body colour is recessive to wild type body colour. You have three sets of flies with wild type bodies designated A, B and C. You crossed A x B and obtained 112 wild type flies; A x C gave 78 wild type and 30 black-bodied; whilst B x C gave 86 wild type flies. Explain the expected genotypic and phenotypic ratios when flies A, B and C are crossed with flies having black bodies.

Genetics

11. In a species of plant petal colour is determined by one pair of alleles and stem length by another. The following experimental crosses were carried out.

exp. A. A yellow flowered plant was crossed with several white flowered plants. The F1 were all yellow flowered plants.

exp. B. A short stemmed plant was crossed with several long stemmed plants. The F 1 were all short stemmed

exp. C. A different yellow flowered, short stemmed plant of the same species was crossed with several white flowered long stemmed plants.
The following F1 were grown.
38 yellow flowered short stemmed
35 white flowered short stemmed
40 white flowered long stemmed
37 yellow flowered long stemmed

a. What are the dominant alleles?
b. Give all genotypes in exp. A and B.
c. Explain exp. C using a Punnett square.

12. In mice, coat colour is controlled by several pairs of alleles, including the following:

A, wildtype colour, dominant over a, black
C, coloured (i.e. pigmented), dominant over c, albino

a. If a black individual, CCaa, is crossed with an albino, ccAA, what will be the appearance of the F1 generation?
b. If an individual of the F1 generation is bred with another individual of the same genotype, what will be the appearance of the F2 generation?

Explain your answer using a Punnett square.

13. A baby has blood type B, his mother has blood type A, his paternal grandfather has blood type A, and his paternal grandmother has blood type B. Determine

a. the genotype of the baby
b. all possible genotypes

14. In shorthorn cattle the coat colour can be red, white or in the heterozygous condition, roan. In addition, the polled (without horns) condition is dominant to the horned condition.

a. What will be the phenotypes and the genotypes of the F1 and F2 generations if homozygous polled red cattle are crossed with white horned cattle?
b. How would you establish a pure breeding strain of red polled shorthorns from your F2 generation?

15. A boy called Mohammed and his sister Latifa have a brother who suffers from haemophilia. Their parents are normal.

    a.   What are Mohammed's chances of having a haemophiliac child?
    b.   What are Latifa's chances of having a haemophiliac child?

16. In Drosophila melanogaster, a gene influencing eye colour has wildtype (dominant) and purple (recessive) alleles. Linked to this gene is another that determines wing length. A dominant allele at this second gene locus produces wild type (long) wings; a recessive allele produces vestigial (short) wings. Suppose a completely homozygous dominant female having wildtype eyes and long wings mates with a male having purple eyes and vestigial wings. First-generation females are then crossed with purple-eyed, vestigial-winged males. From this second cross, offspring with the following characteristics are obtained:

252 wild type eyes, wild type wings
276 purple eyes, vestigial wings
42 wild type eyes, vestigial wings
30 purple eyes, wild type wings
----
600 offspring total.

What is the distance between the genes?

Genetics

ECOLOGY

4

Chapter contents

- Communities and ecosystems
- Photosynthesis, respiration and energy
- Populations, natural selection and evolution.
- Human Impact
- Ecological techniques

Ecology

4.1 COMMUNITIES AND ECOSYSTEMS

In the study of ecology, you will find that many of the words used are new to you. It is important that you know their meaning very well and that you can distinguish between terms that refer to similar (but not identical) concepts. It is equally important that you can give named examples where appropriate. The named examples given in this section are kept as simple as possible, so that they should be easy to remember. Try to watch as many nature documentaries as you can; they will often illustrate the issues discussed below.

4.1.1-7
Define ecology, ecosystem, population, community, species, habitat. Explain what is meant by the biosphere.

© IBO 1996

Ecology: the study of relationships between living organisms and between them and their environment.

Ecosystem: a community and its abiotic (non-biological) environment.

Population: a group of organisms of the same species living in the same area at the same time and capable of interbreeding.

Community: a group of populations living and interacting with each other in a habitat.

Species: a group of organisms which look alike, can interbreed and produce fertile offspring.

Habitat: the physical area in which individuals of a certain species can usually be found.

Biosphere: total of all areas where living things are found; including deep ocean and part of the atmosphere.

4.1.8
Describe what is meant by a food chain giving three examples, each with at least three linkages (four organisms).

© IBO 1996

A food chain is the transfer of the sun's energy from producers to consumers as organisms feed on one another.

Example 1a: seaweed → limpet → starfish → seagull
Example 1b: seaweed → limpet → crab → seagull
Example 1c: seaweed → periwinkle → crab → seal
Example 1d: seaweed → periwinkle → octopus → seal

Example 2:   oakleaves → caterpillars → robin → sparrowhawk
Example 3:   phytoplankton → shrimp → smelt (fish) → heron
Example 4:   grass → grasshopper → frog → rat

4.1.9 & 4.1.12
Describe what is meant by a food web.
Draw a food web given appropriate information, containing up to 10 organisms.

© IBO 1996

A food web is the interconnected pattern of food chains in an area.

If you look at example 1a - 1d from the previous section, you will realise that these food chains are interconnected, forming a simple **food web** in the following way:

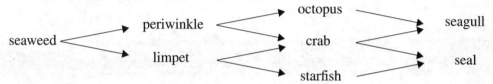

Another example is:

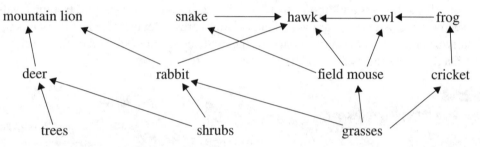

An even more complex example is:

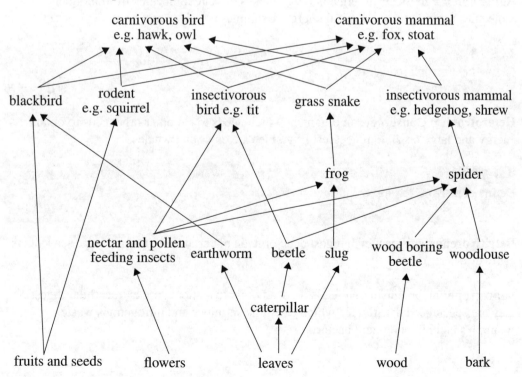

Ecology

Studying the above, you may be able to suggest some additional arrows which were overlooked. If you try to label the animals to their trophic level (see 4.1.10 below), you will see that a secondary consumer can also be a tertiary consumer at other times.

4.1.10
Define trophic level.

© IBO 1996

Trophic level: the level of the food chain at which an organism is found.

4.1.11
Deduce the trophic level(s) of organisms in a food chain and a food web.

© IBO 1996

The first organism in a food chain or food web is known as a **producer**. In other words: producers are the first trophic level. The second trophic level consists of primary consumers (herbivores and/or omnivores) eating the producers. The third trophic levels are the secondary consumers (carnivores and/or omnivores). Beyond this level, organisms can usually be considered to be secondary, tertiary, etc. consumers since they are rarely restricted to one species for food. (see section 4.1.10)

4.1.13
Define autotroph (producer).

© IBO 1996

Autotroph = Producer: an organism that makes organic molecules from inorganic molecules using light or chemical energy. Examples are algae and grass.

4.1.14
Define heterotroph (consumer).

© IBO 1996

Heterotroph = Consumer: an organism which needs to eat other organisms to obtain energy and large organic molecules. Examples are cows and whales.

4.1.15 & 16
Define detritivore & saprotroph (decomposer)

© IBO 1996

Detritivore: organisms which ingest dead organic material. For example large scavengers, eathworms.

Saprotroph: an organism that feeds on dead organic matter using extracellular digestion; they are specialised detritivores which consume cellulose and nitrogenous waste. Examples include, fungi and bacteria.

4.2 PHOTOSYNTHESIS, RESPIRATION AND ENERGY RELATIONSHIPS
PHOTOSYNTHESIS

4.2.1
State that light is the initial energy source for almost all communities.

© IBO 1996

Light is the initial energy source for almost all communities. While materials cycle through the ecosystem, energy enters (autotrophs) and is used or changed into other forms of energy and leaves the ecosystem again.

4.2.2
Describe the fact that photosynthesis involves an energy conversion in which light energy is converted to chemical energy.

© IBO 1996

In the process of **photosynthesis**, light energy is converted into chemical energy. The substances needed for photosynthesis are **carbon dioxide** and **water**, which, in the presence of **sunlight** and **chlorophyll**, can produce **glucose** and **oxygen**. This is far from a one-step reaction as you will see in section 9.2. Some of the energy in the light will be converted into chemical energy in glucose.

4.2.3
State that white light from the sun is composed of a range of wavelengths (colours).

© IBO 1996

The most usual light for photosynthesis is sunlight. Sunlight is white light, made of all colours together. Different colours are actually different wavelengths of light. On one side of the spectrum, you find violet light with the shortest wavelength and the most energy, on the other side red light with the longest wavelength and the least energy. To one side of the visible spectrum (shorter wavelength), the electromagnetic radiation continues as ultraviolet, X rays, etc. To the other side (longer wavelength) we find infrared, radio waves etc.

4.2.4
State that chlorophylls are the main photosynthetic pigments.

© IBO 1996

Most plants are green. The green colour is caused by the presence of the pigment chlorophyll. Chlorophyll is the main photosynthetic pigment. Chlorophyll is green, i.e. it reflects green light and absorbs all other colours. Several different kinds of chlorophyll exist, each with their own specific absorption spectrum.

Ecology

4.2.5
Outline the differences in absorption of red, blue and green light by chlorophylls.

© IBO 1996

Since chlorophyll appears to be green, green light is reflected. From this you can conclude that green light is not absorbed very well. An absorption spectrum can be determined in the following way:

You shine white light through a chlorophyll solution. Some frequencies will be absorbed, others will not. If the remaining light is directed through a prism, you will get the usual spectrum but with some colours 'missing'. These are the colours which were absorbed by the chlorophyll solution. The pattern obtained will look like the diagram below.

Absorption spectrum of chlorophyll

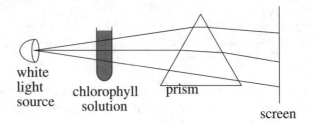

This shows that orange-red light and blue light are mostly absorbed while green light is mostly reflected/transmitted. There is some variation between the different kinds of chlorophyll but the above is true for all.

Not all wavelengths absorbed are used for photosynthesis equally efficiently. As well as creating an absorption spectrum, it is possible to have an action spectrum which shows the amount of photosynthesis at different wavelengths. An action spectrum actually does resemble an absorption spectrum but represents the effectiveness of each wavelength in the process of photosynthesis. The differences found are caused by the fact that not all pigments which contribute to the absorption spectrum, contribute equally efficiently to photosynthesis and hence the action spectrum.

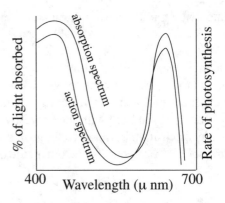

4.2.6
State that light energy is used to split water molecules to produce oxygen and hydrogen, and to produce ATP.

© IBO 1996

In the process of photosynthesis, light energy is used to split water molecules into hydrogen ions, H^+, and oxygen, O_2, also producing adenosine triphosphate, **ATP**. This is called the **light dependent stage**.

4.2.7
State that ATP and hydrogen, (derived from the photolysis of water) are used to fix carbon dioxide to make organic molecules.

© IBO 1996

The H⁺ and ATP thus produced, then drive the formation of glucose from carbon dioxide. Water is formed in this process. This occurs in the **light independent stage**.

4.2.6 and 4.2.7 are summarised in the diagram below.

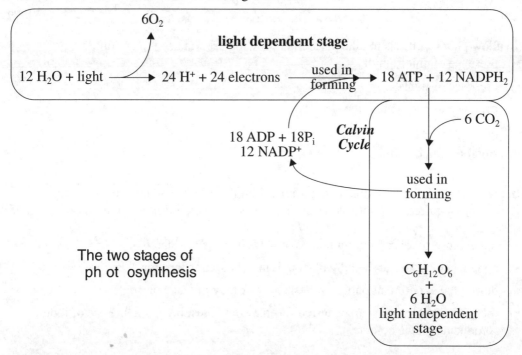

The two stages of photosynthesis

4.2.8
Explain that photosynthesis can be monitored by the production of oxygen, the uptake of carbon dioxide or the increase in biomass.

© IBO 1996

If you want to measure how much photosynthesis is taking place per unit time (minute, hour, decade), you can measure one of the following factors.

1. Since photosynthesis utilised carbon dioxide, it is theoretically possible to place a plant in an enclosed space, measure the available carbon dioxide before and after the experiment. This will tell you how much carbon dioxide was used for photosynthesis. This is a complicated method. It is also possible to allow the carbon dioxide to interact with water, producing bicarbonate and hydrogen. Hence the acidity of the resulting solution will indicate the amount of carbon dioxide present.

2. It is easier to look at the other side of the equation. Photosynthesis produces oxygen and glucose. It is possible to measure how much oxygen a plant produces over time. It

Ecology

is also possible to measure how much heavier a plant is after photosynthesis. We need to make sure that we measure the change in organic matter and not, for example, the change in water content. Therefore, we need to determine the biomass by completely dehydrating (drying) the plant before weighing it. Needless to say, no organism survives this treatment.

4.2.9
Outline the effects of temperature, light intensity and carbon dioxide concentration on the rate of photosynthesis.

© IBO 1996

To allow photosynthesis to take place the following criteria need to be fulfilled:
- presence of chlorophyll
- presence of light
- presence of carbon dioxide
- presence of water
- suitable temperature

In practical work, it is possible to determine if all the above (and maybe some other factors) are required for photosynthesis. The basis of the experiment is the following set of observations:
- the glucose produced during photosynthesis is turned to starch;
- a plant can be 'destarched' by placing it in a dark cupboard for 2 days;
- during the experiment photosynthesis may or may not take place;
- the presence of starch can be tested for in a simple test involving the use of iodine (producing a blue colour).

CO_2 requirement
Two destarched plants are placed in the light but covered with a transparent plastic bag. Under the bag (with one of the plants), a small beaker of soda-lime (NaOH) is placed (this absorbs carbon dioxide from the air). When the plants are tested for starch the next day, the plant with the soda-lime will have had no photosynthesis, while the other plant will show the presence of starch indicating the need for carbon dioxide to allow photosynthesis.

Light requirement
A destarched plant left in a dark cupboard will contain no starch a day later while a similar plant placed in the light will have photosynthesised and starch will be found. Alternatively we can cover part of a leaf of a green destarched plant which is placed in the light and show the absence of starch the next day. So, light is essential for photosynthesis.

Chlorophyll requirement
Using a destarched plant with variegated leaves (green and white) and placing it in the light will show that the white parts contain no starch a day later, showing the need for chlorophyll in the process of photosynthesis.

The factor the furthest away from its optimum value will limit the amount of photosynthesis. This is then the **limiting factor**. If you improve this factor, the rate of photosynthesis will increase until another factor becomes the limiting factor. If you plot a graph of the rate of photosynthesis versus light intensity, the graph will go up until light is no longer the limiting factor. Then, the amount of photosynthesis will remain constant.

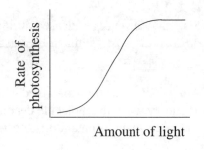

An example of this is found in greenhouses. In the Netherlands, people growing vegetables often do this in greenhouses. They will heat these greenhouses to provide the plants with the optimum temperature. They may switch on lights at night to provide optimum lighting conditions.

Of course, water is readily available in that country and will be provided to the plants. It has been found that adding carbon dioxide to the air in the greenhouse will increase the yield since all other factors are near optimum and carbon dioxide has become the limiting factor.

4.2.10
State that respiration involves the breakdown of organic molecules to release energy stored by photosynthesis.

© IBO 1996

Cellular respiration: breakdown of organic molecules to release energy stored by photosynthesis.

4.2.11
State that the carbon dioxide fixed by photosynthesis is released by respiration.

© IBO 1996

In the process of photosynthesis, carbon dioxide from the air is fixed in glucose molecules. This glucose can be turned into other organic molecules. In the process of cellular respiration, the reverse happens and the carbon dioxide is released again.

One of the causes of global warming is the **greenhouse effect** (the atmosphere behaves like the glass of a greenhouse). Certain gases in the atmosphere will allow light to pass through but stop the heat from radiating out again. This is what the glass panels of a greenhouse do. One of the most significant greenhouse gases is carbon dioxide. Using fossil fuels, we are releasing carbon dioxide into the atmosphere which was taken from it by plants living a long time ago and has been fixed as organic molecules since, for example, as coal and oil.

Ecology

4.2.12
State that the energy released during breakdown/respiration of complex compounds in an organism is used within an organism to do work or is lost as heat.

© IBO 1996

Almost all the energy contained in the chemical bonds of, for example glucose, is released as energy when the molecule decomposes. The products of respiration are carbon dioxide and water, which contain little energy.

Some of the energy released is captured by the reaction ADP + Pi → ATP and the rest is released as heat. The ATP will, in turn, have the energy to do the work needed by the organism.

4.2.13
Define biomass.

© IBO 1996

Biomass: the total mass of organic matter in organisms or ecosystems. Water is not organic matter and so is not included. Synonymous to 'standing crop'.

4.2.14 & 4.2.15
Explain biomass and energy transfer in a food chain in terms of growth, respiration, cell activities and waste.
State that when energy transformations take place, including those in living organisms, the process is never 100% efficient, commonly being 10-20%.

© IBO 1996

In general, when energy is transferred from one trophic level to another, only approximately 10% is used for growth. The rest is used for other metabolic activities and waste.

Example: 100 kg grass → 10 kg antelope → 1 kg cheetah

The efficiency of this process depends on a number of factors. For example a snake eating a mouse is going to be fairly efficient. The mouse is easy to digest and the snake is not active and maintains body temperature by behavioural rather than physiological means.

4.2.16
Explain what is meant by a pyramid of energy and reasons for its shape.

© IBO 1996

If we want to represent some data about the trophic levels in a certain area we can make a pyramid of numbers. Required is information about the number of organisms in each trophic level in the area. For this you will need to decide the trophic level of each species. The width of each level represents the number of organisms at that level. The ideal pyramid of numbers looks like this:

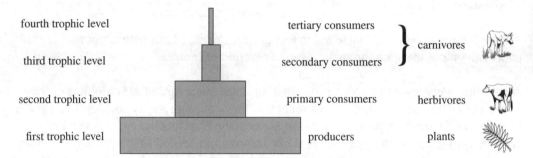

However, if we are studying an insect population in a tree, there will be one producer (the tree) and thousands of primary consumers (the insects). So the pyramid of numbers then looks like this:

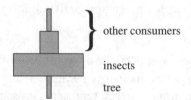

The range of numbers in a pyramid of numbers can be so great that it is difficult to draw the pyramid to scale (one tree, thousands of insects, a few dozen spiders and a few birds). A pyramid of biomass can then give a more meaningful representation of the figures. For a pyramid of biomass, the (dry) mass of a trophic level needs to be measured or estimated. A drawback is that an accurate measurement of the dry mass usually means complete destruction of the organisms involved.

A pyramid of biomass will usually have a shape close to that of the ideal model. Seasonal changes will usually explain any deviations from this shape. For example, a pyramid of biomass in an Italian lake in summer and winter:

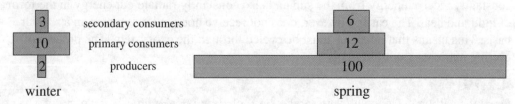

When the producers are small, for example algae, it is likely that their turnover rate is high. This means that their production can equal that of a tree but the amount of biomass is

Ecology

much smaller. This also means that the biomass of the producers can be smaller than that of the primary consumers. A typical example of this is found in the English Channel, where the biomass of the phytoplankton is about one-fifth of the biomass of the zooplankton.

If you find this a difficult concept, see if you can follow this scenario:
Let us assume that one individual of zooplankton needs to eat 2 individuals of phytoplankton every day. Yet every phytoplankton individual can produce 10 offspring a day. So if all offspring are eaten, 1 individual of phytoplankton can sustain 5 individuals of zooplankton.

Although the numbers are 'made up' in this example, it may have helped your understanding of the situation of an inverted pyramid of biomass.

A pyramid of energy will never be 'inverted' in shape since this employs figures of the amount of energy per unit area (or volume) that flows through the trophic level in a given time period. So the high production of the algae in the English Channel is taken into account. A pyramid of energy requires all the information needed for a pyramid of biomass as well as the energy values for the different organisms. To obtain these values, samples of these organisms are destroyed.

4.2.17
Design a pyramid of energy given appropriate information.

© IBO 1996

To construct a pyramid, you need the appropriate information. For a pyramid of energy the values need to be expressed in energy per unit area per unit time, e.g. $kJ/m^2/month$. The lowest bar of the pyramid of energy represents gross primary production. The next bar expresses the energy ingested as food by primary consumers and so on.

4.2.18
Explain that energy enters and leaves an ecosystem, but nutrients must be recycled.

© IBO 1996

As you saw above, the energy stored in products of photosynthesis by autotrophs, is gradually used as these products make their way through the trophic levels. Finally all energy captured in chemical bonds will have been changed into other forms of energy which cannot be passed on. So energy is NOT recycled through the ecosystem. We constantly receive energy from the sun and also constantly radiate out energy in the form of light and heat. The earth, however, does not receive nor send out matter on a regular basis. This means that nutrients must be cycled through the ecosystem. The process of cycling nutrients requires energy.

4.2.19
Draw the carbon cycle to show the processes including photosynthesis, respiration, combustion and fossilization.

© IBO 1996

In outline, plants use carbon dioxide (CO_2) for photosynthesis and produce it again in respiration. Animals eat plants and produce carbon dioxide in respiration.

The action of humans is mainly found in the area of adding CO_2 to the environment by burning fossil fuels. The CO_2 stored in plants and animals hundreds of millions of years ago are now used rapidly to provide us with energy. The CO_2 which has been removed from the cycle for a long time is now rapidly being returned to it. This increases the CO_2 level in the atmosphere and causes the '**greenhouse effect**'. This means that the Earth reflects less heat and absorbs more. The average temperature will go up and result in melting of the ice caps at the poles which increases the level of the seawater and floods areas of land as well as speeding up the process of desertification.

All this is not helped by our practice of burning tropical rain forests for agricultural land. The crops have less photosynthesis than the natural vegetation and will remove less CO_2 from the atmosphere. After a few years the yield will be too low and a new section will be selected. The used land is so depleted of nutrients that very few plants will grow there.

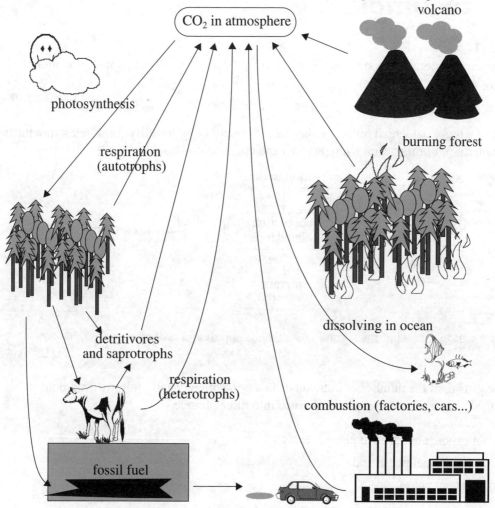

Ecology

4.2.20
Explain the role of saprotrophs (decomposers) in returning elements to the environment in inorganic form.

© IBO 1996

All living things die sooner or later. For some this may be sooner than anticipated if they are eaten, others may live out their full life span. Even then, they produce organic waste during their life. Somehow, the organic nutrients need to be returned to the soil as inorganic minerals to allow the cycle to start again.

Various names are used relating to the group of organisms involved in this process. **Detritivores** (e.g. earthworms), **decomposers** and **saprotrophs** or **saprobes** (e.g. fungi) are generally all used to refer to organisms returning elements to the environment in inorganic form.

4.3 POPULATIONS, NATURAL SELECTION AND EVOLUTION.

4.3.1
Outline how population size can be affected by natality, immigration, mortality and emigration.

© IBO 1996

Four factors exist which influence the size of a population: **natality** (birth rate), **mortality** (death rate), **immigration** ('moving in') and **emigration** ('moving out').

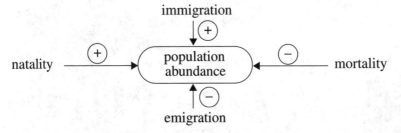

4.3.2.
Draw a graph showing the sigmoid (S-shaped) population growth curve.

© IBO 1996

The growth of the number of individuals in a population usually follows a sigmoid (S-shaped) curve. This curve can be divided into three phases:

- an exponential growth phase,
- a transitional phase and
- a plateau phase.

The exponential growth phase.

All organisms have the ability to increase their numbers rapidly if there are no factors which control them. Without controlling factors, elephants, which are considered to be very slow breeders, can go from one pair to 19 million elephants alive in less than 750 years, even calculating minimal reproduction and life span.

Typically, a bacterium can undergo division (mitosis, asexual reproduction) every 20 minutes. In 2 hours the following happens:

Time (hours:minutes)	0	0:20	0:40	1:00	1:20	1:40	2:00
Number of bacteria	1	2	4	8	16	32	64

In the first 20 minutes, the number of bacteria was increased by 1. In the last 20 minutes of the 2 hour interval, it increased by 32. The graph of this population growth looks like this:

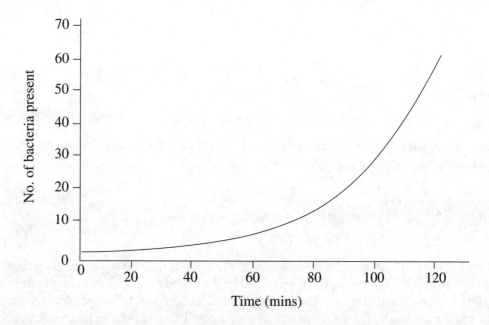

This is an example of the innate capacity for increase of a species where the increase in population size is as rapid as its members can reproduce.

This kind of curve is called a **geometric curve**. Sometimes the growth is called **exponential** (2^n where n is the number of generations).

The transitional and plateau phases.

Of course this situation of exponential growth only occurs occasionally, for example when the environment suddenly changes and only a few individuals survive. The increase in numbers will not continue to be geometric and at a certain point the increase will slow down until it is zero. The number of individuals in the population has then reached the maximum which can be supported by the environment. It has reached its carrying capacity.

Ecology

The result will be intraspecific competition for food, shelter, nesting space, etc. This leads to differential mortality where individuals which are best fitted to the environment survive and reproduce ('survival of the fittest').

So the growth curve of a population is S-shaped or sigmoidal (see diagram below).

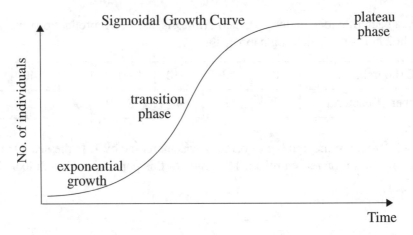

4.3.4.
Define carrying capacity.

© IBO 1996

Carrying capacity: the maximum number of individuals of a species that can be sustainably supported by the environment.

4.3.5
List three factors which set limits to population increase.

© IBO 1996

Many factors can limit population increase. There are several methods of dividing these into categories. Two more common ways will be discussed here. One of them divides the factors limiting population size into density dependent and density independent factors. The other way divides these factors into extrinsic population regulating mechanisms and intrinsic population regulating mechanisms.

The division of factors into density dependent or density independent is fairly simple as can be seen below.

Density dependent factors affect a larger proportion of the population as the density increases. Examples are mortality due to predation and disease, and (intra-specific) competition.

Density independent factors affect a proportion of the population regardless of its density. Examples are mortality due to weather conditions or earthquakes. However, density independent disasters tend to have a greater effect on a population when the density of that population is higher. For example a drought will kill some individuals in a desert population of any size, but the effect will be far worse if the density were greater.

As a result, another system of classifying population-regulating mechanisms can be used. This classification utilises the criterion of whether the mechanism originates outside or inside the population, as can be seen below.

Extrinsic population-regulating mechanisms originate outside the population and include biotic as well as physical factors. For example food supplies, natural enemies, disease, weather.

Intrinsic population-regulating mechanisms originate in an organism's anatomy, physiology or behaviour. Some species (e.g. kangaroos and koalas) can resorb the developing embryo when conditions are crowded and resources limited. Lemmings will migrate (in large groups) when local resources are depleted. Competition is the ultimate intrinsic population-regulating mechanism.

4.3.6
State that populations tend to produce more offspring than the environment can support.
© IBO 1996

Populations tend to produce more offspring that the environment can support.

4.3.7
Explain that the consequences of the potential overproduction of offspring is a struggle for survival.
© IBO 1996

This over-production of offspring leads to intraspecific competition and survival of the individuals best suited to that particular environment.

4.3.8
State that the members of a species show variation (cross reference 3.3.3).
© IBO 1996

Variation between individuals of a species arises by sexual reproduction and/or mutations. Variation is non-directional or random. The selection process is dictated by the environment and leads to **differential survival**.

4.3.9
Explain how, by natural selection, the best adapted will survive to breed.
© IBO 1996

The result is that the individuals best adapted to a particular environment will survive. They will be able to get the most food, find the best shelter, find a mate, reproduce and care for their offspring as well as not be eaten by other species. Please remember that if the environments are different, the 'best adapted' may be different too. Also, environments

Ecology

may change. This can happen gradually or suddenly for example due to natural disasters. As a result the criteria for 'best adapted' will also change.

4.3.10
Discuss the theory that species evolve by natural selection.

© IBO 1996

This process can lead to changes in the species. It can also lead to speciation. When two groups of a species are in different environments and they cannot interbreed, selection pressure will be different and eventually they will become different species (e.g. many of the species of finches in the Galapagos islands).

4.3.11
Discuss the need for evolution in response to environmental change.

© IBO 1996

If a species cannot adapt to the changing environment, then the species will die out. As the dinosaurs did not find a way to deal with the climate becoming colder, they did not survive. Their place was taken by the (homeothermic-warm blooded) mammals.

4.4 HUMAN IMPACT
4.4.1-3
Outline two examples of local or global issues of human impact causing damage to an ecosystem or the biosphere, one of which must be the increased greenhouse effect.
Explain the causes and effects of the two issues in 4.4.1, supported with data.
Discuss measures which could be taken to contain or reduce the impact of these issues, with reference to the functioning of the ecosystem.

© IBO 1996

Humans can make a very large impact on their environment. Some other species also do this, for example beavers, but we have the dubious honour of being the only species to create global environmental issues. Three examples of global issues are discussed here: 'the greenhouse effect', 'the hole in the ozone layer' and 'acid rain'. Each issue will be discussed completely, including what it is (section 4.4.1), its causes and effects (section 4.4.2) and measures to reduce the impact (section 4.4.3). These aspects are discussed in this section for each issue. (See also Section G.6)

The greenhouse effect.
When you want to grow plants in a place or at a time when it really is too cold to grow them, you may decide to build to build a greenhouse. This essentially is a glass house in which to grow plants. During a bright but cold day, the sunlight will enter your greenhouse through the panes of glass. Inside, some of it is changed into heat. The heat is caught inside the greenhouse because its radiation cannot travel out through the glass, instead it is reflected back into the greenhouse. Therefore, on a bright winter day, your greenhouse will be a lot warmer than the outside temperature, even without additional heating systems.

This is also happening in the Earth's atmosphere and has led to the name 'greenhouse effect'.

The greenhouse effect is the combination of factors suspected to cause global warming. The Earth receives energy from the Sun in the form of visible light and UV radiation. This energy passes through the atmosphere like light through glass. The Earth sends out energy, mainly in the form of heat. Some of this heat will not leave the atmosphere but instead be reflected back. Greenhouse gases act like the glass in the greenhouse and are responsible for reflecting the heat back to Earth. The higher the concentration of greenhouse gases, the more heat is prevented from leaving Earth.

Examples of greenhouse gases are:
- carbon dioxide,
- carbon monoxide
- water vapour,
- oxides of nitrogen
- ozone (in the troposphere) and
- CFCs.

Carbon monoxide is a result of the incomplete burning of organic materials (fuel). Ozone is extremely useful in the higher layers of the atmosphere in reflecting UV light but acts as a greenhouse gas in the troposphere. Methane is produced by bacteria living on manure. Oxides of nitrogen are also a result of combustion, as well as resulting from animal waste. Chlorofluorocarbons are found in aerosol sprays, refrigerators and airconditioners.

The greenhouse effect is necessary to maintain life on Earth. If all greenhouse gases were absent, the Earth would be far too cold. The global issue of the greenhouse effect is that human actions increase the greenhouse gases which presumably cause global warming. (Not all scientists agree with this conclusion.)

For millions of years, the lower part of the Earth's atmosphere has contained approximately 0.03% CO_2. The recent problems are largely caused by the rapid increase of CO_2 in the atmosphere. Photosynthesis will reduce the amount of carbon dioxide available but deforestation reduces the amount of photosynthesis. Respiration will add to the amount of carbon dioxide in the atmosphere and so does the burning of fossil fuels. The oceans are

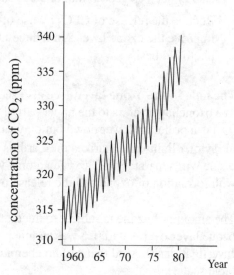

Atmospheric CO_2 concentration at Mauna Loa Observatory (Roberts, 1986)

Ecology

estimated to fix and release 40×10^{12} kg carbon per year.

On land, the plants fix about 35×10^{12} kg per year by photosynthesis. Plants and animals respire 10×10^{12} kg and the decompers release 25×10^{12} kg every year. The burning of fossil fuel adds approximately 5×10^{12} kg yearly. Some scientists estimate that deforestation decreases the carbon fixation by the same amount.

Over the last 30 years, the level of CO_2 has increased rapidly and is expected to be 0.06% around 2050. Together with the increase in the above mentioned gases, this could have a global warming effect of around 3°C. The increase in the amount of water vapour creates its own greenhouse effect and adds to the problems. This could flood, for example, parts of Europe, Bangladesh and many tropical islands as well as causing major shifts in world climate zones. It has been speculated that insects would thrive in this warmer climate and destroy many crops. Yields would be lower anyway because our plants are not adapted to the higher temperatures.

It is well known that the Earth's temperature fluctuates over long periods. The ice ages were times when the average temperature was lower than it is now. Some scientists say that any global warming may well be caused by factors other than greenhouse gases and the recent temperature fluctuations are not over a long enough period to prove anything. However, it seems prudent to try and reduce the release of all of the greenhouse gases as much as possible.

Ways of preventing the release of more greenhouse gases or even the reduction of greenhouse gases are rather obvious. The main issue is to reduce the amount of carbon dioxide as this is the most common greenhouse gas.
- Reduce deforestation especially by burning; attempt reforestation.
- Reduce the use of fossil fuels by increasing the use of alternative energy (renewable energy such as geothermal or solar energy).
- Reduce the release of CFCs by use of alternatives and/or recycling. This will also improve the ozone layer and reduce UV radiation which potentially reduces photosynthesis.

The hole in the ozone layer. (see also section G.6)
The ozone layer refers to the upper part of the Earth's atmosphere containing ozone. Ozone (O_3) is a pollutant lower down but plays a vital role in the upper atmosphere. 99% of the ultraviolet light which strikes the Earth is reflected by ozone. When struck by UV light, ozone will separate into 3 oxygen atoms which are very reactive. They will recombine with formation of ozone and the release of the absorbed energy as heat. ($3\ O_2 \rightarrow 2O_3$).

The amount of ozone is subject to fluctuations over the year. In 1985, it was found that the ozone layer over Antarctica was so much reduced it was called an ozone hole. Investigations showed that certain chemicals were able to destroy ozone molecules on a large scale.

The major cause of the depletion of the ozone layer is **CFCs**. This stands for **chlorofluorocarbons**. These molecules are very useful in refrigerators, airconditioners, fire extinguishers and as propellants in aerosol sprays. They were used in large amounts and were thought to diffuse harmlessly into the stratosphere where they were broken down by sunlight.

This is partly true except that in the process of breaking them down, a chlorine atom is produced. The effect of chlorine on the ozone layer is very serious. The chlorine will react with an ozone molecule and break it apart so that it will not reform. One chlorine atom can destroy 100 000 ozone molecules.

The effect of the reduction of ozone is an increase in the amount of UV radiation which reaches the Earth's surface.

The main effect of UV radiation on living tissues is found in the harm it does to DNA. High levels of UV radiation alter the structure of DNA which eventually may cause (skin) cancer. UV light can be absorbed by organic molecules causing them to dissociate forming atoms or groups with unpaired electrons. These substances are very reactive and can cause unusual reactions to take place.

Phytoplankton are very sensitive to UV light. These organisms at the base of aquatic food chains function less well in even only moderate levels of UV light. Higher levels cause them to die. Since phytoplankton also contribute significantly to the total oxygen production on Earth, damage to these organisms will have far reaching consequences.

Terrestrial plants have been shown to have a lower yield when UV levels are increased. Nitrogen fixing bacteria have been shown to be killed by high levels of UV light.

Again the solutions to this problem are rather obvious. First we must stop the increase in the release of CFCs. Then we can attempt to reduce the amount of CFCs released until it has become very small. The ozone layer will be able to repair itself when the CFCs are removed but it will take time.

In 1990, 93 nations signed an international agreement to phase out the use of CFCs by the year 2000. It will take a century before the ozone layer will be complete again since CFCs are broken down only slowly.

When the refrigerator, freezer or airconditioner is disposed of, it is usually not because the CFCs have disappeared. So it is possible to collect these unwanted items and recycle the CFCs. In many countries, systems have been set up to do this.

Alternatively, producers are looking for alternatives to CFCs and 'green' refrigerators are now available. Although the alternatives tend to be more costly, it does not seem to have a very large effect on the total price of the refrigerator or airconditioner. In many countries, CFCs are no longer allowed to be used in aerosol sprays and alternatives have been found. This is usually indicated on the aerosol can so that the consumers can choose to buy the product for that reason.

Some scientists have reported that the increase in the hole in the ozone layer is slowing down. It is expected to stop growing soon and they are hopeful to be able to show it getting smaller in a next decade.

Acid rain (see also section G.6)
Rain is normally slightly acidic (pH 5.6) due to the presence of carbonic acid (from dissolved CO_2). Over the last few decades, the pH of the rain has become lower than this and a pH between 4.0 and 4.5 has unfortunately become usual. In very severe cases, levels as low as pH 2.1 have been measured.

The presence of sulfuric acid (H_2SO_4) and nitric acid (HNO_3) have caused this increase in acidity. These gaseous oxides will dissolve in the water droplets and come down as precipitation. Sulfur oxides are produced by volcanic eruptions, combustion of high sulfur coal and oil and the smelting of sulfur containing ores. Nitrogen oxides are also produced by volcanic eruptions, petrol combustion in cars and by generating electricity by burning coal, oil or gas.

Many plants cannot tolerate acid rain. Coniferous forests will be damaged, especially the highest parts of each plant, by the direct impact of the acid rain. Plants living on limestone ($CaCO_3$) have some advantage in that the limestone will somewhat buffer the pH of the water and nutrients will leach less and remain available to the plants. Those less fortunate may show reduced germination, decrease in seedling survival, reduced growth and less resistance to disease.

Lakes and streams can be severely affected by this. Especially lakes at high elevations will suffer the consequences. A study in 1977 showed that many high altitude lakes had a pH of 5.0 or less and that most of these no longer sustained life. Pictures of healthy trout and those coming from acidic streams show a clear difference in size and general appearance with those from acidic lakes being smaller.

Reducing emission of gases that promote acid precipitation can be done by, for example, reducing flue-gas (emissions resulting from the burning materials) as well as fuel desulfurisation (removal of sulfur) and the use of alternative energy sources.

4.5 ECOLOGICAL TECHNIQUES

In the IB examination, you are allowed to use your calculator in paper 2 and 3 but NOT in paper 1. So do practise the statistical calculations below using your calculator but also remember that you also have to understand what you are doing (rather than just knowing which buttons to press).

4.5.1
Describe one method used to measure each of three abiotic characteristics of a habitat including light.

© IBO 1996

The environment determines which species can survive in an certain area. Abiotic factors play an important role of this. In our study of ecology, we often want to gather information about abiotic factors such as light intensity, temperature and the availability of water.

Light intensity can be measured with a light meter. It can tell you how much light a certain place has at a certain time. When considering the light as an abiotic factor, you would also want to know the length of the day (duration of light) as well as the wavelengths/colours of light.

Temperature is measured with a thermometer. Other aspects to include in this study would be the differences between day and night temperature and the differences between summer and winter temperature.

The availability of water depends on the amount of rain. You would be interested in the annual rainfall (which could be measured with a rain meter or a funnel and a measuring cylinder) as well as the spread over the year. Also the structure of the soil will determine how much water will run off into streams or will be retained in the soil.

4.5.2
Define random sample.

© IBO 1996

Since statistics are a useful tool in ecology, we will introduce some simple aspects here.

If we want to study some individuals to draw some conclusions about the entire population, we want to take a random sample.

random sample: a method to ensure that every individual in a population has an equal chance of being selected.

4.5.3
Describe one technique used to estimate the population size of one animal species based on a capture-mark-release-recapture method.

© IBO 1996

Ecology

If we want to know the size of a certain population, we can use the techniques known as 'capture - mark - release - recapture'.

This sounds a lot more complicated that it really is. What you will do is you will capture a number of organisms (random sample) and mark them (without harming them or changing their behaviour). You then release them back into the original population. The assumption is that they will mix with the unmarked individuals in a random way.

After a suitable time you will again capture a random sample. A certain proportion of your second sample will be marked. This is the same proportion as the original first (marked) sample was to the entire population.

This techniques assumes that natality, mortality, immigration and emigration are zero.

There are 3 ways of understanding this. They are all related and you may understand all three. If not, you should choose the one that you find easy to remember.

In every case:

n_1 = number in the first sample captured and marked

n_2 = total number in the second sample

n_3 = number of marked individuals in the second sample

1. The proportion that n_3 is to n_2 is equal to that of n_1 to the entire population. In an equation this becomes: $\dfrac{n_3}{n_2} = \dfrac{n_1}{\text{total population}}$

2. From the above, you get the following equation. If you simply memorise this equation, you can work with data of the capture-mark-release-recapture method,

    population size = $\dfrac{n_1 \times n_2}{n_3}$

3. A visual approach can also be used:

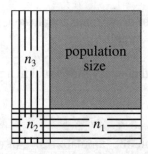

You can see that this is really no different from 1. The first sample (n_1) is a proportion of the entire population but at the time of sampling, you do not what proportion. When you have taken the second sample (n_2), you can see that the marked section of this sample (n_3)

is a certain proportion of the second sample (n_2). You assume that this proportion is the same for n_1 to the entire population.

EXAMPLE
In a capture/release study on a small island, 100 squirrels were captured, tagged and released. Three weeks later another 150 squirrels were captured. Of these, 37 were tagged. Estimate the population of squirrels on the island.

SOLUTION

n_1 = number in the first sample captured and marked = 100

n_2 = total number in the second sample = 150

n_3 = number of marked individuals in the second sample = 37

$$\text{population size} = \frac{n_1 \times n_2}{n_3} = \frac{100 \times 150}{37} = 405.4$$

There are about 400 squirrels on the island.

4.5.4
Describe one method of random sampling used to compare the population numbers of two plant species based on quadrat methods.

© IBO 1996

A very simple tool which is often used in estimating plant populations is the **quadrat**. A quadrat is a wire, shaped into a square of a known size, for example 0.25 m².

When you want to know the population size of one or more plant species in a uniform area, you can take random samples of this area by throwing the quadrat and investigating the area in the quadrat.

When the area shows variation (slope or seashore), then the quadrat can be used together with a **line transect**. You take some string to make a line across the area (across the variation). You can either investigate all plants touching the line or you can place your quadrat at regular intervals and investigate the area within the quadrat.

4.5.5
Evaluate graphical presentations of ecological data.

© IBO 1996

Graphical presentations of ecological data need to be studied from practical work. See the Biology portfolios.

Ecology

4.5.6
Define mean.

© IBO 1996

The **mean** is the 'average' value obtained by dividing the total of a set of values by the number of values. So a series of test grades could be:
$$72\%, 63\%, 98\%, 87\%, 95\%$$
$$\text{mean: } (72\% + 63\% + 98\% + 87\% + 95\%)/5 = 415\%/5 = 83\%$$

4.5.7
Define mode.

© IBO 1996

The **mode** is the most frequent value in a set of observations. So a series of marks for practical work could be:
$$2, 3, 3, 2, 1, 1, 2, 1, 3, 3, 3$$
The mode here is 3.

4.5.8
Define median.

© IBO 1996

The **median** is the central value in a set of observations arranged in order.
Using the above examples the median in the test grades would be 87% and in the practical marks the median would be 2.

$$1\ 1\ 1\ 2\ 2\ \boxed{2}\ 3\ 3\ 3\ 3\ 3$$

4.5.9
State that the term standard deviation is used to summarise the spread of variables around the mean and that 68% of the values fall within one standard deviation of the mean (plus and minus).

© IBO 1996

The **standard deviation** is used to summarise the spread of variables around the mean. 68% of the values (of a normal distribution) fall within one standard deviation of the mean (plus and minus).

The mean is calculated as follows: $\bar{x} = \dfrac{\sum fx}{n}$

where
- $\bar{x}$ = mean
- x = class value
- f = frequency
- n = number of values

Since the standard deviation is a measure for the spread of the variables, we start the calculations by subtracting the mean from each value.

Then we square these numbers, add them and divide by the total number of values −1. The number acquired this way is called the variance.

Using sample standard deviation assuming we have a sample and want to make assumptions about the entire population. Since you are most likely to have missed the extremes of the population in your sample, your are dividing by $n-1$ rather than n.

$$\text{variance} = \frac{\sum (x - \bar{x})^2}{n - 1}$$

The standard deviation is the square root of the variance.

$$\text{Standard deviation} = \sqrt{\frac{\sum (x - \bar{x})^2}{n - 1}}$$

Another way of calculating the standard deviation more directly is using the following formula:

$$\text{Standard deviation:} \sqrt{\frac{\sum fx^2 - \frac{(\sum fx)^2}{n}}{n - 1}}$$

EXAMPLE
The number of eggs in the nests of a sample of a species of bird is shown below. Find the mean and sample standard deviation of these numbers of eggs.

| 5 | 3 | 5 | 3 | 4 | 2 | 0 | 2 | 1 | 2 |

SOLUTION
The mean is found by adding the data to get 27. The mean is now found by dividing by the number of items of data (10) to get a mean of 2.7. Note that, whilst no nests had 2.7 eggs, the decimal place would, for example, allow us to predict that 100 nests should contain $100 \times 2.7 = 270$ eggs.

Direct entry of the data into a graphic calculator (in this case a Texas TI-83) gives an immediate answer:

First use the STAT mode and enter the data as a list.

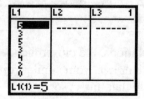

Mean

Standard deviation

The median can be found by scrolling down.

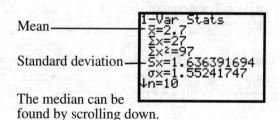

Ecology

To calculate the standard deviation using the formula $\sqrt{\dfrac{\sum(x-\bar{x})^2}{n-1}}$, we could use a table:

Number of eggs	$x - \bar{x}$	$(x - \bar{x})^2$
5	2.3	5.29
3	0.3	0.09
5	2.3	5.29
3	0.3	0.09
4	1.3	1.69
2	-0.7	0.49
0	-2.7	7.29
2	-0.7	0.49
1	-1.7	2.89
2	-0.7	0.49

The sum of the third column gives $\sum(x - \bar{x})^2 = 24.1$

The sample standard deviation $= \sqrt{\dfrac{\sum(x - \bar{x})^2}{n - 1}} = \sqrt{\dfrac{24.1}{10 - 1}} = 1.6363917$

4.5.11
Describe how standard deviation is useful in comparing the means and spread of ecological data between two or more sites.

© IBO 1996

Since 68% of all values lie within the range of the mean plus or minus one standard deviation (for normally distributed data). This rises to 95% for the mean plus or minus two standard deviations.

A small standard deviation indicates that the data is clustered closely around the mean value. A large standard deviation indicates a wider spread around the mean.

This group of Moorish idols (reef fish) are very similar to each other. Their weights could be expected to have a small standard deviation.

When comparing two samples from different populations, the closer the means and standard deviations, the more likely the samples are drawn from similar (or the same) population. The bigger the difference, the less likely that this is so.

Smaller samples will create a variation simply by random factors in the individual values. A sample of population A of 3 plants may give the values of 10 cm, 20 cm and 30 cm high. A sample of population B may give the values of 20 cm, 20 cm, 20 cm. The mean of the sample size will be the same for the populations but the standard deviations will differ. However, if one of the plants in population B had measured 40 cm, the results would have been very different.

Two conclusions drawn from the above are:
1. considering only the means may give a distorted picture
2. small samples are unreliable

EXERCISE

1. A community and its abiotic environment is known as:

    A   a population.

    B   an ecosystem.

    C   a habitat.

    D   a community.

2. A species is a group of organisms that:

    A   live in a community.

    B   do not eat each other.

    C   share the same habitat and look similar to one another.

    D   can interbreed, producing fertile offspring.

3. The food chain: leaf litter → earth worm → sparrow implies that:

    A   earthworms eat leaf litter.

    B   sparrows feed on leaf litter.

    C   leaf litter is essential food for earthworms.

    D   sparrows eat only earthworms.

4. The biosphere is best defined as:

    A   a habitat containing a group of unrelated species.

    B   a community and its abiotic environment.

    C   the physical area in which individuals of a certain species can usually be found.

    D   everywhere living creatures are found.

Ecology

Questions 5 & 6 refer to the following diagram:

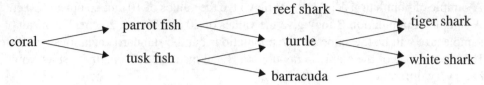

5. The diagram is best described as:

    A   a food pyramid.

    B   a food chain.

    C   a food web.

    D   a species diagram.

6. The diagram indicates that:

    A   turtles are occasionally eaten by tiger sharks.

    B   barracuda eat coral.

    C   tusk fish eat noting but coral.

    D   white sharks will eat anything.

7. The level of the food chain at which an organism is found is known as its:

    A   consumer level.

    B   producer level.

    C   web level.

    D   trophic level.

8. Seaweed and grass are known as

    A   autotrophs.

    B   saprotrophs.

    C   heterotrophs.

    D   detritivores.

9. Parrots and eagles are known as:

    A   autotrophs.

    B   saprotrophs.

    C   heterotrophs.

    D   detritivores.

10. Which one of the following graphs best represents the rate of photosynthesis in a green plant as light levels are altered?

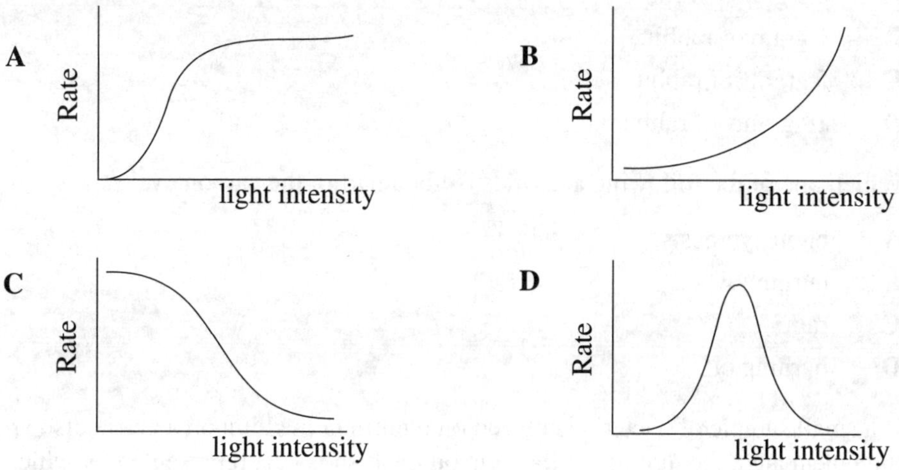

A

Rate vs light intensity

B

Rate vs light intensity

C

Rate vs light intensity

D

Rate vs light intensity

11. Which one of the following is not required for effective photosynthesis?

    A   chlorophyll.
    B   light.
    C   starch.
    D   carbon dioxide.

12. The main products of respiration are:

    A   energy, carbon dioxide and water.
    B   starch and oxygen.
    C   energy and glucose.
    D   ATP and energy.

Questions 13 & 14 refer to the following diagram:

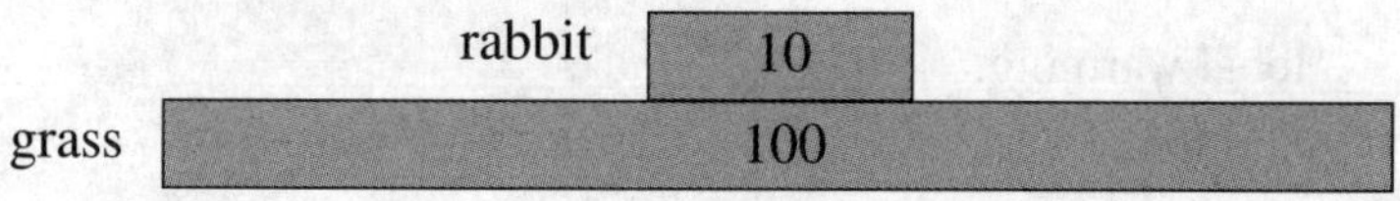

rabbit	10
grass	100

13. The diagram is known as

    A   a food pyramid.
    B   a food chain.
    C   a food web.
    D   a species diagram.

Ecology

14. The diagram indicates if a rabbit eats 50grams of grass, this will be turned into:

    A   0.5 grams of rabbit.
    B   1 gram of rabbit.
    C   5 grams of rabbit.
    D   10 grams of rabbit.

15. Which one of the following activities is not a part of the carbon cycle?

    A   photosynthesis.
    B   burning wood.
    C   rain.
    D   burning oil.

16. If a small sample of bacteria is placed on a nutrient medium on a petrie dish and then incubated, the number of bacteria on the dish is best represented by which of the following graphs?

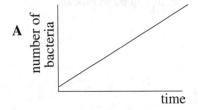
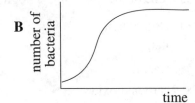
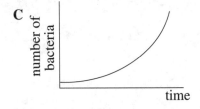
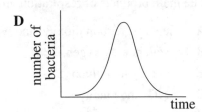

17. The main cause of the greenhouse effect is:

    A   global warming.
    B   pollution.
    C   burning of fossil fuel.
    D   increased human population.

18. The main cause of the 'ozone hole' is thought to be:

    A   the greenhouse effect.
    B   the release of chlorofluorocarbons into the atmosphere.
    C   the release of carbon dioxide into the atmosphere.
    D   the release of ozone into the atmosphere.

19. In a study of a bird population on a small island, 500 birds were captured, ringed and released. Later, a sample of 200 birds were captured. Of this second sample, 14 were found to be ringed. These results suggest that the population of birds on the island is closest to:

A   35
B   7100
C   12 600
D   1 400 000

20. The numbers of eggs produced in each laying by a group of loggerhead turtles had a mean of 90 and a standard deviation of 7. These data mean that:

A   nests containing 130 eggs would be very common.
B   nests containing 130 eggs would be quite common.
C   nests containing 130 eggs would be very rare.
D   nests containing 130 eggs would never be observed.

21. Find the mean and standard deviations of the following data sets. Give your answers correct to one decimal place:

a.
54	51	57	57	58	57
59	55	55	61	52	55
54	56	61	52	60	56
55	53	54	58	56	60

b.
136	135	132	140	134	130
130	131	133	132	140	136
144	140	138	138	128	131
132	129	138	144	131	135

c.
1318	1325	1395	1278	1316	1381
1377	1387	1360	1298	1379	1296
1394	1345	1310	1344	1311	1392
1396	1310	1326	1350	1289	1407

22. The capture-recapture method requires that a random sample of animal is captured and marked (the 'initial capture'). Later a second group is captured (the 'second capture'). Of these, some will be marked (the 'number marked'). For the following capture data, estimate the total population of animals.

	initial capture	second capture	number marked
a.	5000	4000	1256
b.	350	300	88
c.	250	500	123
d.	800	1000	35

HUMAN HEALTH AND PHYSIOLOGY

5

Chapter contents

- Digestion and nutrition
- The transport system
- Defense against infectious disease
- Gas exchange
- Homeostasis
- Reproduction

5.1 DIGESTION AND NUTRITION

5.1.1
Explain why digestion of large food molecules is essential (cross reference topic 2).

© IBO 1996

Large food molecules need to be digested before being absorbed. The process of absorption of food molecules requires them to pass into a cell lining the gut. To do so, molecules must be small and soluble. Therefore large molecules like polysaccharides, proteins and lipids need to be broken down into their building blocks before they can be absorbed.

5.1.2
Explain the need for enzymes in digestion (cross reference topic 2).

© IBO 1996

Like many other biological reactions, the breakdown of these large molecules is a slow process. Enzymes lower the required activation energy and make this process sufficiently fast.

5.1.3
State the source, substrate, products and optimum pH conditions for one amylase, one protease and one lipase.

© IBO 1996

The three kinds of macronutrients are carbohydrates, lipids and proteins. Since enzymes are highly specific, each molecules has at least one enzyme to break it down into its components. But as we can recognise three kinds of macronutrients, we can recognise three groups of enzymes:

- the amylases digest carbohydrates,
- the proteases digest proteins and
- the lipases digest lipids.

Examples of each are given below.

enzyme	salivary amylase	pepsin	phospholipase A2
source	saliva	gastric juice	pancreatic juice
substrate	starch	proteins	phospholipid
product	maltose	polypeptide	glycerol, phosphate, fatty acids
optimum pH	7-8	2-3	8

Saliva is produced by the salivary glands.

Gastric juice is produced by cells in the wall of the stomach; the pepsin producing cells are called **chief cells**.

5.1.4
Draw a diagram of the digestive system including mouth, oesophagus, stomach, small intestine, large intestine, anus, liver, pancreas, gall bladder.

© IBO 1996

Schematic diagram of the human digestive system (alimentary canal and associated glands).

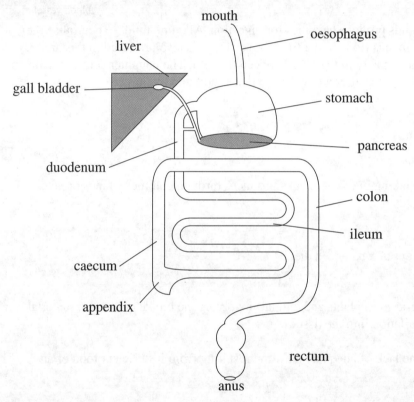

5.1.5
Draw a villus in vertical section.

© IBO 1996

The **small intestine** is a very important place for chemical digestion but it is also the major site of absorption. For this reason the wall of the small intestine is not just simply a smooth inner surface of a tube, as you may have imagined from the schematic and simplified diagrams, but covered with **villi** (sing. villus), which are small fingerlike projections made of many cells. The cells of the villus often also have projections into the lumen of the gut: **microvilli** (compare with a plant root hairs). The purpose of villi and microvilli is to increase the surface area of the small intestine.

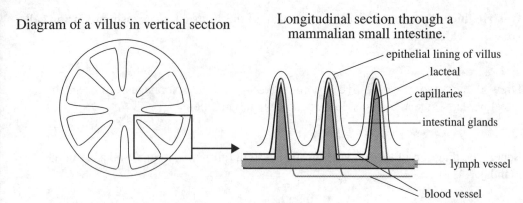

The intestinal glands produce enzymes for digestion. Alkaline fluid (to neutralise the gastric acid) and mucus (to help the food move along) is also produced in this area. Mitosis takes place mostly in the cells between the villi. The daughter cells move up and reach the tip of the villus after a few days and are shed into the lumen.

5.1.6
Define absorption.

© IBO 1996

Absorption: the taking in of chemical substances through cell membranes or layers of cells.

5.1.7
Explain the concept of a balanced diet.

© IBO 1996

Nutrition: the process in plants and animals involving the intake of nutrient materials and subsequent assimilation into the tissues.

Malnutrition: the lack of adequate nutrition resulting from insufficient food or an unbalanced diet.

A **balanced diet** is one which contains the right amounts of macronutrients and micronutrients to satisfy all the body's need.

A balanced diet (or complete diet) contains the following groups of substances:
- carbohydrates
- proteins
- fats
- vitamins
- minerals

Carbohydrates, proteins and fats are needed in bulk and called **macronutrients**. **Micronutrients** are needed in smaller quantities and are minerals and vitamins. Water and roughage (dietary fibre) are sometimes considered to be part of a complete diet but have no nutritional value.

A balanced diet can be obtained by combining energy rich foods (carbohydrates and fats), body building foods (proteins) and protective foods (vitamins and minerals) in each meal. Of course the quantities of each should be sensible and the total energy intake per day should be no more than the body uses.

Children up to approximately 12 years need less food than adults since their bodies are smaller. From 12 to 15-18 years, young people need to supply their growing bodies with more food than they will need later. Old people are usually less active and need less food. Generally the higher the level of activity, the more food the person needs. Also in general, men need more food than women. Pregnant and lactating (breast feeding) women need up to 25% more food than they would at other times.

5.1.8
Explain the general importance of vitamins and minerals.

© IBO 1996

Vitamins are a group of molecules which cannot be made by the body. Vitamins often function as coenzymes. Some minerals are a structural part of the body (Ca and P for bones/teeth), others are part of other molecules (I in thyroxine, Fe in haemoglobin/cytochromes)

5.1.9
Outline one health problem concerned with malnutrition.

© IBO 1996

Anorexia nervosa is a condition where the person refuses to eat adequate amounts of food. It usually is brought on by dieting. The person often continues to perceive herself/himself as fat, even when this is not the case. Complex psychological motives can be involved.

After almost all of the available reserve food has been used up, the body will start to use protein as a source of energy. Eventually, the heart muscles will waste away which results in death. Obviously severe weight loss is a symptom of anorexia, in females the absence of a monthly cycle can also be a symptom.

Too much or too little food will create a non-balanced diet. But also the wrong combination of food can be a non balanced diet. In a western society, people generally eat too much fat and salt and not enough fresh fruit and vegetables (fibre). Problems with blood cholesterol levels and cancer of the colon are suspected to be related to diet.

Human Health and Physiology

Other problems associated with malnutrition are:

nightblindness	lack of retinol (vit A)	rods lack visual pigment
beri-beri	lack of thiamine (vit B_1)	general weakness
anaemia	lack of cyanocobalamin (vit B_{12})	malformed erythrocytes
scurvy	lack of ascorbic acid (vit C)	bleeding gums
rickets	lack of calciferol (vit D)	deformed bones
	lack of Ca/P	
goitre	lack of iodine	enlarged thyroid
hypothyrodism	lack of thryroxine	lethargy
kwashiorkor	lack of proteins	apathy, impaired growth, skin ulcers
constipation	lack of dietary fibre	.....

5.2 THE TRANSPORT SYSTEM

5.2.1
Describe the action of the heart in terms of collecting of blood, pumping of blood and opening and closing of valves.

© IBO 1996

A major component of the heart is muscle tissue. The heart collects blood (from the body and the lungs) and pumps it under pressure through the body and the lungs. The movement of the blood is regulated by valves in the heart.

5.2.2
Draw a diagram of the heart showing all four chambers, associated blood vessels and valves.

© IBO 1996

Mammals have a double circulation, i.e. the blood coming from the body goes through the heart, to the lungs, back to the heart and then to the body. The heart is therefore divided into a right and a left side. Each side has an atrium and a ventricle. The atria collect the blood and their walls are thin; the walls of the ventricles are thick and muscular since they pump the blood into the arteries. The left ventricle has a thicker wall than the right ventricle since it has to pump the blood through the entire body.

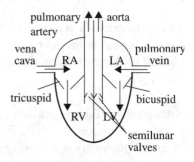

If you look at a more detailed diagram of the heart, you will find that the vena cava really exists of two vessels: the vena cava superior, from above the heart, and the vena cava inferior, coming from below the heart. The pulmonary vein is really 4 veins, 2 from each lung.

The direction of the blood flow is controlled by the valves. Valves are flaps of tissue without muscle. Valves are opened and closed by the flow of blood. As long as the blood flows in the right directions, the valves remain pushed open. As soon as the blood starts to flow in the wrong direction, they are pushed shut.

It may help to consider valves as doors. If a group of people are pushing against an outward opening door, the door will be pushed open. However, if these people suddenly turn around and try to go back, they will only push the door shut.

The atrio ventricular valve between the right atrium and the right ventricle is called the tricuspid valve; the one between the left atrium and left ventricle is called the bicuspid valve. The valves in the pulmonary artery and in the aorta are semilunar valves. The valves have some strings of connective tissue attached to them. This prevents them from opening to the wrong side.

5.2.3
Outline the way the heart beats and is regulated in terms of its myogenic nature, nerve and hormone stimulation (cross reference 5.5).

© IBO 1996

The contractions of the cardiac muscles are brought about by nerve impulses which originate not from the brain but from a specific region of the right atrium: the **Sino Atrial Node (SAN)**. The SA node is made from specialised muscle cells. The SA node releases an impulse at regular intervals which spreads across the walls of the atria, causing simultaneous contractions. The impulse cannot spread to the muscles of the ventricles except in the region of the **Atrio Ventricular Node (AVN)**. The AV node is connected to the bundle of His (specialised cardiac fibres) which branches out into the **Purkinje tissue**. From the AV node, the impulse travels through the **bundle of His** down to the apex of the heart and from there spreads up through the Purkinje tissue. This causes the ventricular contractions to start at the apex and push the blood up into the arteries.

Although the heart is largely autonomous in its contractions, the brain and some hormones can influence the frequency of the heartbeats. Impulses from a nerve from the sympathetic nervous system will increase the heart rate, messages from the vagus nerve (part of the parasympathetic system) will decrease the cardiac frequency.

Some hormones also have an effect on the heart rate: **adrenalin** (epinephrine) increases cardiac frequency.

If the SA node does not function properly, it is quite easy to implant an artificial pacemaker to carry out this function. With a well adjusted pacemaker, a person with a malfunctioning SA node can live a long and active life.

5.2.4
Explain the relationship between the structure and function of arteries, capillaries and veins.

© IBO 1996

Human Health and Physiology

Blood is circulated through the body by the contractions of the heart. Mammals (contrary to for example insects) have a closed circulatory system, i.e. the blood is confined to bloodvessels. There are three kinds of bloodvessels:

arteries: move blood away from the heart.
have thick muscular walls but no valves.
move blood at high speed (10-40 cm/s).
contain blood at high pressure (80-120 mm Hg).

capillaries: small numerous bloodvessels in tissue.
very thin walls.
move blood at low speed (< 0.1 cm/s).
contain blood at moderate pressure (15 mm Hg).

veins: move blood towards the heart (except hepatic portal vein).
have quite thin walls and valves.
move blood at moderate speed (5-20 cm/s).
contain blood at low pressure (< 10 mm Hg).
have valves to prevent blood from flowing back.

5.2.5
State that blood is composed of plasma, erythrocytes, leucocytes and platelets.

© IBO 1996

By weight, about 8% of the human body is blood. It has the following composition: plasma (50 - 55%) and cells (45 - 50%).

The plasma is 90% water and dissolved in it are:
- proteins (e.g. fibrinogen, globulins, albumins).
- dissolved food.
- hormones.
- waste materials.

The cells are:
- erythrocytes (red blood cells) (90%).
- leucocytes (white blood cells).
- thrombocytes (platelets).

5.2.6
State that the following are transported by the blood: heat, nutrients, oxygen, carbon dioxide, hormones, antibodies, waste products.

© IBO 1996

The function of the circulatory system is transport. The following are transported by the blood:

heat, nutrients, oxygen, carbon dioxide, hormones, antibodies and waste products.

Transport of oxygen and carbon dioxide is associated with the eryothrocytes (section H.6); all the others are dissolved in the blood.

5.2.7
Outline one health problem concerned with disorders of the transport system.

© IBO 1996

Several health problems can be associated with the transport system:
Coronary heart disease and heart pacemakers (see section 5.2.3) are the best known.

Coronary heart disease is a partial or complete blockage of some of the coronary arteries. Since the coronary arteries supply the muscle tissue of the heart with food and oxygen, a blockage may cause damage to part of the muscle tissue which results in improper functioning of the heart. Angioplasmy and bypass operations are two medical solutions to coronary heart disease.

5.3 DEFENSE AGAINST INFECTIOUS DISEASE

5.3.1
State that disease can be caused by a variety of organisms.

© IBO 1996

Disease: any change, other than injury, which interferes with the normal functioning of the body.

A variety of organisms can cause disease. Disease causing organisms are called pathogens. Some are listed below.

Viral diseases:
Chicken pox, Common cold, German measles = Rubella, Mononucleosis, Influenza, Measles, Mumps, Viral pneumonia, Poliomyelitis.

Bacterial diseases:
Diphteria, Meningitis, Pneumonia, Scarlet Fever, Syphilis, Tetanus, Tuberculosis, Whooping cough.

Rickettsiae: (prokaryotes which grow only in living cells)
Typhus (carried by fleas and lice).

Protozoa: (unicellular eukaryotic heterotrophs)
Plasmodium causes Malaria.
Entamoebe histolytica causes Amoebic dysentery.
Trypanosoma causes African sleeping sickness.

Human Health and Physiology

5.3.2
Explain how skin and mucous membranes act as barriers against microbes.

© IBO 1996

The best way to prevent disease is to prevent the pathogens from entering the body. The skin plays a major role in this. When unbroken, it is almost impossible for any microorganism to penetrate. Weak points are those where we are not protected by skin. Most of these areas have defenses of their own. Mucus is an often used barrier. It traps microorganisms and prevents further entry.

Lungs: mucus and cilia which transport the mucus to the throat.
Stomach: very acid environment.
Eyes: tears contain lysozymes (enzymes which destroy bacterial cell walls).
Vagina: mucous and acidic environment.

5.3.3
Outline that phagocytic leucocytes ingest disease causing organisms in the blood and in body tissues.

© IBO 1996

Despite all of the above measures to keep pathogens out of the body, many of them do manage to get in. This is called infection (a successful invasion of the body by pathogens) but does not always lead to disease. Leucocytes (white blood cells) are the body's defense against pathogens after they have entered. They can be found in the blood but also in the body's tissues, e.g. lungs. Several different kinds of leucocytes exist, some of which are phagocytic, i.e. they simply will 'eat' (**phagocytosis**) any cell which is not recognised as 'body own' (determined by the 'code' on the outside of the cell surface membrane)

5.3.4
State the difference between antigen and antibody.

© IBO 1996

Antigen: a molecule recognised as foreign by the immune system.
Antibody: a globular protein that recognises an antigen (also known as 'immunoglobulin')

5.3.5
Explain antibody production.

© IBO 1996

Antibodies have several ways of dealing with antigens. One common method is to gather several antigens and lump them together. This interferes with the microorganism's usual life cycle and makes them a better target for the phagocytic leucocytes.

Antibodies are produced by the **B-lymphocytes** or **B-cells**. These are one kind of leucocytes (white blood cells). All blood cells are produced in the bone marrow. B-lymphocytes also differentiate in the bone marrow before moving to the lymph nodes.

When an antigen has entered the body, a phagocytic leucocyte will ingest the invader and travel to a B-cell in the lymph nodes. Here the phagocytic leucocyte will present the antigen to the B-cell. The presence of another leucocyte (a T helper cell) will then cause the B-cell to clone itself many times. A few of these cloned cells will remain as memory cells but the majority differentiate into plasma cells. Plasma cells secrete large amounts of (only one kind of) antibodies which is released into the lymph, which drains into the blood.

Memory cells are kept so that the antibody production will be faster at the next invasion of the same antigen.

A B-cell, and its subsequent clones, will be able to produce only one kind of antigen.

T-cells also originate from the bone marrow but travel to the thymus in an immature state and mature there. Two different kinds of T-cells develop: T helper cells and cytotoxic T cells. The T helper cells play a role in the production of antibodies by B-cells. T helper cells interact with the phagocytes and the B-cells to come to the production of the correct antibodies. Cytotoxic T cells directly kill pathogens in what is known as the **cell-mediated response**.

5.3.6
Outline the effects of HIV on the immune system.

© IBO 1996

Acquired Immune Deficiency Syndrome (**AIDS**) is caused by the Human Immunodeficiency Virus (**HIV**). People can become infected with the HIV virus in several ways:
- sexual contact.
- blood contact: contaminated hypodermic syringes.

  transfusion of blood/blood products.

  mother to foetus via placenta.
- mother to child via breast milk.

When the HIV virus infects a person, the virus will specifically infect and destroy the T-helper cells. This interferes with the specific defense (see section 5.3.2) and often leads to a number of 'opportunistic diseases' like a rare form of pneumonia and skin cancer.

Human Health and Physiology

5.4 GAS EXCHANGE

5.4.1
Describe four features of alveoli that allow them to carry out gas exchange efficiently.
© IBO 1996

Four features of **alveoli** that allow efficient gas exchange:
- large surface area.
- thin (short diffusion distance).
- moist (gases need to dissolve before passing membranes).
- good blood supply (to maintain concentration gradient).

5.4.2
Explain the necessity for a ventilation system.
© IBO 1996

One of the characteristics of life is respiration. Respiration in this case has the biological meaning of the release of energy. This means that cellular respiration goes on in every living cell. Since anaerobic respiration releases about 5 - 7% of the energy that aerobic respiration releases (see section 9.1 or C.5), most cells require a constant supply of oxygen to function properly. (A few species of bacteria, living in the mud at the bottom of ponds and lakes, cannot tolerate oxygen; they employ alternative pathways to release energy.)

Unicellular organisms and small multicellular organism have few problems in gaseous exchange. The required gases will diffuse into and out of their system. For larger organisms this is not possible due to their smaller surface area over volume ratio. The decrease of the surface area over volume ratio is quite rapid as the size of an organism increases.

Two examples to convince you:
Example 1: If you study the cubes below, you can see the 4 shaded sides.

This is the amount of surface area which is 'lost' if you change these 3 small cubes into 1 larger structure, whilst no volume is lost.

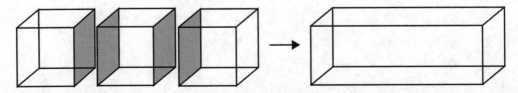

Example 2: Again study the cubes below.

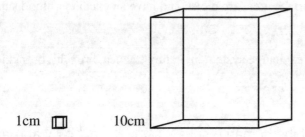

The smaller cube:
The surface area is $6 \times 1 \times 1$ cm^2 = 6 cm^2
The volume is $1 \times 1 \times 1$ cm^3 = 1 cm^3
The surface area over volume ratio is $\frac{6}{1} = 6$

The larger cube:
The surface area is $6 \times 10 \times 10$ cm^2 = 600 cm^2
The volume is $10 \times 10 \times 10$ cm^3 = 1 000 cm^3
The surface area over volume ratio is $\frac{600}{1000} = 0.6$

This means that the smaller cube has 6 cm^2 of surface for every cm^3 of volume, whereas the larger cube has 0.6 cm^2 of surface for every cm^3 of volume. (The mathematical explanation is that which increasing size of factor y, the surface increases with y^2 but the volume increases with y^3).

So, when organisms become larger, there is simply not enough surface for gaseous exchange. An added problem is that the oxygen, once inside the organism, has to travel a long way to reach some cells. Since diffusion in liquids is a fairly slow process, this is unsatisfactory.

When the size of the organism is limited, the above problems can be solved by flattening the body (e.g. flatworms). This increases the surface area and decreases the diffusion distance. However, in larger organisms, even this measure is insufficient and a need for a respiratory surface exists.

Insects have a system of tracheal tubes. These many tubes run from the exoskeleton throughout the insect's body. They are partially air filled, which ensures much faster diffusion.

Larval amphibians often have external gills. These are thin structures with a large surface area especially suited for gaseous exchange. Due to their position they are easily damaged.

Fish have internal gills. These are similar to external gills but safely inside the body. A fine capillary network transports the gases around the body.

Human Health and Physiology

Many terrestrial animals have lungs. A ventilation movement refreshes the air in these sacs, which have a large surface area, are moist, and have an excellent blood supply.

5.4.3
Draw a diagram of the gas exchange system including trachea, bronchi, bronchioles and lungs.

© IBO 1996

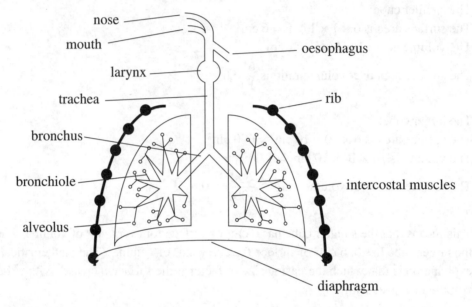

Schematic diagram of the human respiratory apparatus

The lungs are found in the chest cavity or thorax, together with the heart. Around each lung, a pleural membrane is found. Lining the inside of the thorax is another pleural membrane. These membranes surround the pleural fluid.

Structure of the lungs.
The air enters the body through the nose or mouth. From there it passes the trachea, into the bronchi which branch into many smaller bronchioles. Finally the air ends up in the airsacs: the **alveoli**. Alveoli are small thin walled sacs in which most of the gaseous exchange takes place. The oxygen diffuses across the wall of the alveolus, through the capillary cells, across the membrane of the erythrocytes to bind with haemoglobin. The blood then transports it to the tissues. Carbon dioxide from the tissues is carried back to the lungs (see section H.6), diffuses into the alveoli and is breathed out.

Ventilation of the lungs
The air in the lungs constantly needs to be refreshed. Since the lungs only have one connection with the atmosphere, the 'old' air needs to be exhaled before fresh air can be taken in.

Biology

Inspiration
The intercostal muscles contract and move the ribcage up and outward. The diaphragm contracts, flattening it downward. Both actions have the effect of increasing the volume of the chest cavity. If the volume increases, the pressure decreases and as a result the air will flow into the lungs.

Expiration
The relaxation of the intercostal muscles and the diaphragm will bring them back into their original position. The volume decreases in the increase in pressure will make the air leave the lungs. In forced expiration the abdominal muscles contract, which increases the pressure in the abdominal cavity. This pushes the diaphragm up further.

A normal breath will move approximately 500 cm^3 of air into and out of the lungs. This is called the tidal volume. After breathing in normally, you can breathe in an extra 3000 cm^3 of air: the **inspiratory reserve volume** or complemental air. If you then breathe out as much as you can, you expire approximately 4500 cm^3. This is your **vital capacity**. All you have left now is approximately 1200 cm^3 of residual air which you cannot force out or your lungs would collapse. The air which you can exhale after breathing out normally is approximately 1100 cm^3 of **expiratory reserve volume** or supplemental air.

5.4.4
State the difference between breathing and cell respiration.

© IBO 1996

Breathing is gaseous exchange; the intake of oxygen and excretion of carbon dioxide. Cell respiration is the process of releasing energy from food (large organic molecules), often using oxygen as the ultimate electron acceptor.

Breathing involves muscle movement and as such requires energy (released by cell respiration). Cell respiration uses oxygen taken into the body via breathing and produces carbon dioxide which needs to be excreted.

5.4.5
State that exercise improves the functioning of the heart and lungs.

© IBO 1996

Exercise improves the functioning of the heart and lungs. Since exercise increases the demand for oxygen, the ventilation rate and depth will increase during exercise as will the cardiac frequency and stroke volume. Regular exercise will increase stroke volume of the heart and tidal volume of the lungs leading to a lowering of heart beat and ventilation rate.

5.4.6
Explain how and why breathing rate varies with exercise.

© IBO 1996

During exercise, more oxygen is used but also more carbon dioxide is produced. Since carbon dioxide is transported by the blood plasma as HCO_3^-, the pH of the blood will be lowered by large amounts of carbon dioxide. Chemosensors (found in the aorta and carotid arteries) will detect this change and send impulses to the breathing centre in the brain. This centre compares the incoming information with the desired value (the set point) and if the blood pH is too low, impulses will be sent to the intercostal muscles and the diaphragm to increase the rate and depth of lung ventilation. This entire system is under involuntary control although some voluntary control is possible.

5.4.7
Outline one health problem concerned with gas exchange.

© IBO 1996

Various health problems associated with gas exchange exist: lung cancer, tuberculosis, asthma, emphysema, etc.

Asthma is caused by hyper-reactivity of the airways leading to constricted airpassages and attacks of coughing, wheezing and respiratory distress. It is the most common chronic respiratory disease among children, affecting approximately 12% of them (UK).

Due to some irritation, the airways of an asthma patient will contract (muscle spasm), increase mucous production and some tissue swelling (oedema) will occur. All this reduces the airflow to the alveoli and causes coughing (which leads to further irritation). Irritants (allergens) can be house dust mite, pollen, animal fur, moulds, some foods and drugs.

Hyperventilation is another problem associated with gas exchange. By breathing deeply and rapidly, the amount of carbon dioxide in the blood is reduced. The resulting change in pH may cause a positive feedback reaction from the respiratory centre, leading to faster and deeper breathing. This positive feedback cycle can be broken by breathing into a bag and thus rebreathing the released carbon dioxide. Breathing will normalise when pH level has returned to its original value. Hyperventilation is often brought on by anxiety or shock.

5.5 HOMEOSTASIS

5.5.1
Define homeostasis.

© IBO 1996

Homeostasis: the maintenance of a constant internal environment despite possible fluctuations in the external environment.

An example of homeostasis in a non-biological system

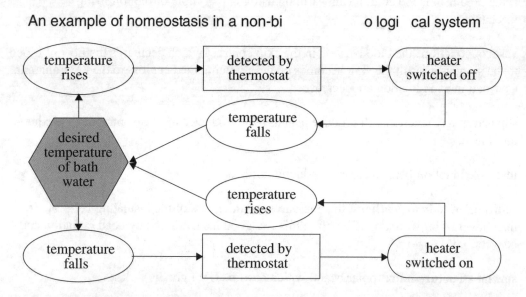

5.5.2
Explain the concept of homeostasis with reference to body temperature and levels of blood glucose.

© IBO 1996

Two examples of homeostasis are the maintenance of body temperature in birds/mammals and the maintenance of bloodsugar levels.

Thermoregulation:
The body of a mammal/bird has thermoreceptors in the skin and in the heat centre in the brain. This way, it monitors temperature changes in the environment as well as changes in the blood temperature.

If the organism is too hot, it can cool down using one or more of the following mechanisms:

vasodilation: the blood vessels in the skin become wider which increases the flow of blood to the skin; as a result the skin becomes warmer which increases heat loss to the environment. Convection and radiation are increased.

sweating: evaporation of fluid from the skin; change of phase (liquid to gas) requires energy which is taken from the body. Panting has the same effect.

decreased metabolism: any reaction produces heat as a by product.

behaviour adaptations: for example birds: bathing - desert rodent: retreat into humid burrows -dogs: dig holes and allow cool earth to absorb heat from belly.

If the organism is too cold, it can warm up using one or more of the following mechanisms:

vasoconstriction: the bloodvessels in the skin contract which decreases the flow of blood to the skin; as a result the skin becomes colder reducing the heat loss to the environment. Convection and radiation are decreased.

shivering: any reaction will produce heat as a by product. Muscles contractions produce a lot of heat.

increased metabolism: increase production of heat.

'fluffing' of hair or feathers: this increases the thickness of the insulating layer of air. thick layer of brown fat or of blubber: this is a good insulator and reduced radiation and convection

special structure hair: (polar bears) which absorbs UV light

The main methods of heat transfer are:
Conduction: transfer of thermal energy without any net movement of the material itself. For example when one end of a metal rod is heated, the other end gets warm. This is relatively slow.

Convection: transfer of thermal energy through the mass motion or flow of some fluid. For example warm air flowing about a room. This is faster.

Radiation: transfer of thermal energy requiring no contact or mass flow. For example, the sun's energy, heat from a stove. This is very fast.

Regulation of bloodsugar levels.
Cells in the pancreas have chemoreceptors which are sensitive to levels of glucose in the blood. Glucose is absorbed from the digested food and is used in cellular respiration. It can also be converted to glycogen and stored. Levels of glucose could go up after a meal and down after exercise if not carefully regulated.

If blood glucose levels are too low, the cells in the pancreas will secrete glucagon. Glucagon is a protein hormone and is secreted into the blood. It will travel to all parts of the body but the liver is the main target organ. Hepatocytes (cells in the liver) will respond to the presence of glucagon by converting glycogen to glucose and releasing it to the blood. They also convert amino acids into glucose (indirectly).

If the blood glucose levels are too high, the cells in the pancreas will secrete insulin. Insulin is a protein hormone and is secreted into the blood. It will travel to all parts of the body. The presence of insulin will make the muscle cells absorb more glucose and the muscle cells and hepatocytes convert glucose into glycogen. In adipose tissue (fat tissue), glucose is converted to fat in the presence of the hormone insulin.

5.5.3
Explain the concept of negative feedback (cross reference 5.6.4 and 5.6.9).

© IBO 1996

Negative feedback: the control of a process by the result or effect of the process in such a way that an increase or decrease in the results or effects is always reversed.

The process of negative feedback requires certain elements to be present. It requires sensors to measure the current situation. The sensors need to pass on the information to a centre which knows the desired value (the norm) and compares the current situation to the norm. If these two are not the same, then the centre activates a mechanism to bring the current value closer to the norm. When this has happened, the centre will turn of the mechanism.

5.5.4
State that the nerve and the endocrine systems are involved in homeostasis.

© IBO 1996

Both the nervous system and the endocrine system are involved in homeostasis. Thermoregulation is done mainly via nerves (see above) but the maintenance of blood glucose levels is mainly carried out via hormones (see above).

5.5.5
State that the nervous system consists of the central nervous system and peripheral nerves composed of special cells called neurons that can carry electrical impulses rapidly.

© IBO 1996

The nervous system can be divided into the **Central Nervous System** (CNS) and the peripheral nerves. The CNS is the brain and spinal cord, everything else is peripheral. The peripheral nerve cells are called **neurons**. Their function is to transport messages in the form of electrical impulses to specific sites. This is done very quickly by local depolarisations of the cell membrane of the neuron.

5.5.6
State that the endocrine system consists of glands which release hormones that are transported in the blood.

© IBO 1996

Human Health and Physiology

The **endocrine system** consists of endocrine glands which produce hormones to the blood. Endocrine glands are **ductless glands**; they do not put their product into a duct, as exocrine glands, for example sweat glands do. Instead, endocrine glands secrete their product (hormones) into the blood which transports it around the body. As the hormone passes cells, only those with special receptors will react to the presence of the hormone. These cells are called **target cells**.

5.5.7
Define excretion.

© IBO 1996

Excretion: the removal from an organism of the toxic waste products of metabolism

5.5.8
Define osmoregulation.

© IBO 1996

Osmoregulation: the maintenance of the osmotic and water potential in a cell or inside a living organism.
[Osmoregulation is the maintenance of a constant internal salt and water concentration in an organism.]

5.5.9
State that the functions of the kidney are excretion and osmoregulation.

© IBO 1996

The functions of the kidney are to regulate the amounts of salt and water in the body. Surplus of either, as well as of undesirable materials, will be excreted.

5.6 REPRODUCTION

5.6.1
Explain that sexual reproduction promotes variation in a species (cross reference 3.3.3).
© IBO 1996

Sexual reproduction promotes variation in a species (see section 3.3.3). Creating gametes by meiosis involves separation of the homologous pairs of chromosomes. Since this process is random, a gamete has a mixture of paternal and maternal chromosomes. Two gametes from different individuals fuse to create a new organism. Since gametes from one individual differ, this mixing will lead to further variation.

5.6.2
Draw diagrams of the adult male and female urinogenital systems.
© IBO 1996

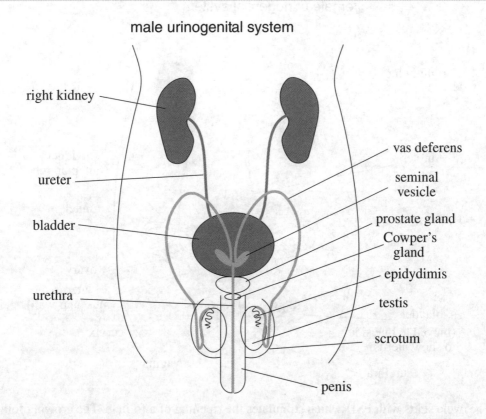

5.6.3
Explain the role of hormones in regulating the changes of puberty (testosterone, oestrogen) in boys and girls, and in the menstrual cycle (FSH, LH, oestrogen and progesterone).
© IBO 1996

Male

Although FSH and LH are usually considered to be female hormones are they also present in the male. FSH is present in the male where it promotes spermatogenesis (production of male gametes). LH is also found in males where it is also known as ICSH (interstitial cell stimulating hormone) and it stimulates secretions of testosterone.

Testosterone (and other hormones, together known as androgens) are produced by the Leydig cells which are found near the bloodvessels in the interstitial tissue of the testes (in between the seminiferous tubules). Testosterone promotes the male secondary sexual characteristics as well as growth and activity of the male reproductive organs.

Female

In the female there are 4 hormones involved in the control of the monthly cycle. They are Follicle Stimulating Hormone and Luteinising Hormone, both produced in the anterior lobe of the pituitary gland, and progesterone and oestrogen, produced in the ovaries.

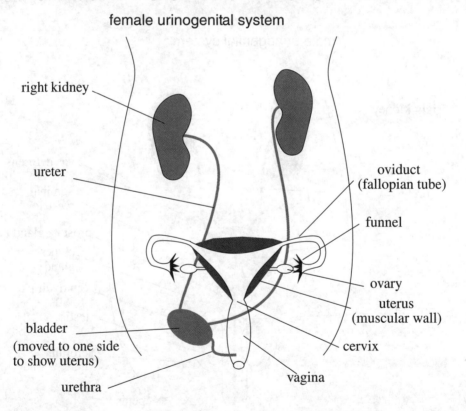

female urinogenital system

The cycle starts with FSH which stimulates the ripening of a follicle. The growing follicle releases oestrogen. Oestrogen increases the thickness of the endometrium and inhibits the production of FSH but stimulates the production of LH. LH stimulates ovulation and the formation of the **corpus luteum**. The corpus luteum produces progesterone which keeps the endometrium intact and inhibits both FSH and LH. If fertilisation does not occur, the corpus luteus degenerates and the pituitary will start producing FSH again to stimulate another follicle.

The functions of oestrogen is also to promote the development of female secondary sexual characteristics and the inhibition of milk secretion. High levels of oestrogen can cause nausea. Progesterone will inhibit ovulation and milk secretion.

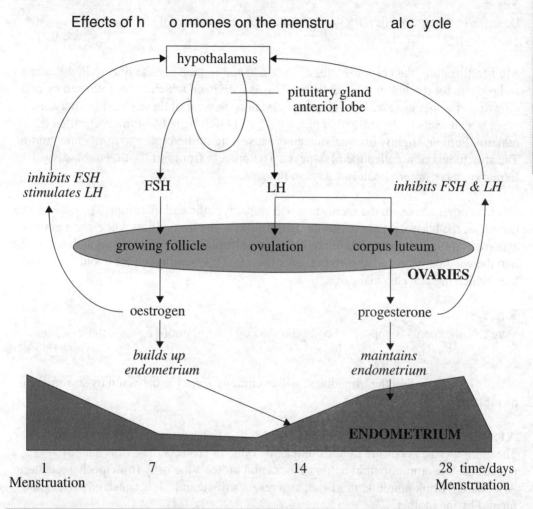

5.6.4
List the secondary sexual characteristics in both sexes.

© IBO 1996

Secondary sexual characteristics are those which are unique to males or females but not present at birth.
Examples are:
males: facial hair, hair on chest, larger voice box, more muscle.
females: breasts, rounder hips.

5.6.5
State the difference between copulation and fertilisation.

© IBO 1996

Human Health and Physiology

Copulation: the act of coupling of male and female animals in sexual intercourse.
Fertilisation: the fusion of male and female gametes.

5.6.6
Describe early embryo development up to the implantation of the blastocyst.

© IBO 1996

After fertilisation, the chromosomes of the male and female gamete line up in the equator and prepare for the first mitotic division. This is a division which is not followed by cell growth and is therefore called a **cleavage division**. Several of these cleavage divisions occur which leads to the formation of a solid ball of cells (the **blastomeres**) called the **morula**. Further, slightly unequal divisions cause a fluid filled space to form in the middle. The structure is now called the **blastocyst**. The process from fertilisation to blastocyst formation takes approximately 4 days in humans.

The blastocyst arrives in the uterus and will implant in the endometrium. The cells on the outside of the blastocyst are called the **trophoblast**. The trophoblast will embed in the endometrium in the process of implantation. The trophoblast will grow trophoblastic villi into the endometrium and absorb nutrients from it. This is sufficient for about 2 weeks after which the placenta takes over.

5.6.7
State that the foetus is supported and protected by the amniotic sac and amniotic fluid.

© IBO 1996

A mass of cells within the trophoblast will eventually form the baby and its surrounding membranes.

A **foetus** is surrounded by amniotic fluid which in turn is surrounded by the **amniotic sac**. The amniotic sac keeps the fluid from leaking out and protects the foetus against infections. The amniotic fluid buffers shocks and protects the baby from mechanical harm. Also, babies drink amniotic fluid and, as a result, urinate in it. It is constantly made and filtered by the mother.

5.6.8
State that the materials are exchanged between the maternal and foetal blood in the placenta.

© IBO 1996

This fast growing foetus needs a good supply of nutrients as well as needing to excrete waste products. Both of these functions are carried out by the **placenta**. The placenta is foetal tissue which invades maternal uterine tissue. The baby's blood runs through bloodvessels which go through through blood spaces filled with maternal blood. An exchange of substances takes place (diffusion). The foetal blood returns to the foetus enriched with nutrients and oxygen, the maternal blood has taken up the carbon dioxide and other waste products from the foetal metabolism, which it will excrete.

Mother and foetus each have their own blood and circulation. Materials are exchanged but the blood does not mix. Many babies have blood groups different from their mothers.

5.6.9
Outline the process of birth and its hormonal control (cross reference 5.5.3), including progesterone and oxytocin.

© IBO 1996

The foetus will (normally approximately 38 weeks after conception) be ready to be born. It will send a signal to the extra-embryonic membranes (like the amniotic sac) which leads to a secretion of **prostaglandins**. These locally produced hormones will initiate contractions of the uterine wall. These contractions push the baby's head against the cervix, making it dilate. Nerve endings in the uterus and cervix will report the contractions to the brain and the posterior lobe of the pituitary gland will release oxytocin. Oxytocin, prostaglandins and a positive feedback system will cause the contractions to become longer and stronger.

When the cervix is fully dilated, the first stage of birth (labour) is over and the second stage (expulsion) begins. While the first stage can take many hours, the second stage usually does not take more than 1 hour. Powerful contractions push the baby out of the uterus.

Now that the uterus is no longer pushing against something, the positive feedback loop is broken and contractions soon reduce. However, a few contractions are needed to expel the placenta from the uterine wall. This is the last stage of birth.

5.6.10
Describe four methods of family planning and contraception.

© IBO 1996

Contraceptives are aimed at preventing conception. Most methods do this by preventing the meeting of sperm and egg.

Some examples are given below:

Permanent methods:
Sterilisation: Cutting or tying the Fallopian tube (oviduct) which prevents the egg cells from reaching the uterus.

Vasectomy: Cutting or tying the vas deferens which prevents the sperm cells from leaving the body.

Mechanical methods:
Condoms: A thin sheath of rubber rolled over the penis before ejaculation will ensure that sperm cells do not enter the female.

Cap / Diaphragm: A firmer rubber cap placed over the female's cervix before intercourse will prevent sperm cells from entering the uterus.

Human Health and Physiology

Chemical methods:
Spermicides: Creams or sponges placed inside the female's vagina which kill sperm cells.

The pill: A tablet containing female hormones. This prevents the release of FSH and thus the formation of the follicle.

Other methods:
IUD: The intra uterine device is placed inside the uterus and prevents implantation of the fertilised egg cell.

Withdrawal: If the man withdraws his penis from the vagina before ejaculation, most of the sperm cells will not enter the female's body.

Rhythm method / Calendar method: Using the calendar or a temperature chart, sexual intercourse is avoided during the fertile period.

The safest are vasectomy/sterilisation and the pill (if used correctly). Diaphragms and condoms are also fairly safe but again need to be used correctly.

Condoms also reduce the chance of Sexually Transmitted Diseases (STD), such as HIV. IUD's do not prevent conception but prevent implantation.

5.6.11
Describe the ethical issues of family planning and contraception.
© IBO 1996

Ethical considerations on the use of contraceptives may include:
considerations about overpopulation, considerations about the future of the child(ren), religious considerations.

5.6.12
Outline the techniques of amniosentesis and chorionic villus sampling
© IBO 1996

Two main techniques are used to detect genetic disorders in unborn babies.

The first one is **chorionic villus sampling**. This is a relatively new technique, newer than amniocentesis. In chorionic villus sampling, a thin tube is inserted into the uterus via the vagina. A sample of cells from the chorion (one of the extra embryonic membranes) is taken and studied.

The second possible technique is **amniocentesis**. This involves inserting a needle into the uterus via a small incision in the abdominal wall. Amniotic fluid, which contains cells from the embryo, is taken. The cells are cultivated and studied.

Chorionic villus sampling can be done at an earlier stage than amniocentesis and does not require cultivation. However, it might have an increased chance of miscarriage or birth defects, although data on this are not clear yet.

More than 400 genetic defects that can be shown by these antenatal techniques. Reasons for doing a survey might be the age of the mother or either parent being related to a person with a genetic disorder.

5.6.13
Outline the process of in vitro fertilisation (IVF).

© IBO 1996

IVF is **in vitro fertilisation** ('in vitro' - 'in glass'). In this procedure the egg and sperm cells 'meet' outside the female's body. A successful procedure results in a so called 'test tube baby'. The first test tube baby was Louise Brown, born in 1978.

IVF is used for women who have blocked oviducts (usually due to an earlier infection) or cannot sustain a pregnancy. The woman donating the egg cells is treated with hormones so that more than one follicle will ripen. The follicles are harvested by sucking them out of the ovaries, using a syringe. They are then mixed with sperm from the prospective father. If the quality of the sperm is poor, a sperm cell may be injected directly into the egg cell. The fertilised egg cells are cultivated to a blastocyst stage and then frozen or placed into the uterus of the female who will carry the children. Usually no more than 4 embryos are placed into the uterus at the same time. If one or more embryos implant successfully, a pregnancy will result.

5.6.14
Discuss the ethical issues of IVF.

© IBO 1996

Ethical issues around IVF can involve the biological/social parenthood, the property of frozen embryos after the parents divorce or die, elderly mothers (e.g. in Italy), 'superbabies'. Do not confuse IVF with donor insemination where the sperm of a man is placed inside a woman and fertilisation may occur in vivo ('in life').

5.6.15
Outline three developments from human embryo research including embryo storage and early detection of chromosome abnormalities.

© IBO 1996

Embryos produced during IVF are stored for later use. The rate of successful implantation is not high while the process of harvesting follicles is unpleasant. So a maximum number of embryos are produced each time but not all are placed into the uterus. The remainder are stored for later use.

If one or more babies are born, the parents do not always concern themselves with the remaining frozen embryos. A clinic in the UK recently had to destroy human embryos which had been stored for the agreed time and the parents could not be traced.

Another issue already mentioned is who owns these embryos and do they have rights? Divorced parents, widows and widowers, heirs all have a stake in this issue. Legally, solutions have not yet been found.

Early detection of chromosome abnormalities can lead to an increased rate of abortion of disabled foetuses. Do only 'perfect' babies have the right to be born? Are you obliged to have an abortion after having had an expensive amniocentesis which shows a disabled child? Abnormalities other than chromosomal can also be detected before birth. This gives rise to the same issues as described above. If embryos could be used to grow selected organs, for example a heart, should 'surplus' embryos be used to grow 'replacement' organs?

EXERCISE

1. Salivary amylase works best at a pH of:

    A   5-6
    B   6-7
    C   7-8
    D   8-9

The diagram opposite shows a schematic representation of the digestive tract. Questions 2-4 refer to the labels on this diagram.

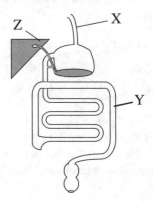

2. Label X points to:

    A   liver.
    B   oesophagus.
    C   stomach.
    D   colon.

3. Label Y points to:

    A   liver.
    B   oesophagus.
    C   stomach.
    D   colon.

4. Label Z points to:

    A   liver.
    B   oesophagus.
    C   stomach.
    D   colon.

Biology

5. The process by which the body takes up the substances that it needs is known as:

    A   digestion.
    B   absorption.
    C   excretion.
    D   respiration.

6. Carbohydrates and proteins are known as:

    A   micronutrients.
    B   macronutrients.
    C   dietary fibre.
    D   minerals.

7. Scurvy is a consequence of dietary deficiency in:

    A   Vitamin A.
    B   Vitamin B_{12}.
    C   Vitamin C.
    D   Vitamin D.

8. The part of the heart labelled X in the schematic diagram is the:

    A   tricuspid valve.
    B   aorta.
    C   right ventricle.
    D   right atrium.

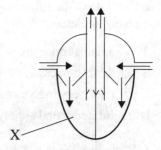

9. The electrical impulses that trigger that trigger the contractions of the heart originate in the:

    A   sino atrial node.
    B   brain.
    C   spinal chord.
    D   atrio ventricular node

10. The main blood vessels are the arteries, veins and capillaries. In order of decreasing blood pressure, these should be arranged:

    A   arteries, veins, capillaries.
    B   veins, capillaries, arteries.
    C   capillaries, arteries, veins.
    D   arteries, capillaries, veins.

Human Health and Physiology

11. The cells of the blood are approximately 90%:

    A   erythrocytes.
    B   leucocytes.
    C   thrombocytes.
    D   plasma.

12. The main mechanism by which the body restricts infections of the stomach is its:

    A   alkalinity.
    B   acidity.
    C   large surface area.
    D   mucous secretions.

13. A globular protein that recognises an antigen is known as:

    A   a virus.
    B   a bacterium.
    C   an antigen.
    D   an antibody.

14. Which one of the following are features of alveoli that allow them to carry out gas exchange efficiently?

    A   small surface area.
    B   their thinness.
    C   their very high moisture content.
    D   limited blood supply.

15. Very small organisms such as algae do not need gills or lungs because:

    A   they have a comparatively small surface area to volume ratio that allows them to exchange nutients and waste products with their environment.
    B   they have a comparatively large surface area to volume ratio that allows them to exchange nutients and waste products with their environment.
    C   they have a comparatively small surface area to volume ratio which means that they need very little food to survive.
    D   they have a comparatively small mass which allows them to exchange nutrients and waste products with their environment.

16. The ribs are moved by the:

    A   intercostal muscles.
    B   diaphragm.
    C   alveoli.
    D   thorax.

17. Cell respiration differs from breathing in that it:

    A   involves muscle movement.
    B   is gaseous exchange.
    C   requires energy.
    D   is the process of releasing energy from food.

18. Which one of the following graphs shows the rate at which an athlete would breathe with increasing exercise:

[Graph A: exponential curve of breathing rate vs amount of exercise]
[Graph B: sigmoid curve levelling off]
[Graph C: straight line increasing]
[Graph D: bell-shaped curve]

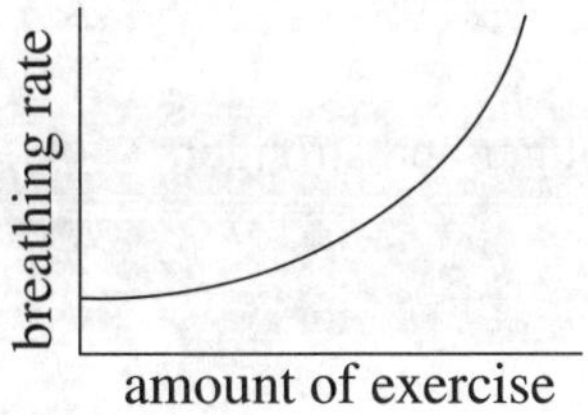

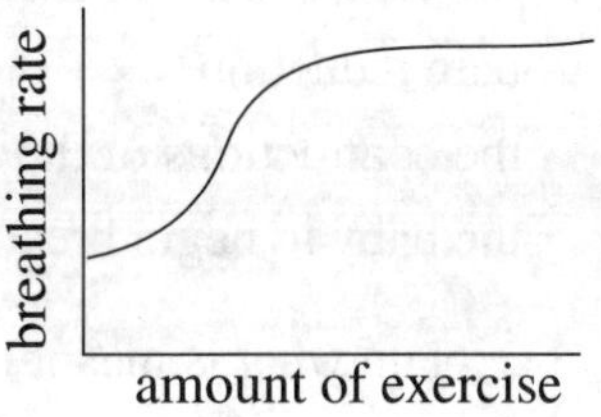

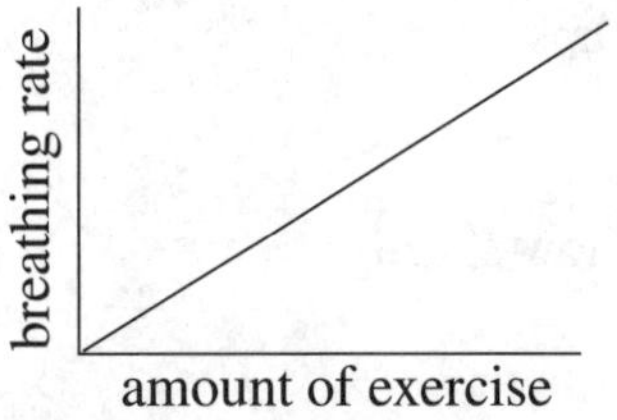

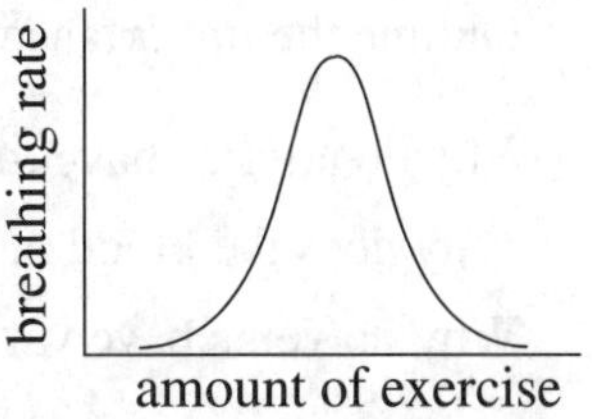

19. The process of temperature regulation in the body is an example of:

    A   homeostasis.
    B   respiration.
    C   excretion.
    D   dilation.

20. The control of a process by the result or effect of the process in such a way that an increase or decrease in the results or effects is always reversed is known as:

    A   convection.
    B   negative feedback.
    C   positive feedback.
    D   shivering.

21. The main function(s) of the kidney is/are:

    A   vitamin synthesis.
    B   filtration.
    C   regulation of bloodsugar levels.
    D   excretion and osmoregulation.

22. The main function of the amniotic sac and fluid is to:

    A   transport the ovum.
    B   protect the foetus
    C   assist in fertilisation.
    D   prevent conception.

23. During birth, oxytocin causes:

    A   the contractions of the uterus to begin.
    B   pain reduction.
    C   the contractions of the uterus to become longer and stronger.
    D   the baby to begin breathing.

24.
    a.   Explain what is meant by 'a balanced diet'.
    b.   List the groups of substances that are part of a balanced diet.
    c.   Outline the importance of each of the groups.

25.
    a.   Why do arteries have thick walls?
    b.   Why does the blood in capillaries flow at low speed?
    c.   Why do veins have valves?

26.
    a.   Outline the effect of the HIV virus on the immune system.
    b.   Outline normal antibody production.
    c.   Explain why the HIV virus interferes in antibody production.

27.
    a.   Draw a diagram of the human respiratory system.
    b.   What features of alveoli makes them suitable for gaseous exchange?
    c.   Why do lungs need to be ventilated?

28.
    a.   List the elements that are involved in thermoregulation in humans.
    b.   Describe how the organism can reduce heat loss.
    c.   Outline the concepts of conduction, convection and radiation.
    d.   Which one causes most heat loss from a human?

29.
    a.   Describe how sexual reproduction promotes variation.
    b.   Birth control pills contain substances resembling oestrogen and progesterone. Explain how keeping these hormones in the blood at a high level prevents pregnancy.

Biology

PRACTICALS

6

Chapter contents

- Sample: Cell division - mitosis

Contributed by John Gibson and David Greig

Practicals

6.1 PRACTICAL WORK

Practical work is a very important part of the study of Biology. An example of a practical taken from 'Biology: A Portfolio of Investigations' by John Gibson and David Greig (IBID Press, ISBN: 0 9585686 3 4, follows.

6.1.1 A SAMPLE PRACTICAL: CELL DIVISION - MITOSIS

RELEVANT IB TOPICS: SSC 1.5   AHL 7.2

BACKGROUND
In the growing root tip of plants there are many cells undergoing mitotic divisions as they divide to give rise to 2 identical daughter cells. When the growing root tips of a plant for example onion, are placed in a fixative such as 70% ethanol, the cells are killed and fixed at the stage of the cell cycle they are in.

When an appropriate stain such as aceto-orcein is used, the DNA is stained and the nucleus or the chromosomes will become clearly visible. It is then possible to identify the stage of cell division that the cell was in at the time of fixing.

AIMS
1. To make and stain a squash preparation of garlic root tips.
2. To observe prepared slides and identify the stages of cell division.
3. To recognise and describe the sequence of events in mitosis.

MATERIALS REQUIRED

Growing roots on a garlic clove	Petri-dish
Razor blade	Fine-tip forceps
Glass slide and coverslip	Teat pipette
Prepared sides showing mitosis	Lens cleaning tissue
Paper towelling	Compound microscope
Aceto-orcein stain	1M Hydrochloric acid
2 pins	25mL beaker
10mL measuring cylinder	

METHOD

The first two steps will probably be done for you by your teacher or the laboratory assistant.

1. Measure 9mL of 1% aceto-orcein in a 25mL beaker using your 10 mL measuring cylinder. Add 1mL of 1M Hydrochloric acid and mix the two solutions together. Pour the mixture into a petri-dish.

2. Remove several growing roots from a garlic clove about 1cm in length and place in the petri-dish. Leave overnight in a cool area.

3. Remove a root tip from the stain by picking up the root tip at the blunt end using a pair of fine forceps. Place the root tip in the centre of a clean glass slide and add 2 drops of fresh aceto-orcein stain to the tip on the slide.

4. Use 2 pins to tear the root tip along its length. This is done by holding the root tip down with one pin and using the other to shred the root tip into several pieces. This allows more stain to penetrate the tissue and will make spreading the cells into a single layer easier.

5. Place a coverslip over the tissue, being careful to exclude all air bubbles.

6. Place the finger of your left hand on the edge of the coverslip so that it will not move and then allow the rounded end of your pencil or biro to gently bounce on the surface of the coverslip by dropping it from about 1cm height above the slide.

7. Cover the coverslip with several layers of paper towelling and place your thumb on top. Gently roll your thumb from side to side. Do not allow your thumb to slide laterally or the cells will be destroyed.

8. Observe the slide under high power (H.P.) magnification (usually about x400) and find a cell in about mid-anaphase. Make a reference of its position using a pointer or grid on the ocular lens if available.

The cells where mitosis is likely to be occurring prior to fixing are near the tip of the root. These cells are smaller, almost cubic in shape whereas cells from further back have grown by the uptake and synthesis of materials and are larger and rectangular in shape.

ASK YOUR TEACHER TO CHECK YOUR SLIDE AT THIS STAGE FOR ASSESSMENT PURPOSES.

9. Obtain a prepared slide showing the various stages of mitosis. Focus on H.P.under the microscope and identify a cell in each of the following stages of mitosis, Prophase, Metaphase, Anaphase, and Telophase.

10. On the next page draw one cell from each of the stages listed above. Also draw one cell in interphase. These drawings are not to be taken from a text book but from the slide in front of you. Draw these to scale, using H.P. magnification, fully labelled, titled and in the correct order.

Practicals

Biology

CELLS

7

Chapter contents

- Membranes
- Cell division - Mitosis
- Differentiation and specialisation of cells

Cells

7.1 MEMBRANES

7.1.1
Explain the dynamic relationship between the nucleus, rough endoplasmic reticulum (rER), Golgi apparatus and cell surface membrane.

© IBO 1996

As can be seen from the fluid mosaic model, membranes are not static structures. The molecules making up the membrane can move in the plane of the membrane. Also this structure has some flexibility so that small amounts of membrane can be added or removed without tearing the membrane (endo- and exocytosis, see section 1.4). Since the structure of the cell surface membrane is essentially the same as that of the nuclear envelope, the endoplasmic reticulum (ER) and the Golgi apparatus, it is possible to exchange membrane sections between them.

If you remember section 1.3, the function of the Golgi apparatus is to prepare substances for exocytosis. This involves wrapping it in a bit of membrane from the Golgi apparatus. This membrane then joins the cell surface membrane in the process of exocytosis.

Many of the substances which the cell 'exports' are proteins and hence the following organelles are involved:
nucleus: chromosomes contain genes coding for proteins, mRNA is made by transcription.
rER: contains the ribosomes which make protein intended for export by translation. The protein then goes into the lumen of the rER, is surrounded by membrane and leaves through the cell surface membrane by exocytosis.

7.1.2
Describe the ways in which vesicles are used to transport materials within a cell and to the cell surface.

© IBO 1996

Vesicles can be used to transport materials within the cell and to/from cell surface membrane. See section 1.4.

7.1.3
Describe the membrane proteins and their positions within membranes.

© IBO 1996

As said in section 1.4.1, proteins can be extrinsic or intrinsic to the cell membrane. This depends on their polar character: a hydrophilic (polar) protein will be embedded between phophate 'heads' or be found just outside the membrane; proteins which have hydrophilic and hydrophobic regions will arrange themselves in the membrane so as to have their hydrophilic regions meet the hydrophilic part of the membrane (the outside of the membrane, the phosphate 'heads') and situate their hydrophobic sections in the centre of the membrane, surrounded by the hydrophobic lipid 'tails' of the phospholipid molecules of the membrane.

Proteins can serve as 'anchors'. The shape of the cell depends on the organisation of its cytoskeleton. This network of microfilaments interconnects in various ways and is also connected to certain proteins in the cell membrane. How the protein is attached (to a bundle of microfilaments or to a planar network) also decides the shape of that particular segment of membrane.

7.1.4
Outline the functions of membrane proteins as antibody recognition sites, hormone binding sites, catalysts for biochemical reactions and sites of electron carriers.

© IBO 1996

Many of the functions of proteins in membranes have already been mentioned in previous sections. They include:
antibody recognition sites: section 1.4.1 - glycoproteins
hormone binding sites: section 1.4.1 - glycoproteins
catalysts for biochemical reactions and
sites of electron carriers: section 1.4.1 - also sections 4.2, 9.1/C5 and 9.2/C4.

7.2 CELL DIVISION - MITOSIS

7.2.1
Describe the behaviour of the chromosomes in each of the four phases of mitosis (prophase, metaphase, anaphase and telophase).

© IBO 1996

The behaviour of the chromosomes in each of the four phases is described below. The purpose of mitosis is to increase the number of cells without changing the genetic material, i.e. the daughter cells are identical to the parent cell in the number of chromosomes, the genes and alleles. Mitosis can occur in haploid, diploid or polyploid cells.

Mitosis is divided into 4 phases: **prophase**, **metaphase**, **anaphase** and **telophase**. When a nucleus is not dividing it can be said that the cell is in interphase.

Please remember that nuclear division is a continuous process. We have separated the division into 4 stages. It is more important to know what typically happens in each stage than to be able to determine the stage of every cell in a microscope slide.

Interphase:
- DNA replication occurs.

Prophase:
- Chromosomes become visible (supercoiling).
- Centrioles move to opposite poles.

Cells

- Spindle formation.
- Nucleolus becomes invisible.
- Nuclear membrane disappears.

Metaphase:
- Chromosomes move to the equator.
- Centromeres attach to spindle.

Anaphase:
- Chromatids separate and move to opposite poles.

Telophase:
- Chromosomes have arrived at poles.
- Spindle disappears.
- Centrioles replicate (in animal cells).
- Nuclear membrane reappears.
- Nucleolus becomes visible.
- Chromosomes become chromatin.

The division of the cell is sometimes included as the last stage of telophase; strictly speaking, however, cell division is not a part of mitosis.

Mitosis can be summarised as the dispersing of the nuclear material, movement of centrioles (if present) to opposite ends, microtubules developing into a spindle, supercoiling of chromatin, attachment to the spindle fibres at the centromere region, separation and movement of chromatids.

7.2.2
Outline the differences in mitosis and cytokinesis between animal and plant.

© IBO 1996

Differences between plant and animal cells in the process of mitosis and cell division are the following:

Animal cells have centrioles which form the focus of the spindle and form the 'asters'. Plants cells do not have centrioles but still form a spindle (but no asters).

Cytokinesis in animal cells occurs by the pinching in of the cell surface membrane caused by microfilaments which have been formed around the cell's equator and constrict.
In plant cells a cell plate is formed by vesicles from the Golgi body moving to the equator and fusing. The cell surface membrane is formed by the membrane of the vesicles. The material of the cell plate becomes the middle lamella; a cell wall is formed on each side between the cell surface membrane and the middle lamella.

7.3 DIFFERENTIATION AND FUNCTIONAL SPECIALISATION OF CELLS

7.3.1
State that unicellular organisms carry out all the functions of life.

© IBO 1996

Unicellular organisms carry out all the functions of life (reproduction, respiration etc).

7.3.2
State that cells in multicellular organisms differentiate to carry out specialised functions.

© IBO 1996

Cells in multicellular organisms differentiate to carry out specialised functions.

7.3.3
Define tissue.

© IBO 1996

Tissue: a group of similar cells performing a particular task.
Examples are blood cells, epithelial cells, palisade cells. A single cell carrying out the task becomes meaningless.

7.3.4
Define organ.

© IBO 1996

Organ: a structural unit made of a group of tissues which work together to perform a function. Examples are heart, liver.

7.3.5
Define organ system.

© IBO 1996

Organ system: several organs working together to perform a job. Examples are the digestive system, the circulatory system.

7.3.6
Explain the hierarchical relationship between cells, tissues, organs and organ systems in multicellular organisms.

© IBO 1996

Tissues are made of cells, organs contain different tissues and several organs contribute to an organ system.

Cells

7.3.7
Calculate linear magnification of drawings.

© IBO 1996

Diagrams and photographs can be shown larger or smaller than reality. To indicate the real size of the object, the magnification can be indicated next to the diagram or picture or a scale bar can be given.

Magnification

× 2

The magnification is given as × 0.5 or × 250, indicating in the first case that the real structure would be twice the size of the picture, in the second case that is 250 times smaller than the picture. To convert magnification into scale bars, remember what magnification means. If the magnification is × 175, it means that 175 cm on the picture would really be 1 cm. So if you draw a scale bar of 1 cm, you should label it 1/175 cm = 0.57 µm.

A scale bar indicates that in reality the length of the scale bar would measure whatever it says below, e.g. 1 µm. If your scale bar has a length of 1 cm, this means that the magnification is 1 cm/ 1 µm = 10 000 times.

7.3.8
State that cells differentiate by expression of some of their genes and not others.

© IBO 1996

While every cell contains all the genetic information to carry out every function, only a small portion of the genetic material is activated. A cell in your toes has the information on how to make the pigment which colours your eyes, but will not use it. So cells differentiate by expression of some of their genes and not others.

Cells affect each other. The differentiation of any one cell is determined by the cell's position relative to others and by chemical gradients.

7.3.9
State that the pathway of differentiation is determined by the cell's position relative to others and by chemical gradients.

© IBO 1996

The pathway of differentiation is determined by the cell's position relative to others and by chemical gradients. Since all cells of an organism originate from the zygote, something must cause cells to specialise differently. Many cells present during early stages of the organism can differentiate into almost any specialised cell type. At some point during development, the fate of each cell becomes committed during determination.

Determination can involve the inheritance of different cytoplasmic constituents or be based on interactions between neighbouring cells.

Cytoplasmic granules composed of DNA and protein can be distributed unequally through the cytoplasm. Subsequent cytokinesis will give rise to cells with unequal amounts of these granules which will differentiate in different ways.

In many regions of a developing organism, the presence of a particular cell type causes the neighbouring cell to differentiate in a specific way. This is called induction. For example: the presence of the notochord induces the ectoderm cells to form the neural tube. It has been suggested that induction occurs because one cell releases substances which diffuse to the other cell and cause it to differentiate. Evidence for this is contradictory. Cells supposedly separated by a thin membrane preventing cellular contact, will have extensions going through the membrane so conclusions cannot be drawn.

EXERCISE

1. One of the functions of the Golgi apparatus is to:

    A   prepare substances for exocytosis.
    B   excrete waste.
    C   synthesise proteins.
    D   respire.

2. The passive movement of water molecules, across a partially-permeable membrane from a region of lower solute concentration to a region of higher solute concentration is known as:

    A   diffusion.
    B   osmosis.
    C   dilution.
    D   extraction.

Cells

3. The process by which the cell takes up a substance by surrounding it with membrane is known as:

- A   endocytosis.
- B   active transport.
- C   diffusion.
- D   osmosis.

4. DNA replication occurs during the:

- A   prophase.
- B   metaphase.
- C   anaphase.
- D   interphase.

5. The nuclear membrane disappears during the:

- A   prophase.
- B   telophase.
- C   anaphase.
- D   interphase.

6. The chromatids separate and move to opposite poles during the:

- A   prophase.
- B   metaphase.
- C   anaphase.
- D   interphase.

7. The heart, liver and kidneys are examples of:

- A   organelles.
- B   organs.
- C   systems.
- D   single tissues.

8. The diagram shows a scaled schematic picture of a cell and a transformation of that picture. The scale of the transformation is closest to:

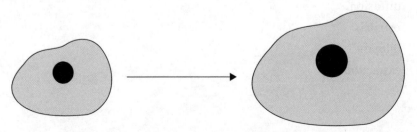

A   1:2
B   3:2
C   2:3
D   1:3

9. Differentiation results in cells

A   transcribing some of their DNA.
B   using all of their DNA during replication.
C   changing type when they replicate.
D   mutating.

10. Lysosomes are vesicles containing digestive enzymes. Explain the role of the nucleus, rER and Golgi apparatus in producing lysosomes.

11. A student investigates a section of a rapidly dividing tissue. The student notes the following :

112   nuclei in Interphase
75   nuclei in Prophase
22   nuclei in Metaphase
27   nuclei in Anaphase
36   nuclei in Telophase

According to these observations :

a.   Which stage in the cell cycle last longest? Explain.
b.   If the total cell cycle lasts 1 hour, how long is each stage?
c.   What aspect of this experiment makes it unsuitable to predict the relative duration of Interphase vs. Mitosis? Would Interphase in reality be longer or shorter? Explain.

12. Refer to the diagram of the eukaryotic cell on page 9.

Remember that a typical eukaryotic cell measures around 20 µm.

a.   What is the magnification of this drawing?
b.   If, on the diagram, you drew a scale bar of 3 cm long, what size should you write with it?

NUCLEIC ACIDS & PROTEINS

8

Chapter contents

- 8.1 DNA structure
- 8.2 DNA replication
- 8.3 Transcription
- 8.4 Translation

Nucleic Acids and Proteins

8.1 DNA STRUCTURE

8.1.1
Outline the structure of nucleosomes including histone proteins and DNA.
© IBO 1996

Analysis of chromosomes has shown that they are made of DNA and protein and a small amount of chromosomal RNA. (The 'chromosomes' of prokaryotes like bacteria are made of only DNA.)

DNA has negative charges along the strand and positively charged proteins are bonded to this by electromagnetic forces. These basic proteins are called **histones**. The complex of DNA and protein is known as **chromatin**.

The total length of DNA in a human nucleus is approximately 2.2 m. To pack all this into a cell means that the total length is shortened to 276 µm = 0.276 mm. This means that the length has to be reduced by a factor 8000 which needs a good organisation if you want to be able to unwind it again. The histone proteins form the skeleton for this.

Recent structures have shown that the genetic material looks like beads on a string. This is caused by the DNA helix combining with 8 small histone molecules. These structures are called **nucleosomes**.

SCHEMATIC DIAGRAM OF THE STRUCTURE OF A NUCLEOSOME
DNA WOUND AROUND CLUSTERS OF HISTONES

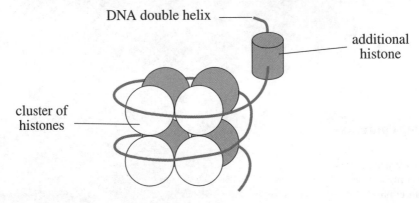

8.1.2
State that only a small proportion of the DNA in the nucleus constitutes genes and that the majority consists of repetitive sequences (cross reference 8.3.4).
© IBO 1996

The units of genetic information are called **genes**. The order of the nucleotides is the 'genetic code', containing the hereditary information.

One gene contains the information to produce one polypeptide (see section 2.6 and 8.3: the 'one gene one polypeptide hypothesis'). Some genes do not code for polypeptides but regulate how other genes are expressed.

Every cell in an organism contains all the genetic information. Not every cell expresses all the genes. A cell in your toe will not express the information for the colour of your eyes. Cells differentiate by having certain genes inactivated. When staining the chromatin, it shows up as lightly staining **euchromatin** and darker staining **heterochromatin**. The euchromatin is less condensed and can be transcribed during interphase. Heterochromatin is more condensed and is rarely transcribed. In eukaryotic cells, less than 10% of the DNA is known to code for polypeptides. Apart from the regulator genes, the function of the remainder of the DNA is as yet unknown.

Most genes have only very few copies. Some genes however, for example those for ribosomal RNA or histones, are present many times. This is rather useful since the cell needs a lot of rRNA and histones. Having many copies of the corresponding genes means that transcription can take place simultaneously in several places.

Part of the non-coding section of DNA is in the form of **repetitive sequences**. The function of these is unknown but humans have found a use for it in the process of DNA profiling.

DNA profiling is the same as DNA fingerprinting. The repetitive sequences are useful in this process See section 2.7.

8.1.3
Explain the structure of DNA including the antiparallel strands, 3'-5' linkages and hydrogen bonding between purines and pyrimidines.

© IBO 1996

As you saw in section 2.4.5, the DNA double helix is made of 2 anti-parallel strands, kept together by hydrogen bonds between the organic bases.

It is possible to assign numbers to the Carbon atoms in the ribose. By convention, we start numbering the C-atoms on the right hand side of the molecule. So, the ribose is numbered as shown in the diagram:

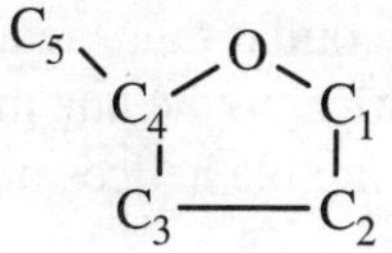

In this way the nucleotide becomes as shown. The phosphate is attached to Carbon # 5 of the ribose. In a single strand of DNA, the phosphate of the next nucleotide will attach to Carbon # 3. This is called a 3' - 5' linkage.

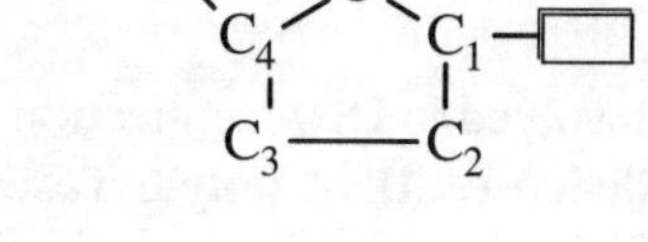

Nucleic Acids and Proteins

A DNA strand therefore, ends on one side with a phosphate on the 5' end of the ribose and a ribose at the 3' end. Since the two strands are anti-parallel, the result is as indicated below.

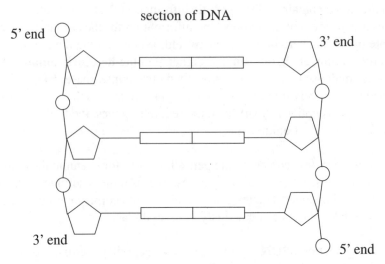

section of DNA

The sequence of the nucleotides is customarily given in the 5' to 3' direction.

8.2 DNA REPLICATION

8.2.1
State that DNA replication occurs in a 5' → 3' direction.

© IBO 1996

As you saw in section 2.5.1 DNA replication has been shown to be a **semi-conservative mechanism** by Meselsohn and Stahl in 1957.

This essentially means that in the process of DNA replication, the DNA double helix 'unzips' and new hydrogen bonds are formed with the organic bases of DNA nucleotides 'floating around' in the cell. **DNA polymerase** creates the 3'-5' linkages between the nucleotides thus creating a new DNA strand, complementary to the original one and identical to the one that was 'unzipped'. The other strand does the same and the result is two double helices of DNA, each with one 'old' and one 'new' strand.

8.2.2
Explain the process of DNA replication in eukaryotes including the role of enzymes (helicase, DNA polymerase III, RNA primase, DNA polymerase I and DNA ligase), Okazaki fragments and deoxynucleoside triphosphates.

© IBO 1996

Involved in DNA replication are the following enzymes:
helicases, DNA polymerase III, RNA primase, DNA polymerase I and **DNA ligase**.
The process of DNA replication is explained in the diagram and text below.

The replication of DNA always occurs in the 5' to 3' direction (of the new growing strand). The hydrogen bonds keeping the 'old' DNA strands together will be broken by the enzymes helicases and the DNA will partially unzip. In order to synthesise a complete new strand, a beginning of the new strand must be present. This beginning in eukaryotic cells is a short strand of RNA called a **primer**. **RNA primase** catalyses the formation of this primer on the exposed single strand of DNA. DNA replication in one strand will take place continuously. This is the **leading strand** and the new DNA will grow in the 5' - 3' direction. The other template strand is antiparallel and DNA synthesis will take place in the opposite direction of the growing fork. The short sequences of DNA formed this way are called **Okazaki fragments**. DNA ligase will join these segments together to form the **lagging strand**. (The lagging strand requires a new primer for each Okazaki fragment.)

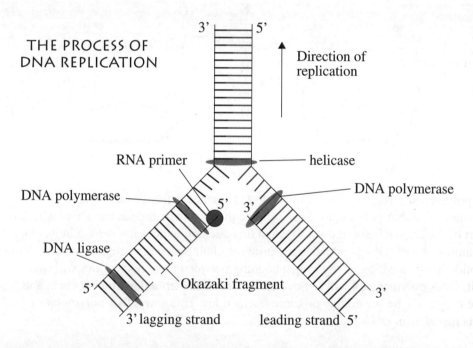

DNA nucleotides are in the form of **deoxyribonucleoside triphosphates** when they form hydrogen bonds with the complementary nucleotide of the DNA strand. A nucleoside is an organic base coupled to a pentose sugar. Deoxyribonucleoside triphosphate consists of an organic base (A, T, C, or G), deoxyribose and three phosphates. The deoxyribonucleoside triphosphates are sometimes known as dATP, dTTP, dCTP and dGTP. The second and third phosphate group are removed as the nucleotide is attached to the growing DNA strand by DNA polymerase. This provides the energy to drive the reaction. (The bond between the 2 phosphate groups will be broken immediately, changing it into 2 inorganic phosphate groups).

DNA polymerase I is responsible for the step by step addition of deoxynucleoside triphosphates (dNTP). It adds the dNTP to the 3' hydroxyl terminus of a pre-existing DNA (or RNA) strand. The phosphate group of the dNTP attached to pentose sugar will form a bond with the Oxygen of the OH group of the pentose of the pre-existing strand, while breaking the bond between this phosphate and the two subsequent phosphates.

Nucleic Acids and Proteins

This is shown in the diagram below.

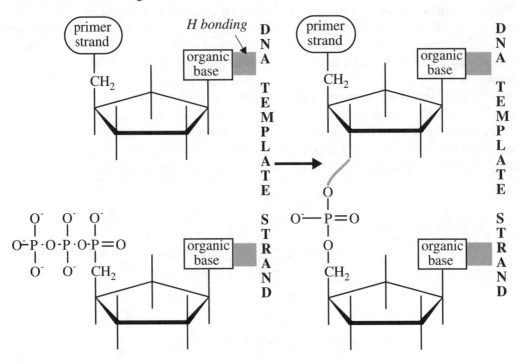

DNA polymerase I is also a 3'-5' exonuclease. This means that (under certain circumstances), DNA polymerase I can remove the last nucleotide at the 3' end which is not part of the double helix. It does so when this nucleotide does not belong in its place. For example, if a dCTP was added to the growing chain when the template has an adenine nucleotide in this position, no hydrogen bonding between these nucleotides will occur. As a result, DNA polymerase I will remove the C nucleotide from the growing chain and replace it with a T before adding on more nucleotides. This way, DNA polymerase I corrects mistakes in DNA replication.

DNA polymerase III was discovered many years after DNA polymerase I. It acts like DNA polymerase I in that it also adds deoxynucleoside triphosphate to a free 3'-OH group. It also is a 3'-5' exonuclease. In addition, unlike DNA polymerase I, DNA polymerase III can act as a 5'-3' exonuclease, removing nucleotides from the 5' end.

DNA polymerase I will erase the primers and fill the gaps thus created. DNA polymerase III is the main enzyme for DNA synthesis.

DNA polymerase III is involved in the synthesis of the DNA of the lagging strand. DNA polymerase III can only catalyse DNA synthesis in a 5'-3' direction (of the new chain). This means that the lagging strand cannot simply add on new nucleotides as the helicase unzips the double strand. To do so would mean adding them on to the 5' end of the growing strand (which would be 3'-5' direction of replication). This is not possible as DNA polymerase III can only work in a 5'-3' direction.

Biology

8.2.3
State that in an eukaryotic chromosome, replication is initiated at many points.

© IBO 1996

In eukaryotic cells, DNA replication is initiated at many points.

8.3 TRANSCRIPTION.

8.3.1
State that transcription is carried out in a 5' → 3' direction.

© IBO 1996

Transcription is the process of synthesizing RNA from a DNA template. It is carried out in a 5' to 3' direction (of the new RNA strand) and involves RNA polymerase. mRNA, tRNA and rRNA all need to be transcribed for protein synthesis to take place.

8.3.2
Outline the Lac Operon model as an example of the control of gene expression in prokaryotes.

© IBO 1996

Gene expression can be studied in prokaryotes more easily than in eukaryotes. In 1961, Jacob and Monod studied the control of protein synthesis in a now classic study involving *E. coli* bacteria and lactose digesting enzymes. They found that *E. coli* does not produce lactose digesting enzymes when grown on a medium without lactose. However, the lactose digesting enzyme is present only minutes after lactose had been added.

Jacob and Monod discovered the that genes can be 'switched on' or 'switched off' as needed, simply by transcribing or not transcribing them into mRNA. A gene that is 'switched on' will be transcribed, producing mRNA. The mRNA will be translated and a protein will be produced. In the *E. coli* study, the protein produced will be the enzyme lactase.

The model for switching genes on and off, proposed by Jacob and Monod, is the **operon model**. An operon includes the promoter, the operator and the structural gene(s). The operon involved in their studies is known as the **lac operon**, an example of an inducible system in prokaryotes.

To transcribe the structural gene for the lactose digesting enzyme into the required mRNA, the RNA polymerase needs to attach to the promoter section of the DNA so that it can make the primer. However, the operator section is situated next to the promoter section. The operator section is the binding site for a repressor protein produced by another gene (the regulator gene) elsewhere on the bacterial DNA (see diagram below).

If the repressor protein is present, it will attach to the operator site and prevent the RNA polymerase from binding to the gene and hence from transcribing the structural gene into mRNA.

However if lactose (or a related sugar) is present, it will bind to the repressor protein. This changes the shape of the repressor protein which can then no longer bind to the operator. Hence now the promoter is no longer blocked, RNA polymerase can come in and the structural gene can be transcribed. mRNA will be formed and the lactose digesting enzyme will be produced.

This is represented in the schematic diagram below.

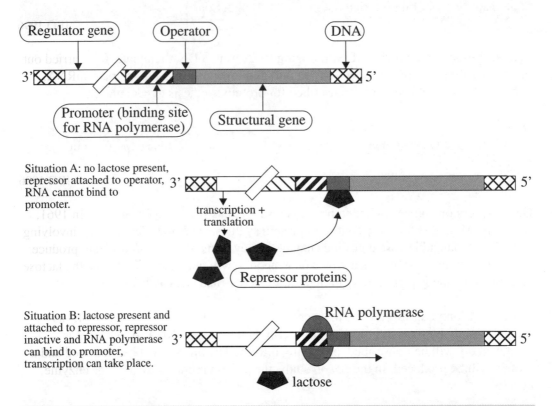

8.3.3
Explain the process of transcription in eukaryotes including the role of promoter region, RNA polymerase, ATP, and terminator.

© IBO 1996

Transcription in eukaryotic cells involves a **promoter region** as in the example seen above. The promoter is the site for binding RNA polymerase which will attach the individual nucleotides together to a single strand of RNA. At the (5') end of the structural gene, a terminator site is found which will stop the transcription process.

RNA polymerase is similar to DNA polymerase and works in almost the same way. RNA nucleotides in the form of ribonucleoside triphosphates form hydrogen bonds with the complementary nucleotide of the DNA strand.

Biology

The only difference between ribonucleoside triphosphate and deoxynucleoside triphosphate which you saw in section 8.2 is one hydroxyl group (-OH group) on Carbon number 2 in the pentose sugar. This changes the deoxyribose into ribose.

As in DNA replication, the ribonucleoside triphosphate will attach (by covalent bond) to the 3' hydroxyl group of the growing strand. The second and third phosphate groups will be removed, providing the energy required to drive this reaction.

8.3.4
State that eukaryotic chromosomes contain far more DNA than is needed to code for their protein products (cross reference 8.1.2).

© IBO 1996

In eukaryotic cells, less than 10% of the DNA present codes for proteins, in humans it is probably even less than 1%. This contrasts sharply with prokaryotes and viruses who express almost all their genes.

Almost half the DNA of a eukaryotic cell consists of nucleotide sequences that are repeated many, many times. This contrasts sharply with the principle in Mendelian genetics that a gene should only be found twice in a eukaryotic cell.

8.3.5
Outline the difference between introns and exons.

© IBO 1996

While in prokaryotes, genes are uninterrupted sections of DNA, it appears that in eukaryotic cells, coding sections of DNA are interrupted with long non-coding intervening sequences. In other words, many genes are discontinuous. The intervening sequences are called **introns**, while the coding sequences are called **exons** (since they are expressed).

8.3.6
State that eukaryotic RNA needs the removal of introns to form mature mRNA and that this process is called splicing.

© IBO 1996

It appears that the entire section of the DNA (introns and exons) is transcribed into RNA. Before the RNA leaves the nucleus, a 'cap' is added to the 5' end (the front) which will protect the mRNA from phosphatases and nucleases and assist in protein synthesis. A 'tail' is added to the 3' end. The 'tail' will increase the stability of the mRNA but is not required for translation or for transport to the cytoplasm.

Enzymes will very precisely cut the bond between exon and intron nucleotides and attach the remaining exon nucleotides to each other. (The introns are taken out.) This process is called **splicing**. The RNA is now ready to travel to the cytoplasm.

Nucleic Acids and Proteins

8.3.7
State that a small group of viruses, known as retroviruses, cause host cells to synthesise viral reverse transcriptase (cross reference 5.3.6 and 12.1.5).

© IBO 1996

A small group of viruses, known as **retroviruses**, cause their host cells to produce a special enzyme called **reverse transcriptase**. The viruses that have RNA for their genetic material, usually use this enzyme for replication. The RNA is the genetic material but can also be used directly as mRNA for protein synthesis. By causing the host cell to produce reverse transcriptase, the retroviruses have their viral RNA transcribed into DNA which then becomes part of the host cell's DNA. The host cell is then directed to transcribe this DNA into many RNA copies and to produce many protein coats so that many new viruses can be assembled. Cancer-causing viruses and HIV are examples of retroviruses.

8.3.8
State that reverse transcriptase catalyses the production of single-stranded 'novel' DNA from RNA.

© IBO 1996

The reverse transcriptase causes the cell to produce a single strand of DNA from the viral RNA which will form a double helix with the viral RNA. The enzyme will then digest the viral RNA strand, leaving the single DNA strand. This DNA will then use the reverse transcriptase again to produce the complementary DNA strand and thus a double stranded DNA molecules is produced, based on the information of the viral RNA.
So reverse transcriptase catalyses the formation of the first DNA strand from RNA. It then hydrolyses the RNA strand and then directs the formation of the DNA strand complementary to the one made earlier.

The new piece of (viral) DNA will be incorporated into a chromosome. In the case of HIV, after infecting a T-helper cell (leucocyte), it may then become dormant and when the cell divides, the viral DNA is simply copied along with the cell's own DNA. Many years later, the viral DNA suddenly becomes active and causes the cell to produce many copies of the virus, harming the T-helper cell. The disease AIDS is diagnosed when many T-helper cells have been destroyed and the proper functioning of the immune system is no longer possible.

8.3.9
Explain why reverse transcriptase is a useful tool for molecular biologists.

© IBO 1996

Although cancer causing viruses and the HIV virus are retroviruses using reverse transcriptase, not all aspects of this enzyme are negative. Molecular biologists have found it extremely useful in producing DNA from mRNA. Extracting insulin from the pancreas of slaughtered cattle used to be the only way of obtaining this hormone required daily by diabetics. Unfortunately, cow's insulin and human insulin are not identical and a few diabetics could not tolerate this difference.

Using reverse transcriptase to produce the gene from mRNA meant that the DNA only carried the exons for insulin since the introns were not in the mRNA and hence not in the reverse transcribed DNA. This made the gene smaller and easier to work with. It was then inserted into *E. coli* bacteria which obligingly started to produce insulin. This insulin is identical to human insulin since it was based on human mRNA.

8.4 TRANSLATION

8.4.1
Outline that the structure of a tRNA allows recognition by a tRNA activating enzyme that binds a specific amino acid to it using ATP for energy.

© IBO 1996

Translation is the synthesis of polypeptide chains. The information that was contained by the sequence of the nucleotides in the mRNA is now translated into a sequence of amino acids.

The process of translation involves all 3 kinds of RNA. They must have been transcribed earlier from different genes. The information about the desired sequence of amino acids is found in the nucleotide sequence (codon sequence) of the mRNA. The ribosomes provide the site for the anticodon of tRNA to bind to mRNA. There are 64 different codons possible on the mRNA and all (except the 3 'stop' codons) have corresponding tRNA which altogether only carry 20 different amino acids. Here we also see that the genetic code is 'degenerate'.

A tRNA molecule is a single strand of RNA which will fold back upon itself because of base pairing between nucleotides of the strand. tRNA has a typical clover leaf shape as shown below.

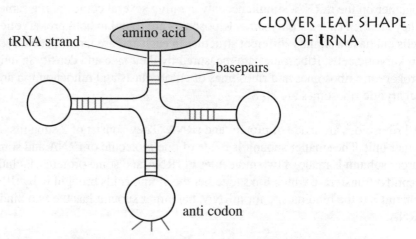

CLOVER LEAF SHAPE OF tRNA

One side of the tRNA is the amino acid attachment site; this always contains the nucleotide sequence CCA. The adenine is found on the 3' end of the RNA strand. On the other side of the tRNA molecule, 3 nucleotides are sticking out. This is the anti-codon which, in the presence of a ribosome, can bind to the codon of the mRNA.

Nucleic Acids and Proteins

The amino acid will bind to the attachment site of the tRNA by a condensation reaction. The -OH of the -C(O)OH end of the amino acid will react with the -OH of the adenosine at the 3' end of the tRNA nucleotide. Water will be formed and the remainder of the amino acid will have formed an oxygen bridge with the tRNA. This is shown in the diagram below.

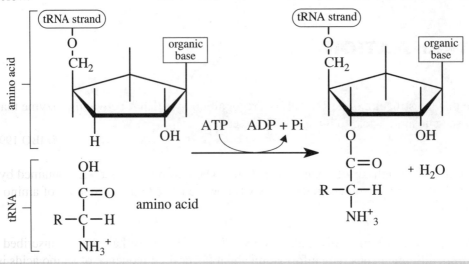

8.4.2
Outline the structure of ribosomes including protein and RNA composition, large and small subunits, two tRNA binding sites and mRNA binding sites.

© IBO 1996

tRNA will not bind to mRNA, unless ribosomes are present. Ribosomes are the most numerous cell organelle. Cells can have thousands of ribosomes, eukaryotic cells even tens of thousands. The nucleolus contains (many copies of) the information on how to make rRNA. Ribosomes measure approximately 25 nm. Ribosomes sometimes are clustered together on the mRNA, simultaneously creating several copies of the same polypeptide. They are then called a **polysome**. Ribosomes are found in both prokaryotic and eukaryotic cells but have a slightly different structure in each and are larger in eukaryotic cells than in prokaryotic cells. Ribosomes are measured by their size and density in units called 'S'. Prokaryotic ribosomes and ribosomes of chloroplasts and mitochondria are 70 S, whereas eukaryotic ribosomes are 80 S.

All ribosomes are made of protein and rRNA. They consist of 2 subunits: a smaller and a larger unit. The smaller subunit is made of one molecule of rRNA and some proteins, the larger subunit is made of two molecules of rRNA and some proteins, including the enzyme peptidyl transferase which links together the amino acids brought in by tRNA. The smaller subunit has the binding site for mRNA, the larger subunit has the two binding sites for tRNA.

8.4.3
State that translation consists of initiation, elongation and termination.

© IBO 1996

The actual process of translation consists of 3 stages: **initiation**, **elongation** and **termination**.

8.4.4
State that translation occurs in a 5' → 3' direction.

© IBO 1996

As stated earlier, the 'start' codon is found at the 5' end of the mRNA. This means that translation takes place in a 5' to 3' direction (of the mRNA).

8.4.5
Explain in detail the process of translation including GTP, ribosomes (including peptidyl transferase), polysomes, start codon and stop codons.

© IBO 1996

Initiation. The smaller subunit of the ribosome will attach to the mRNA 5' end (in prokaryotes sometimes even before transcription is completed). The first codon is the 'start' codon AUG, also coding for the amino acid methionine (met). The tRNA with the anticodon UAC, carrying methionine will attach to the mRNA. This causes the larger subunit of the ribosome to bind to the smaller one and the tRNA fits into one of the two binding sites (the P site). The energy needed for attaching the larger subunit to the smaller one is provided by GTP.

Elongation. At the end of initiation, the second codon of the mRNA is lined up with the second binding site of the larger subunit of the ribosome (the A site). A tRNA (carrying an amino acid) with an anticodon complementary to the mRNA codon will fit into this slot. Peptidyl transferase will form a bond between the two amino acids. The ribosome moves on the next codon, releasing the first tRNA. The second tRNA now occupies the P site and the A site is available for another tRNA. This process continues until the ribosome reaches a 'stop' codon on the mRNA.

Termination. When the ribosome reaches a 'stop' codon (UAG, UAA or UGA), there is no tRNA available which has the correct anticodon. The A site will not be occupied, translation stops and the ribosmal subunits separate.

8.4.6
State that free ribosomes synthesise proteins for use primarily within the cell itself and that bound ribosomes synthesise proteins primarily for secretion and lysosomes.

© IBO 1996

The distribution of ribosomes depends on the function of the protein they make: if the proteins are to be used inside the cell, the ribosomes tend to be found throughout the cytoplasm; if the protein is to be exported (secretion) or used by lysosomes, the ribosomes are generally stuck to the endoplasmic reticulum. These protein will then enter the lumen of the rER as they are produced and they will be processed from there.

8.5 PROTEINS

8.5.1 & 8.5.2
Explain the four levels of structure of proteins, indicating their significance.
Outline the difference between fibrous and globular proteins, with reference to two examples of each type.

© IBO 1996

Four 'levels' are distinguished in the structure of a protein:

- **primary structure:** the sequence of the amino acids in the chain. The linear sequence of amino acids with peptide linkages affects all the subsequent levels of structure since these are the consequence of interactions between the R group. Each amino acid is characterised by its R group. Polar R groups will interact with other polar R groups further down the chain and the same goes for non-polar R groups.

- **secondary structure:** the coils of the chain, for example α helix and β pleated sheath. α helix structures are found in hair, wool, horn, feathers; β pleated sheath is found in silk. Hydrogen bonds are responsible for the secondary structure. Fibrous proteins like collagen and keratin are in helix or pleated-sheet form caused by a regular repeated sequence of amino acids. They are structural proteins.

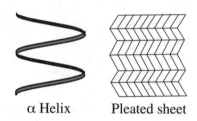

α Helix   Pleated sheet

- **tertiary structure:** the way the helix chain is folded caused by the interactions of the R groups. Hydrophobic groups cluster together on the inside (away from the water) as do hydrophilic groups which are found on the outside (near the water). Some amino acids have a sulfur molecule in their R group. Two of these molecules may come together and the sulfur atoms will form a covalent bond: the disulfide bridge. Hydrogen bonds are also involved in the formation of the tertiary structure. Bonds between an ion serving as a cofactor (see section 8.6) and the R group of a certain amino acid may also be responsible for the folds in the polypeptide. Proteins which have a globular (folded) shape are globular proteins for example haemoglobin. Microtubules are globular proteins which are structural. Enzymes are globular proteins. The folding of the polypeptide creates the 'active site', i.e. the location where the substrate binds to the enzyme so that the reaction can take place. (see section 8.6).

- **quaternary structure:** many proteins (especially large globular proteins) are made of more than one polypeptide chain. Together with the greater variety in amino acids, this causes a greater range of biological activity. The different polypeptide chains are kept together by hydrogen bonds, attraction between positive and negative charges, hydrophobic forces and disulfide bridges or any combination of the above. Cofactors may also assist in the quaternary structure.

Biology

8.5.3
Explain the significance of polar and non-polar amino acids (cross reference 7.1.3, 1.4.1 and 1.4.2).

© IBO 1996

Of the 20 amino acids commonly used to build proteins, 8 have non-polar (hydrophobic) R groups. The others have polar R groups and are soluble in water. The non-polar amino acids in the polypeptide chain will cluster together in the centre of the molecule and contribute to the tertiary structure of the molecule. In general, the more non-polar amino acids a protein contains, the less soluble it is in water. In membranes, proteins are found in between phospholipids. The phospholipid layer is polar on the outside and non-polar in the centre. The protein will often arrange itself so that it exposes hydrophilic portions to the outside of the membrane and hydrophobic sections to the centre. (see section 1.4)

8.5.4
State six functions of proteins, giving a named example of each.

© IBO 1996

Functions of proteins:
- **enzymes:** all enzymes are (globular) proteins, for example amylase which catalyses the reaction: starch → maltose.
- **hormones:** some hormones are proteins, others are steroids. An example of a protein hormone is insulin.
- **antibodies:** antibodies or immunoglobulins are globular proteins assisting in the defense against foreign particles.
- **structural proteins:** collagen is a fibrous, structural protein which build tendons and is an important part of your skin.
- proteins are part of the cell membrane, playing a role in the passage of substances into and out of the cell.
- haemoglobin is a protein which easily binds to oxygen due to the haem group attached to it.

8.6 ENZYMES

8.6.1
State that metabolic pathways consist of chains and cycles of enzyme catalysed reactions.

© IBO 1996

Very few, if any, chemical changes in a cell result from a single reaction. **Metabolic pathways** are chains and cycles of enzyme catalysed reactions.

The reactions will rarely occur spontaneously at reasonable speed at room temperature. Therefore enzymes are used to speed up a reaction. You may safely assume that biological reactions only occur in the presence of enzymes.

8.6.3
Explain that enzymes lower the activation energy of the chemical reactions that they catalyse.

© IBO 1996

How do enzymes work?

The presence of an enzyme will speed up the reaction because the active site will facilitate the chemical change. This happens by a means of lowering the activation energy. Every reaction requires a certain amount of **activation energy**. If two molecules are going to react with each other, they need to collide with a certain speed. The higher the activation energy, the higher the speed required. At a low temperature, only few molecules will have this speed which means that the rate of reaction is low.

The active site of the enzyme assists in the chemical reaction by lowering the required activation energy. This means that more molecules are able to react and the rate of reaction will increase.

The diagram shows what happens to the energy level during an exothermic reaction.

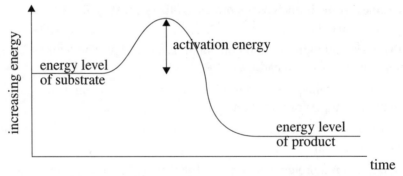

Now if we added enzyme to this reaction, we could plot the following diagram:

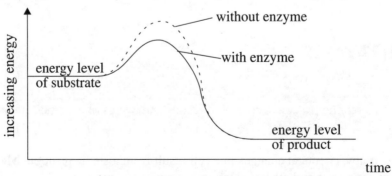

As you can see, the enzyme has reduced the required activation energy.

Biology

8.6.2
Describe the 'induced fit' model.

© IBO 1996

As we discussed earlier (2.3.3) a lock and key model exists to explain the specificity of enzymes. The lock and key model was first suggested by Emil Fischer in 1894. Soon after, it appeared that certain enzymes can catalyse several (similar) reactions. The **induced fit model** suggests the following: The active site may not be as rigid as orginally was thought. Its shape will adapt somewhat to allow several slightly different substrates to fit. The active site will interact with the substrate and adapt to make the perfect fit. It is like a glove which will fit on several hands but not on a foot.

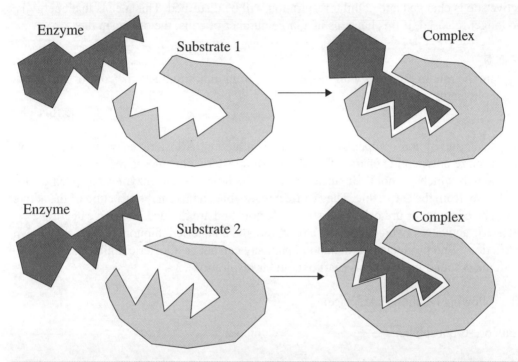

8.6.4
Explain the difference between competitive and non-competitive inhibition with reference to one example of each type.

© IBO 1996

A number of molecules exist which can reduce the rate of an enzyme controlled reaction. These molecules are called **inhibitors**. There are two kinds of inhibitors:

- **competitive inhibitors:** the inhibiting molecule is so similar to the substrate molecule that it binds to the active site of the enzyme and prevents the substrate from binding. Adding more substrate will reduce the effect of the inhibitor.

example:
Prontosil (an antibiotic) which inhibits synthesis of folic acid (vit B, which acts as a coenzyme) in bacteria. The drug will bind to the enzyme which makes folic acid. The folic

acid will no longer be made and the bacterial cell dies. The animal cells are not damaged since they do not make folic acid but absorb it from food. The animal cells therefore lack the enzyme and the drug has no effect.

- **non-competitive inhibitors:** the inhibiting molecule binds to the enzyme in a place which is NOT its active site. As a result, the shape of the active site of the enzyme changes and the substrate molecule will no longer fit. Adding more substrate will have no effect on the reaction rate.

example:
Cyanide (CN^-) will attach itself to the -SH groups in an enzyme. It thereby destroys the disulfide bridges (-S-S-)and changes the tertiary structure of the enzyme. The shape of the active site is changed and cellular respiration will be disturbed. This means that energy is no longer released. If this happens in a large number of cells, the organism dies.

8.6.5
Explain the role of allostery with respect to feedback inhibition and the control of metabolic pathways.

© IBO 1996

A special kind of non-competitive inhibition is **allostery**. Allosteric enzymes are made of two or more polypeptide chains. The activity of allosteric enzymes is regulated by compounds which are not their substrates and which bind to the enzyme at a specific site well away from the active site. They cause a reversible change in the structure of the active site. The compounds are called allosteric effectors and are divided into two categories: allosteric activators (which speed up a reaction) and allosteric inhibitors (which slow down a reaction). End products of a metabolic pathway can act as allosteric inhibitors. An example is found in glycolysis (part of cellular respiration).

The following reactions take place:

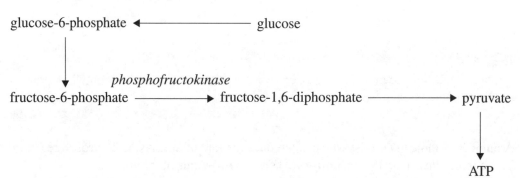

Phosphofructokinase catalyses the reaction of fructose-6-phophate to fructose-1,6-diphosphate. The chain of reactions will continue and eventually form ATP. If ATP is already present, it will bind to the phosphofructokinase and change the shape of the active site, decreasing the activity. However, when ATP is not present, phosphofructokinase is in its active form and ATP will be produced. So ATP is the allosteric inhibitor of phosphofructokinase. This is an example of negative feedback.

EXERCISE

1. The complex of DNA and protein is known as:

    A   a histone.
    B   a nucleosome.
    C   genes.
    D   chromatin.

2. The units of genetic information are called:

    A   histones.
    B   nucleosomes.
    C   genes.
    D   chromatin.

3. The diagram shows a section of DNA. The point labelled X is the:

    A   3' end.
    B   5' end.
    C   gene.
    D   chromosome.

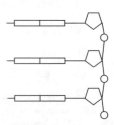

4. State that DNA replication occurs in a

    A   random direction.
    B   helical direction.
    C   3' → 5' direction.
    D   5' → 3' direction.

5. The diagram illustrates DNA replication. The point marked X is the:

    A   lagging strand.
    B   leading strand.
    C   DNA polymerase.
    D   Okazaki fragment.

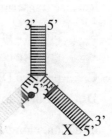

6. Which of the following is not an enzyme involved in DNA replication?

    A   helicase.
    B   DNA ligase.
    C   DNA polymerase.
    D   amylase.

Nucleic Acids and Proteins

7. Transcription is carried out in a:

    A random direction.

    B helical direction.

    C 3' → 5' direction.

    D 5' → 3' direction.

8. In the Lac Operon model, genes can be

    A 'switched on' or 'switched off' as needed, by enzymes.

    B 'switched on' or 'switched off' as needed, by transcribing or not transcribing them into mRNA.

    C 'switched on' or 'switched off' as needed, by transcribing or not transcribing them into DNA.

    D 'switched on' or 'switched off' as needed, by translation.

9. The enzyme called reverse transcriptase is produced by:

    A all viruses.

    B retroviruses.

    C parasites.

    D bacteria.

10. HIV is caused by:

    A a retrovirus.

    B lifestyle.

    C diet.

    D a bacterium.

11. Reverse transcriptase is a useful tool because it can be used to:

    A produce proteins.

    B produce enzymes.

    C produce DNA from mRNA.

    D produce mRNA from DNA.

12. Initiation, elongation and termination are the main stages in:

    A transcription.

    B translation.

    C replication.

    D decomposition.

13. Explain the structure of DNA by drawing a segment of 6 linked nucleotides.

14. What is the function of each of the following enzymes in DNA replication?
    a. helicase
    b. RNA primase
    c. DNA polymerase III
    d. DNA polymerase I
    e. DNA ligase

15. This question is about the lac operon model.
    a. What is the immediate role of the regulator gene?
    b. What is the role of the promotor region?
    c. Where do repressor proteins bind?
    d. How does the repressor stop production of the lactose digesting protein?
    e. What happens when lactose is present?

16. What is the role of the following in translation?
    a. GTP
    b. ribosomes
    c. polysomes
    d. codons

17. Which types of bonds are involved in each of the four levels of protein structure?

18. a. Where does the competitive inhibitor bind to the enzyme?
    b. Where does the non-competitive inhibitor bind to the enzyme?
    c. Give an example of each.

CELL RESPIRATION AND PHOTOSYNTHESIS

9

Chapter contents
- Cell respiration
- Photosynthesis

Cell respiration and photosynthesis

9.1 CELL RESPIRATION

9.1.5
Draw the structure of a mitochondrion as seen in electronmicrographs.

© IBO 1996

As you saw in Section 4.2.10, cell respiration is the release of energy from organic molecules made in photosynthesis. The process by which this energy is released will be studied in more detail in this section.

Below you find a schematic diagram of the structure of a mitochondrion, based on EM data.

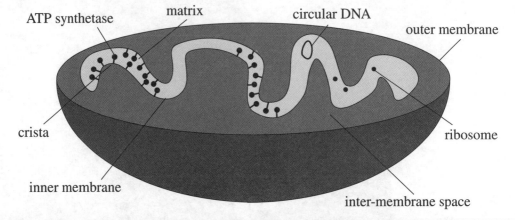

9.1.1
Outline that oxidation involves the loss of electrons from an element whereas reduction involves gain in electrons, and that oxidation frequently involves gaining oxygen or losing hydrogen; whereas reduction frequently involves loss of oxygen or gain in hydrogen.

© IBO 1996

In cell respiration, as in photosynthesis (see Section 9.2), reactions often involve the movement of electrons. This kind of reaction is called **redox reaction**. In these reduction-oxidation reactions, one compound loses some electrons and the other compound gains them.

OIL RIG: **O**xidation **I**s **L**oss (of electrons), **R**eduction **I**s **G**ain (of electrons).

The process of oxidation often involves gaining oxygen (hence its name) or losing hydrogen while reduction often involves the loss of oxygen or gain in hydrogen.

A substance which has been reduced, now has the power to reduce others (and become oxidised in the process); e.g. NADH and NADPH.

9.1.2
Outline what is achieved by the process of glycolysis including phosphorylation, lysis, oxidation and ATP formation.

© IBO 1996

Glycolysis takes place in the cytoplasm and produces 2 pyruvate molecules from every glucose in the following reaction:

Glucose + 2ADP + 2P_i + 2NAD+ → 2Pyruvate + 2ATP + 2NADH + 2H^+ + 2H_2O
No oxygen is needed in this step of the reaction.

The structural formula of pryruvate is:

$$\begin{array}{c} O=C-O^- \\ | \\ C=O \\ | \\ CH_3 \end{array}$$

The following key steps take place in glycolysis (a more detailed diagram can be found in option C section C.5.2, chapter 19. You should consult this diagram as well.):

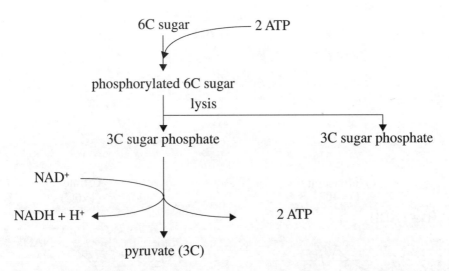

(you do not need to memorise this but you must remember the principles)

What happens is that in the cytoplasm one 6-C sugar is converted into two 3-C compounds (pyruvate) with a net gain of 2 ATP + 2 NADH + H^+.

To achieve this it is necessary to change glucose into fructose-1,6-diphosphate which is then split into two 3-C compounds (**lysis**). This process requires energy. Subsequently, the 3-C compounds are oxidised into pyruvate. During this process, energy is released (ATP is formed) and NAD is reduced into NADH.

Phosphorylation: the process of making ATP in vivo.

Cell respiration and photosynthesis

9.1.3
Outline aerobic respiration including oxidative decarboxylation of 2-oxopropanoate (pyruvate), Krebs cycle, NADH + H⁺ and electron transport chain.

© IBO 1996

If oxygen is present, pyruvate is transported to the mitochondrial matrix and the reactions continues in the following way:

$$\text{Pyruvate} + \text{CoA} + \text{NAD}^+ \rightarrow \text{Acetyl CoA} + CO_2 + \text{NADH} + H^+$$

or

$$CH_3.CO.COOH + CoAS\text{-}H + NAD^+ \rightarrow CO_2 + NADH + H^+ + CH_3CO\text{-}S\text{-}CoA$$

The reaction is known as the **link reaction** because it forms the link between glycolysis and the Krebs cycle. This process is known as oxidative decarboxylation of pyruvate (2-oxopropanoate).

Krebs cycle

The **Krebs cycle** occurs in the matrix of the mitochondria and produces CO_2, NADH + H⁺, $FADH_2$ and ATP.

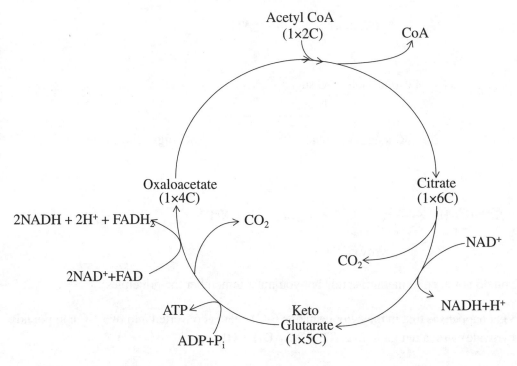

SUMMARY OF THE KREBS CYCLE

As you can see, one turn of the Krebs cycle yields:
    2 × CO_2 3 × NADH + H⁺   1 × $FADH_2$ 1 × ATP

Remember that one of the key roles of the co-enzymes NAD⁺ and FAD are to collect and remove H⁺.

The last step of aerobic respiration is the Electron Transport Chain. The ETC passes two hydrogens (and two electrons) from NADH or $FADH_2$ from one electron carrier to another (found in the inner membrane of the mitochondrion) by a series of redox reactions. The final acceptor is oxygen (producing water). This process produces 3 ATP for every 2 hydrogens (and 2 electrons) from NADH + H^+ and 2 ATP for every 2 hydrogens (and 2 electrons) from $FADH_2$.

9.1.4
Describe oxidative phosphorylation in terms of chemiosmosis including proton pumps, a proton gradient and ATP synthetase (cross reference 9.2.4).

© IBO 1996

Oxidative phosphorylation occurs on the inner membrane of the mitochondria. As the electrons are passed down the electron transport chain, protons are being pumped across that membrane. The resulting **proton gradient** drives the production of ATP (from ADP and P_i) by ATP synthetase. This is the chemiosmotic theory of Peter Mitchell.

As was said in the previous section, the net result of this process is that 1 NADH + H^+ supplies enough energy to produce 3 ATP from 3 ADP + 3 P_i and 1 $FADH_2$ supplies enough energy to produce 2 ATP from 2 ADP + 2 P_i. During these reactions NADH + H^+ and $FADH_2$ are returned to the form of NAD^+ and FAD.

The mechanism of this series of reactions is that the energy from NADH + H^+ and $FADH_2$ is transferred to ATP through a series of electron carriers.

This series of electron carries finally yields H^+ and electrons to oxygen (O_2) to form water (H_2O). However if no oxygen is present, this reaction cannot take place. As a consequence, no NAD^+ or FAD is formed and hence the Krebs cycle cannot operate. This will cause acetyl CoA to accumulate and as a result it will no longer be produced from pyruvate. Glycolysis will continue to operate however, since, even without oxygen, it is possible to break down pyruvate and release some energy. This process is less efficient though.

The chemiosmotic theory of Peter Mitchell.
It had already been obvious for some time that a link existed between the electrons being passed down the electron transport chain and the production of ATP. Peter Mitchell discovered that during the passing of the 'high energy' electrons down the electron transport chain, protons are being pumped across the inner mitochondrial membrane.

There is a build up of H^+ ions in the intermembrane space. The diffusion and electromagnetic forces will drive H^+ through the ATP synthetase molecule. As the H^+ ions go through the ATP synthetase molecule, the potential energy they posses will be used to drive ATP synthesis.

Cell respiration and photosynthesis

THE ELECTRON TRANSPORT CHAIN IN CELLULAR RESPIRATION

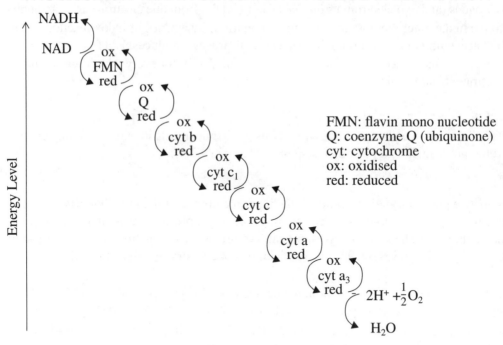

FMN: flavin mono nucleotide
Q: coenzyme Q (ubiquinone)
cyt: cytochrome
ox: oxidised
red: reduced

CHEMIOSMOTIC COUPLING OF ELECTRON TRANSPORT CHAIN AND OXIDATIVE PHOSPHORYLATION

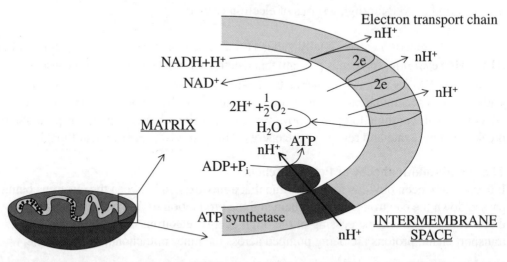

9.1.6
Explain the relationship between the structure of the mitochondrion and its function.
© IBO 1996

Keeping in mind all of the above information, it is useful to return to the structure of the mitochondrion.

The **outer membrane** is a regular membrane, separating the mitochondrion from the cytoplasm. Its structure is based on the fluid mosaic model. It is impermeable to H^+ ions.

The **intermembrane space** has a higher concentration of H^+ ions because of the electron transport chain. It pH is lower.

The **inner membrane** is folded into **cristae** to provide maximum space for the electron carriers and ATP synthetase. It is impermeable to H+ ions. Its structure is based on the fluid mosaic model with the electron carriers and the ATP synthetase embedded among the phospholipid molecules. The ATP synthetase molecules can be seen on the cristae.
The **matrix** contains the enzymes which enable the Krebs cycle to proceed.

Glycolysis takes place in the cytoplasm. Pyruvate is transported to the matrix of the mitochondrion and decarboxylated to acetyl CoA which enters the Krebs cycle. The resulting $NADH + H^+$ and $FADH_2$ give their electrons to the electron carriers in the inner membrane. The electrons move through the membrane as they are passed from one electron carrier to another in a series of redox reactions. During this process, H^+ ions are pumped from the matrix into the intermembrane space, creating a potential difference. Electromagnetic and diffusion forces drive the H^+ ions back to the matrix through the ATP synthetase which uses the energy released to combine ADP and P_i into ATP, which is released into the matrix.

9.1.7
Describe the central role of ethanoyl (acetyl) CoA in carbohydrate and fat metabolism.
© IBO 1996

In the above sections, you have seen the key role in the carbohydrate metabolism, played by acetyl CoA, essentially linking the Krebs cycle to glycolysis. Acetyl CoA plays a similar key role in the metabolism of fatty acids. Since fats contain more energy per gram than carbohydrates or proteins (see Section 2.2.7), an efficient system must exist to break down fatty acids in cellular respiration. The long chains of fatty acids are oxidised, effectively breaking off sections of 2 Carbon molecules. These are changed into acetylCoA and enter the Krebs cycle.

9.1.8
Outline fermentation to 2-hydroxypropanoate (lactate) and to ethanol, and the circumstances in which they occur in cells.
© IBO 1996

When no oxygen is available, cells are capable of anaerobic respiration. Although anaerobic respiration releases some energy, the yield is much lower than that of aerobic respiration because a lot of the energy of the glucose remains 'locked up' in the end product (lactic acid or ethanol).

aerobic respiration:
$$C_6H_{12}O_6 + 6\ O_2 \rightarrow 6\ CO_2 + 6\ H_2O + 2880\ kJ$$

Cell respiration and photosynthesis

anaerobic respiration: (e.g. alcoholic fermentation, in yeast)
$$C_6H_{12}O_6 \rightarrow 2\ C_2H_5OH\ (ethanol) + 2\ CO_2 + 210\ kJ$$

or lactic acid production in muscle cells
$$C_6H_{12}O_6 \rightarrow 2\ C_3H_6O_3\ (lactic\ acid) + 150\ kJ$$

In anaerobic respiration, glycolysis takes place in the cytoplasm as we have seen before. During this process, the net gain is 2 ATP and 2 NADH + H^+. The reaction cannot continue in the manner described above since the ultimate oxygen acceptor is not available. This means that the electron carriers are reduced, and the Krebs cycle is unable to continue. To allow glycolysis to proceed, two conditions must be met. Pyruvate (or pyruvic acid) cannot be allowed to accumulate since this would change the equilibrium of the reaction and a supply of NAD^+ is needed. Both conditions can be satisfied by reducing pyruvate into ethanol or lactic acid. NADH + H^+ is oxidised into NAD^+ in the process.

In the production of ethanol, CO_2 is released and the reaction is therefore not reversible. Lactic acid (or lactate) can be used to form pyruvate again (using energy) when oxygen becomes available.

9.2 PHOTOSYNTHESIS

9.2.1
Draw the structure of a chloroplast as seen in electronmicrographs.
© IBO 1996

Photosynthesis occurs in the chloroplasts. These cell organelles found in cells of green plants are 2 - 10 mm in diameter and ovoid in shape when found in higher plants (in green algae their shape varies).

STRUCTURE OF A CHLOROPLAST FROM EM

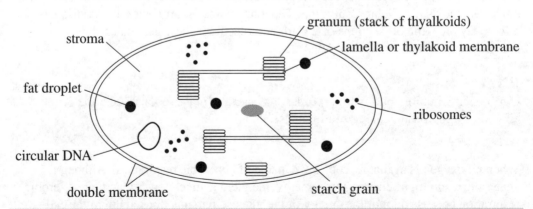

9.2.2
State that photosynthesis consists of light-dependent and light-independent reactions.
© IBO 1996

As you saw in section 4.2.6 and 4.2.7, photosynthesis is NOT a simple one step reaction. It consists of a series of reactions which can be grouped into a **light dependent stage** and a **light independent stage**. The light dependent stage will only take place in the light, the light independent stage can occur at any time, if provided with the required materials. Outside the laboratory, these materials (ATP and NADPH) come from the light dependent stage.

Some texts will still use the terms 'light stage' and 'dark stage'. These are incorrect since they imply that light is required for one stage and darkness for the other. So please do not use them.

9.2.3 & 9.2.4
Explain the light-dependent reactions including the photoactivation of Photosystem II, photolysis of water, electron transport, cyclic and non-cyclic photophosphorylation, photoactivation of Photosystem I and reduction of $NADP^+$.
Explain photophosphorylation in terms of chemiosmosis (cross reference 9.1.4).

© IBO 1996

The diagram in section 4.2.6 gave you some idea of what is happening in the two stages. Now we will look at the light dependent stage in some more detail.

NON-CYCLIC PHOTOPHOSPHORYLATION

Cell respiration and photosynthesis

Non-cyclic photophosphorylation

The light hits the pigments of **photosystem II (PS II)**, which are mainly found in the grana. The pigments involved are mainly **chlorophyll a** and they absorb light at 680 nm and are sometimes called P_{680}. Absorbing this light energy excites some electrons which as a result leave their normal position and move away from the nucleus. They are taken up by an electron acceptor X, resulting in a chlorophyll a molecule with a positive charge. The electrons are then passed through a number of electron carriers in the membrane via oxidation-reduction reactions (see Section 9.1.4) and will end up at PS I.

The presence of Chl a$^+$ will induce the **lysis of water** so that oxygen, H$^+$ and electrons are released. P_{680}^+ is the strongest biological oxidant known.

- The electrons are taken up by Chl a$^+$ (which returns to Chl a).
- The oxygen is released as a waste product.
- The H$^+$ are pumped to the inside of the **grana** (the lumen), they accumulate until the diffusion and electromagnetic forces are enough to drive them through proton channels in the ATP synthetase, driving the chemiosmotic reaction ADP + P$_i$ → ATP (See Section 9.1.4).

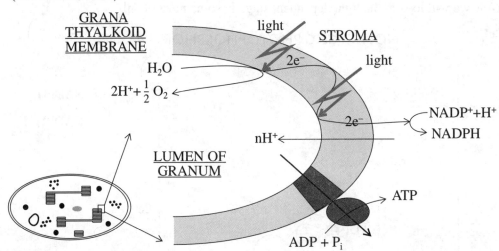

The light also hits **photosystem I (PS I)**, which is mainly found in the intergranal lamella. Due to a slight difference in the protein environment, PS I absorbs light at 700 nm and is also known as P_{700}. Again, the electrons absorb the light energy and move away from the nucleus. They leave the chlorophyll a molecule and are taken up by electron acceptor Y. They are then passed on and taken up by NADP$^+$ which combines with an H$^+$ and is reduced to form NADPH. The Chl a$^+$ receives electrons from the electron carrier chain and becomes an uncharged Chl a molecule.

In **cyclic photophoshorylation**, the electrons from PS I go to electron acceptor Y but instead of being used to produce NADPH, they go through the membrane via several electron carriers (redox reactions) and are returned to PS I. PS II is not involved. This process is cyclic, as its name suggests. It does not produce NADPH but it does produce ATP.

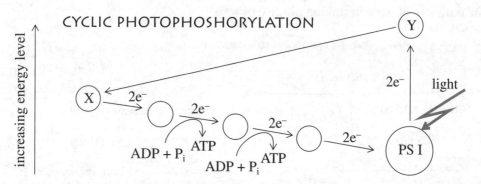

9.2.5
Explain the light-independent reactions including the roles of ribulose bisphosphate (RuBP) carboxylase, reduction of glycerate 3-phosphate (GP) to triose phosphate (TP or GALP), NADPH + H$^+$, ATP, regeneration of RuBP and synthesis of carbohydrate and other products.

© IBO 1996

The light independent stage also has some detail which was not included in 4.2.6. Below you find a diagram of the **Calvin cycle**.

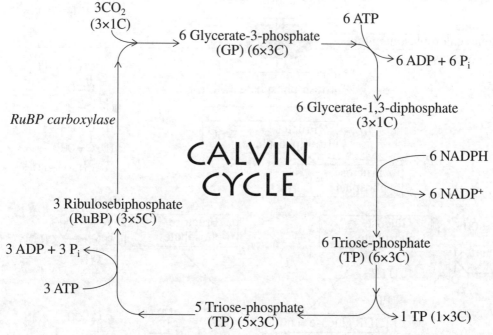

The Calvin cycle takes place in the stroma of the chloroplast. ATP provides the energy and NADPH provides the reducing power needed for biosynthesis using carbon dioxide. **RuBP** is the carbon dioxide acceptor and (catalysed by RuBP carboxylase) will take up CO_2, forming GP. GP will be reduced to TP but this conversion needs energy from ATP and reducing power from NADPH. TP can be converted to glucose, sucrose, starch, fatty acids and amino acids and other products. Of course, TP is also converted into RuBP to keep the cycle going. This process requires energy from ATP.

Cell respiration and photosynthesis

The pathway of carbon during photosynthesis

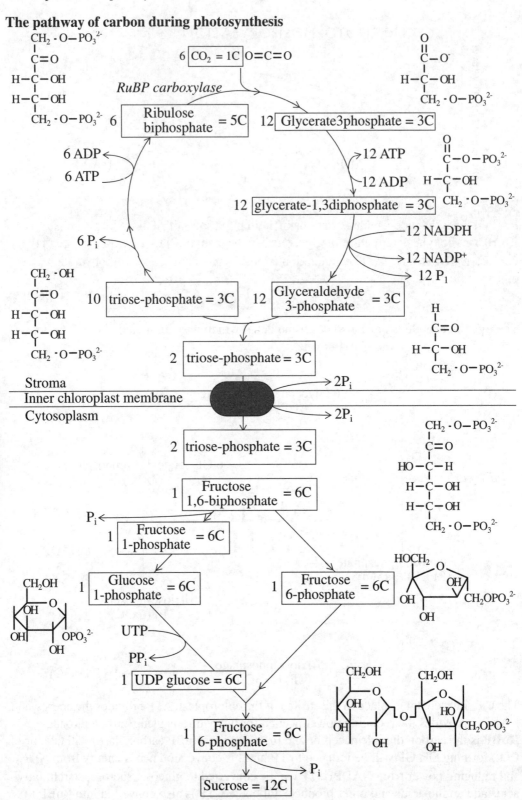

9.2.6
Outline the differences in carbon dioxide fixation between C_3, C_4 and CAM plants, noting their adaptive significance.

© IBO 1996

The pathway of Carbon fixation described in 9.2.5 is called the C_3 pathway because GP (a three carbon compound) is the first recognisable compound after fixation of CO_2. Two alternatives to this exist: a C_4 pathway and a CAM pathway.

The C_4 pathway uses PEP (propenoate-2-phosphate or phospho-enol-pyruvate) to accept CO_2 which will produce malate, a four carbon compound. The enzyme involved is PEP carboxylase. This enzyme has a higher affinity for CO_2 than RuBP carboxylase so that CO_2 can be fixed at lower concentrations. The malate will be moved into the bundle sheath cell where it is returned to pyruvate and CO_2. The CO_2 is then fixed in the normal way by RuBP.

Schematic diagram of a cross section of a C_3 and a C_4 leaf

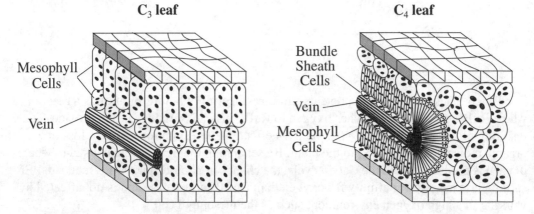

Schematic diagram of the C4 pathway.
a. the anatomy of a C4 leaf and b. electron micrograph of a cross section of a typical C4 leaf.

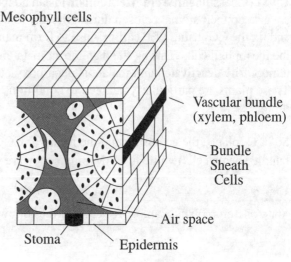

SCHEMATIC DIAGRAM OF THE HATCH-SLACK PATHWAY.

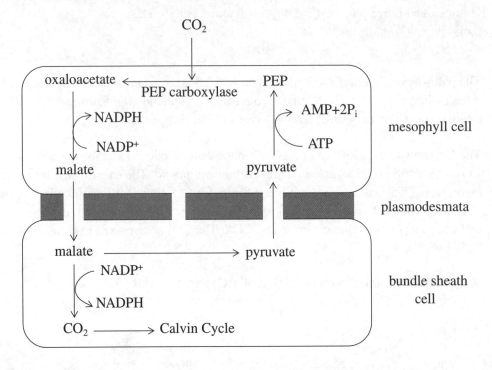

The Hatch Slack pathway also prevents **photorespiration**. Photorespiration is a reaction where RuBP carboxylase attaches oxygen to RuBP instead of fixing carbon dioxide. The resulting product is of no use to the plant. In effect, photorespiration causes a loss of organic molecules (the RuBP) to the plant. By separating the light dependent stage (where oxygen is produced) from the Calvin cycle, the chance of photorespiration is reduced. PEP carboxylase has a lower affinity for oxygen than RuBP carboxylase and is not affected by working in a high oxygen environment such as the mesophyll cell.

CAM (Crassulaean Acid Metabolism) is an adaptation where some plants living in dry areas keep their stomata closed during the day to conserve water. They open them at night and fix the CO_2 using PEP carboxylase to form malate which is stored in a large vacuole in the mesophyll cells. During the day, the stomata are closed, PEP carboxylase is temporarily deactivated and malate is decarboxylated releasing the CO2 for synthesis. These plants are particularly adapted to conditions of high light, high temperature and drought.

So in C_4 plants, the capturing and the fixing of CO_2 are separated in space (mesophyll and bundle sheath cells) while in CAM plants they are separated in time (night and day).

9.2.7
State one crop plant example for each of the following: a C_3, C_4 and CAM plant.

© IBO 1996

Examples of C_3 plants are: rice, wheat, potatoes.
Examples of C_4 plants are: sugar cane, maize (sweet corn).
Examples of CAM plants are: pineapple, prickly pear, vanilla orchid.

9.2.8
Describe how photosynthetic pigments can be separated and identified by means of chromatography.

© IBO 1996

Chromatography is a group of techniques that separates molecules based on differences in the way they become distributed between two phases. Paper chromatography has a sample applied as a small spot to a piece of filter paper. The solvent is allowed to flow across the paper by capillary action. Molecules in the sample will migrate at different rates depending on their relative affinity for the solvent and the paper.

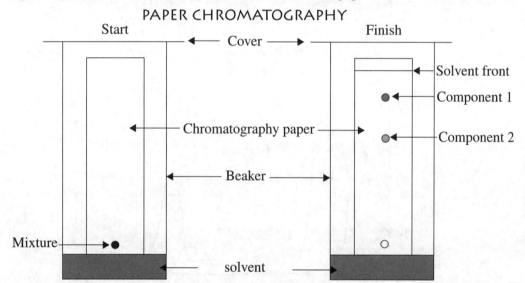

When a sample of photosynthetic pigments are subjected to this technique, you will see the different colours separating.

9.2.9 & 9.2.10
Explain the relationship between the action spectrum and the absorption spectra of photosynthetic pigments.
Explain the concept of limiting factors with reference to light intensity, temperature and concentration of carbon dioxide.

© IBO 1996

The **action spectrum** of photosynthesis is a diagram which tells you how much photosynthesis goes on at any wavelength of light. The action spectrum depends largely upon how much light is absorbed. As you can see in the diagram below, it is closely related to the absorption spectra of several pigments.

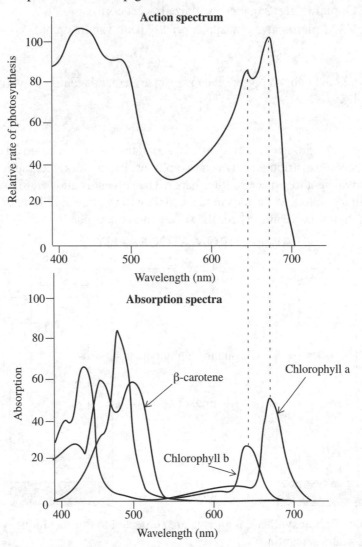

Limiting factors have already been discussed in section 4.2.9.

The factor the furthest away from its optimum value will limit the amount of photosynthesis. This is then the limiting factor. If you improve this factor, the rate of photosynthesis will increase until another factor becomes the limiting factor. If you plot a graph of the amount of photosynthesis versus light intensity, the graph will go up until light is no longer the limiting factor. Then, the amount of photosynthesis will remain constant.

Limiting factors for photosynthesis are: light intensity, temperature, concentration of carbon dioxide.

EXERCISE

Questions 1 & 2 refer to the schematic diagram of a mitochondrion.

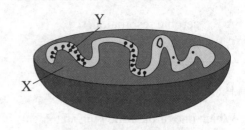

1. X is the:

    A   intermembrane space
    B   DNA
    C   matrix
    D   ATP synthetase

2. Y is the:

    A   crista
    B   DNA
    C   matrix
    D   ATP synthetase

3. In a redox reaction:

    A   both compounds lose electrons.
    B   both compounds gain electrons.
    C   electrons are not invloved in the reaction.
    D   one compound loses electrons and the other compound gains them:

4. Oxidation is:

    A   loss of electrons.
    B   gain of electrons.
    C   removal of oxygen.
    D   removal of hydrogen.

5. Glycolysis is:

    A   the decompsition of glucose to produce pyruvate molecules.
    B   the decompsition of glucose to produce sucrose molecules.
    C   the decompsition of glucose to produce water molecules.
    D   the decompsition of glucose to produce carbon dioxide and water molecules.

6. The process of making ATP in vivo is known as:

    A   hydrolysis.
    B   glycolysis.
    C   phosphorylation.
    D   translation.

Cell respiration and photosynthesis

7. The production of alcohol from sugars by fermentation is an example of:

    A   aerobic respiration.
    (B)   anaerobic respiration.
    C   hydrolysis.
    D   reduction.

8. Which one of the following statements is not true?

    (A)   Light is necessary for every stage of photosynthesis.
    B   Photosynthesis will not occur without light.
    C   Some stages of photosynthesis do not require light.
    D   Photosynthesis is a multi-step reaction.

9. P_{680}^+ is:

    (A)   a part of chlorophyll.
    B   a part of DNA
    C   a biological reducing agent.
    D   an inorganic oxidant.

10. The Calvin cylcle is a part of:

    A   the light dependent stage of photosynthesis.
    (B)   the light independent stage of photosynthesis.
    C   fermentation.
    D   respiration.

11. Rice, wheat and potatoes are examples of:

    (A) C_3 plants.
    B C_4 plants.
    C   CAM plants.
    D   tropical plants.

12. Which one of the following techniques is most suitable for separating the pigments of chlorophyll?

    A   distillation.
    B   crystallisation.
    (C)   paper chromatography.
    D   absorption spectroscopy.

13. Which one of the following is not a limiting factor for photosynthesis?

    A   light intensity.
    (B)   temperature.
    C   concentration of carbon dioxide.
    D   light intensity, temperature, concentration of oxygen.

14.
    a.   Where in the cell does glycosis take place?
    b.   Where in the cell does the Krebs cycle take place?
    c.   Where in the cell is the electron transport chain found?
    d.   Draw a diagram of the structure of a mitochondrion as seem with the electron microscope.
    e.   How does the structure of the site for the Krebs cycle relate to its function?
    f.   How does the structure of the site for the electron transport chain relate to its function?
    g.   What would happen to the all parts of aerobic respiration of the outer membrane of the mitochondrion became permeable to protons (hydrogen ions)?

15.
    a.   What is the function of the ATP and NADPH produced in non-cyclic photophosphorylation?
    b.   What would be the purpose of cyclic photophosphorylation?
    c.   What is the advantage of non-cyclic photophosphorylation over cyclic photophosphorylation for the plant?
    d.   What is the purpose of the Calvin cycle?

16. Compare and contrast the process of ATP production in chloroplasts and mitochondria.

Cell respiration and photosynthesis

GENETICS

10

Chapter contents
- Meiosis
- Dihybrid crosses
- Autosomal gene linkage and gene mapping
- Statistical analysis
- Polygenic inheritance
- Applications of genetics

Genetics

10.1 MEIOSIS

10.1.1
Define homologous chromosomes.

© IBO 1996

Homologous chromosomes: Chromosomes with the same gene loci in the same sequence which are capable of pairing up to form bivalents diring the first prophase of meiosis.

10.1.2
Describe the behaviour of the chromosomes in the phases of meiosis.

© IBO 1996

The behaviour of chromosomes in the various phases of meiosis is described below.

Interphase: * DNA replication.

Prophase I: * Chromosomes condense.
* Nucleolus becomes invisible.
* Spindle formation.
* Synapsis : homologous chromosomes side by side.
(the pair is now called a bivalent, the crossover points are called chiasmata).
* Nuclear membrane disappears (sometimes considered as early metaphase).

Metaphase I: * Bivalents move to the equator.

Anaphase I: * Homologous pairs split up, one chromosome of each pair goes to each pole.

Telophase I: * Chromosomes arrive at poles.
* Spindle disappears.

Prophase II: * New spindle is formed at right angles to the previous spindle.

Metaphase I: * Chromosomes move to the equator.

Anaphase II: * Chromosomes separate, chromatids move to opposite poles.

Telophase II: * Chromosomes have arrived at poles.
* Spindle disappears.
* Nuclear membrane reappears.
* Nucleolus becomes visible.
* Chromosomes become chromatin.

Again cell division (cytokinesis), strictly speaking, is not a part of meiosis but is often considered to be the last stage of telophase II.

10.1.3 CROSSING-OVER.
Outline the process of crossing-over (cross reference 10.3.2).

© IBO 1996

The importance of meiosis as a source of variation is not only found in creating new combinations of the parents chromosomes. New combinations of genes within the chromosome are possible through a process called crossing over. When, during Prophase I, synapsis occurs, the chromatids of the bivalent are close together. It is then possible that parts of two chromatids overlap, break at the chiasmata and re-attach to the other chromatid.

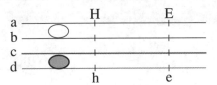 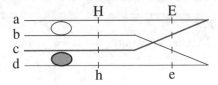

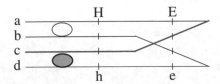 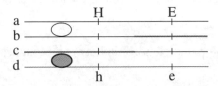

Each of the chromatids a, b, c, d will end up in a gamete. Before crossing over two gametes would have contained H and E (e.g. brown hair and brown eyes) and the other two would have contained genetic information h and e (e.g. blond hair and blue eyes). After crossing over is completed, gamete a will contain H and E (brown hair, brown eyes), gamete b : H and e (brown hair, blue eyes), gamete c : h and E (blond hair and brown eyes) and gamete d : h and e (blond hair, blue eyes). Gametes b and c are therefore new combinations. They are called **recombinants**. See section 10.3.2.

10.1.4
Define chiasma.

© IBO 1996

Chiasma: the points at which homologous chromosomes remain in contact as chromatids move apart during Prophase I of meiosis or a cross shaped structure formed by crossing over between chromosomes or two chromatids.

10.1.5
Explain how meiosis results in an effectively infinite genetic variety in gametes through crossing over in Prophase I and random orientation in Metaphase I (cross reference 3.3.3).

© IBO 1996

Genetics

Meiosis results in an effectively infinite genetic variety in gametes through crossing over in Prophase I and random orientation in Metaphase I.

The number of different types of gametes produced by random orientation alone is 2^n where n = haploid number. Add to this the effect of crossing over and the resulting variation indeed is infinite.

10.1.6
Define recombination.

© IBO 1996

Recombination: the reassortment of genes or characters into different combinations from those of the parents.

10.1.7
State Mendel's Second Law (Law of Independent Assortment).

© IBO 1996

Mendel's Law of Independent Assortment (second law): any one of a pair of characteristics may combine with either one of another pair.

10.1.8
Explain the relationship between Mendel's Laws and meiosis.

© IBO 1996

Mendel's second law applies to traits carried on different chromosomes. Since any combination of chromosomes is possible in Metaphase I, any one of a pair of characteristics may combine with either one of another pair.

E.g. pea plants: gene: shape of pea   alleles: wrinkled, round
                  gene: colour of pea alleles: yellow, green

When crossing two plants which are heterozygous for both traits (genes), the offspring will show all combinations : green-round, green-wrinkled, yellow-round and yellow-wrinkled. This shows that the genes for shape and colour inherit independently.

10.2 DIHYBRID CROSSES

Students are expected to be able to chose letters representing alleles with care to avoid possible confusion between capitals and small letters (upper and lower case). Students are expected to be able to apply their understanding of dihybrid crosses to organisms not familiar to them.

10.2.1
Calculate and predict the genotypic and phenotypic ratios of offspring of dihybrid crosses involving unlinked autosomal genes.

© IBO 1996

Calculation and prediction of genotypic and phenotypic ratios of offspring of dihybrid crosses involving unlinked autosomal genes is discussed below.

Remember the following ratios:

P: AaBb   ×   AaBb
                                      ↓

(genotypic ratios)

F_1:   1   AABB   2   AaBB   1   aaBB
      2   AABb   4   AaBb   2   aaBb
      1   AAbb   2   Aabb   1   aabb

(phenotypic ratios)

F_1:   9   dominant - dominant
         3   dominant - recessive
         3   recessive - dominant
         1   recessive - recessive

See questions at the end of the chapter.

10.2.2
Identify which of the offspring in dihybrid crosses are recombinants.

© IBO 1996

Recombination: the reassortment of genes or characters into different combinations from those of the parents.

Recombination has often been restricted to linked genes but it also applies to non-linked situations. For example, in the cross with tall, white (Ttrr) with short red (ttRr) the F_1 will contain four different phenotypes - tall white (Ttrr) short red (ttRr), tall red (TtRr) and shorte white (ttrr). The tall red and short white are recombinants.

Genetics

10.3 AUTOSOMAL GENE LINKAGE AND GENE MAPPING

10.3.1
State the difference between autosomes and sex chromosomes.

© IBO 1996

Autosomes are all chromosomes which are not sex chromosomes. **Sex chromosomes** are those chromosomes which help in determining the sex of an individual.

10.3.2
Explain how crossing over in Prophase I (between non-sister chromatids of a homologous pair) can result in an exchange of alleles.

© IBO 1996

Crossing over in Prophase I between non-sister chromatids of a homologous pair can result in an exchange of alleles. See section 10.1.3.

10.3.3
Define linkage group.

© IBO 1996

Linkage group: a group of genes whose loci are on the same chromosome.

10.3.4
Explain an example of a cross between two linked genes.

© IBO 1996

An example of a cross between two linked genes. The notation AaBb that was used for the non-linked dihybrid crosses will be modified to represent crosses involving linkage. The letter representing linked genes will be represented as vertical pairs, as shown in the diagram.

$$\frac{A \quad a}{B \quad b}$$

Drosophila (fruit flies) only posses 8 chromosomes and are easy to breed. They mature quickly and have been found useful when studying genetics. When studying dihybrid crosses in fruit flies, the result did not always correspond with the expected ratio. For example : a pure breeding fruit fly with a tan body and long wings was crossed with a mutant having a black body and short wings. The F_1 all had tan bodies and long wings. The F_2 was 75% tan with long wings and 25% black with short wings. The only possible reason for this was that body colour and wing length are found on the same chromosome. They are linked.

b^+: tan body   b: black body
w^+: long wings   w: short wings

Using the above notation, a wild type *Drosophila* with a tan body and long wings can be represented by:

$$\frac{b^+ \quad w^+}{b^+ \quad w^+}$$

A mutant *Drosophila* with a black body and short wings could be represented by:

$$\frac{b \quad w}{b \quad w}$$

So the cross of these two organisms would be represented by:

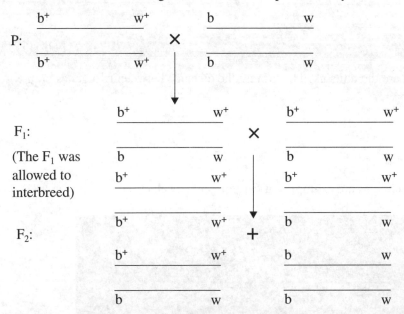

10.3.5
Identify which of the offspring in such dihybrid crosses are recombinants.

© IBO 1996

Identify which of the offspring are recombinants (see also section 10.2.2).

Recombinants in linked genes are those combinations of genes which the parents did not possess. So in the above example, recombinants would be:

$$\frac{b \quad w^+}{b \quad w} \quad \text{or} \quad \frac{b^+ \quad w}{b \quad w}$$

Genetics

10.3.6
Analyse cross over value (COY) data to construct gene maps of up to four genes using two-point testcross data.

© IBO 1996

The **cross over value** (COV) is the percentage of recombinant offspring out of the total ('original' and recombinant). The cross over value is expressed as a percentage. The map distance is then the same number expressed as cM.

10.3.7
Define centimorgan.

© IBO 1996

Centimorgan (cM) are the units used to express the distance between two genes on the same chromosome.

Racehorses are examples of careful genetic selection

10.4 STATISTICAL ANALYSIS

10.4.1
Analyse both monohybrid and dihybrid genetic crosses using the chi-squared test.

© IBO 1996

The chi squared test will tell you how much your data differ from your expected values. If they are different, you then want to know what the probability is that this difference is due to chance. If the probability is less than 5% ($p < 0.05$) than we accept that the data is significantly different from the expectations.

The difference between this test and the Students' t-test is that the t-test is parametric (i.e. deals with numbers e.g. height or size) and that the chi-square test is non-parametric (i.e. data belong to different categories).

Question: Is the height of students in Grade 5 different to that in Grade 9?
Answer: use Students' t-test.

Question: Is ling (a plant) usually present in areas where bilberries are present? **Answer:** use chi-squared test.

The formula for the **chi-squared test** is: $\chi^2 = \sum \dfrac{(O-E)^2}{E}$

where
- O = observed value
- E = expected value
- n = number of classes/categories
- $p < 0.05$ is accepted as being significant.

A table of critical values for the chi-squared test at $p = 0.05$. If the χ^2 you found is bigger than the critical value given in the table, then you can reject H_0 (there no significant difference between the observed and expected value) and accept that your observations are significantly different from your expectations.

Degrees of freedom	Critical value
1	3.84
2	5.99
3	7.81
4	9.49
5	11.07
6	12.59
7	14.07
8	15.51
9	16.92

Genetics

Degrees of freedom	Critical value
10	18.31
11	19.68
12	21.02
13	22.36
14	23.69
15	24.99
16	26.30
17	27.59
18	28.87
19	30.14
20	31.41
21	32.67
22	33.92
23	35.17
24	36.42
25	37.65
26	38.89
27	40.11
28	41.34
29	42.56
30	43.77

EXAMPLE 1:

This investigation studies whether woodlice prefer a dry or a humid atmosphere. It consists of 5 trials, using 10 animals per trial. The results are shown below :

Distribution of ten woodlice after 3 minutes

Trial	Dry atmosphere	Humid atmosphere
1	3	7
2	4	6
3	3	7
4	5	5
5	4	6
Total	$O = 19$	$O = 31$

Does this distribution show that woodlice prefer a humid atmosphere or does it not? Try to answer this question before you continue.

SOLUTION 1:

You need to answer the following questions:
- What is H_0?
- According to H_0, what are the expected values?
- What is χ^2?

- How many degrees of freedom are there?
- What is the critical value ($p = 0.05$)?
- Is χ^2 bigger or smaller than the critical value?
- Do you reject or accept H_0?

H_0 says that there is no difference in the distribution of woodlice.

E would then be 25 for each area.

$$\chi^2 = \sum \frac{(O-E)^2}{E}$$
$$= \frac{(19-25)^2}{25} + \frac{(31-25)^2}{25}$$
$$= 2.88$$

There are 2 categories and therefore the degrees of freedom are $2 - 1 = 1$

The critical value is 3.84

$\chi^2 = 2.88$ which is below the critical value.

Therefore we cannot reject H_0 (that there is no difference in the distribution of woodlice).

EXAMPLE 2:
In a genetics experiment tall pea plants were crossed with short pea plants. The resulting F_1 was self-fertilised and the F_2 consisted of 787 tall plants and 277 short plants. Does this result confirm Mendel's explanation or not?

SOLUTION 2:
You need to answer the following questions:
- What is H_0?
- According to H_0, what are the expected values?
- What is χ^2?
- How many degrees of freedom are there?
- What is the critical value ($p = 0.05$)?
- Is χ^2 bigger or smaller than the critical value?
- Do you reject or accept H_0?

H_0 would be based on Mendel's first law and it would state that the expected ration of tall: short plants = 3 : 1.

The expected values would be : total $F_2 = 787 + 277 = 1064$
based on the expected ratio of tall : short = 3 : 1 we would expect

Genetics

tall : short = 798 : 266

$$\chi^2 = \sum \frac{(O-E)^2}{E}$$

$$= \frac{(787-798)^2}{798} + \frac{(277-266)^2}{266}$$

$$= 0.61$$

There are 2 categories and therefore the degrees of freedom are $2 - 1 = 1$

The critical value is 3.84

$\chi^2 = 0.61$ which is below the critical value.

Therefore we cannot reject H_0.

10.5 POLYGENIC INHERITANCE

10.5.1
Define polygenic inheritance.

© IBO 1996

Polygenic inheritance concerns the inheritance of a characteristic which is controlled by more than one gene.

10.5.2
Explain that polygenic inheritance can contribute to continuous variation, using three examples including human skin colour.

© IBO 1996

Polygenic inheritance can contribute to continuous variation.

EXAMPLE 1: SHAPE OF THE COMB IN POULTRY.
Different shapes of comb exist :

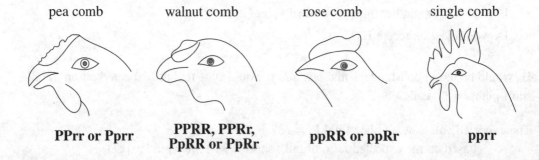

pea comb — PPrr or Pprr

walnut comb — PPRR, PPRr, PpRR or PpRr

rose comb — ppRR or ppRr

single comb — pprr

If a true breeding pea-combed fowl is crossed with a true breeding rose-combed fowl, the F_1 all display a walnut comb. If two of these are bred, the F_2 shows every kind of comb seen so far as well as a single comb.

Study the diagram and draw the appropriate Punnett square.

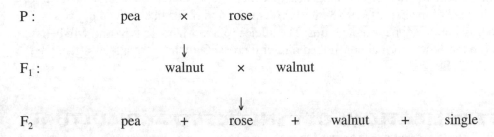

P :   pea   ×   rose

F$_1$:   walnut ×   walnut

F$_2$:   pea + rose + walnut + single

EXAMPLE 2: HUMAN SKIN COLOUR.

Human skin colour involves the interaction of at least 3 independent genes. If we assume that A, B and C will each represent alleles for dark skin, then a, b and c will represent alleles for light skin.

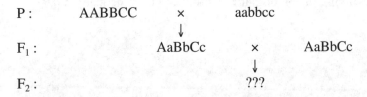

P :   AABBCC   ×   aabbcc

F$_1$:   AaBbCc   ×   AaBbCc

F$_2$:   ???

To find the F_2 genotypes with the corresponding ratios (and from there the phenotypes with the corresponding ratios) make a Punnett square.

Essentially you assume that the alleles are codominant and that the greater the number of dominant alleles, the darker the skin.

Plot a histogram of the phenotypes against the ratios found. Some of Mendel's results for the observed ratios of dominant and recessive characteristics of peas (F_2 generation) are shown opposite. The theoretical values are 3:1.

Other example of polygenic traits are eye colour and height.

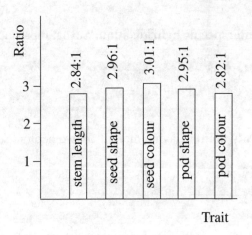

Genetics

10.5.3
Explain how interaction between genes can cause modified Mendelian ratios in dihybrid crosses.

© IBO 1996

Depending on the interactions of the genes, modified Mendelian ratios can be the result of a cross. Instead of the expected 9 : 3 : 3 : 1, you might find ratios like 9 : 3 : 4 or even 9 : 7. These ratios would be phenotypic ratios. The alleles follow Mendel's Law and will behave identical to those in a normal non-linked dihybrid cross. Only how the alleles affect the genotype has changed.

10.6 APPLICATIONS OF GENETICS TO AGRICULTURE AND HORTICULTURE

10.6.1
Define inbreeding.

© IBO 1996

Inbreeding: Reproduction involving fusion of gametes produced by genetically related individuals.

10.6.2
Define outbreeding.

© IBO 1996

Outbreeding: Reproduction involving fusion of gametes produced by genetically unrelated individuals.

10.6.3
Define interspecific hybridisation.

© IBO 1996

Interspecific hybridisation: Sexual reproduction between members of different species.

10.6.4
Define polyploidy.

© IBO 1996

Polyploidy: Having more than two haploid sets of chromosomes.

10.6.5
Define F_1 hybrid vigour.

© IBO 1996

F_1 hybrid vigour: Vigour due to high levels of heterozygosity.

10.6.6
Outline one example for each of the above terms.

© IBO 1996

Inbreeding:
- Self-fertilisation in plants can be considered to be the ultimate form of inbreeding.
- Inbreeding is commonly used in domestic animals such as sheep, cattle and pigs to strengthen desirable characteristics.

Outbreeding:
- outbreeding can be done between breeding lines within the species or even between species.
- within species: a short-horned cattle and a black Angus cattle were crossed to produce offspring with superior beef and rapid growth.
- between species: a horse can be crossed with a donkey to produce a mule which is stronger and more resistant to disease than either parent.
- between species: the Macoun apple was crossed with a variety of crab apple which produced the Liberty apple which is similar in taste to the Macoun apple but more resistant to disease.

Interspecific hybridisation / F_1 hybrid vigour:
- see the examples of outbreeding between species above.

Polyploidy:
- produces plants which are hardier, bigger and more productive
- strawberries, Daylilies, Freesias, many plants bought for their flowers

10.6.7
Describe a total of three examples of the use of transgenic techniques in agriculture and/or horticulture.

© IBO 1996

Examples of the use of transgenic techniques in agriculture and/or horticulture are:

1. Transfer of human genes into bacteria e.g. to produce insulin, factor VIII (clotting factor).

2. Transfer of human genes into mammals e.g. a-1-antitrypsin (emphysema drug) into sheep milk, e.g. factor IX (clotting factor) into sheep milk.

3. Transfer of winter flounder fish gene into tomatoes to make them frost resistant.

4, Transfer of the T toxin gene from Bacillus thuringiensis into tomatoes to make them resistent to the tobacco mosaic virus.

Genetics

10.6.8
Discuss the ethical issues arising from the use of transgenic techniques.

© IBO 1996

Ethical issues arising from the use of transgenic techniques include:
- genetically manipulated soya beans. Do products need to be labelled to inform consumers?
- animal rights movements. Cattle with blond hair and blue eyes may soon be created?

10.6.9
Discuss the need, to maintain the biodiversity of wild plants/ancient farm breeds as a reservoir of alleles which may have future value.

© IBO 1996

The growing human population relies on three plants (rice, corn, wheat) for more than half of its food.

Potato famine in Ireland:
> In the 1800's Ireland relied on only a few varieties of potatoes for almost all its food. None were resistant to a fungus that spread quickly and in the 1840's more than 2 million people died of starvation.

Research is being done into possibilities of using wild plants to become new crop plants or to improve resistance to insects and disease.

In the 1970's a rare species of wild corn was found in Mexico. It is highly resistant to disease and some animal pests. It was found only on a 120 acre plot which was grazed by cows. This rare species was discovered before it was wiped out. How many are not?

EXERCISE

1. During meiosis, DNA replication occurs during:

    A   Prophase I
    B   Anaphase I
    C   Interphase
    D   Telophase I

2. The reassortment of genes or characters into different combinations from those from of the parents is known as:

    A   meiosis.
    B   recombination.
    C   chiasma.
    D   interphase.

3. Two bean plants are crossed. The flowers of this type of bean are either white or yellow. The pods are either short or long. The plant that is crossed could **not** have:

    A white flowers and long pods.

    B white flowers and short pods.

    C red flowers and short pods.

    D yellow flowers and long pods.

4. A group of genes whose loci are on the same chromosome is known as a:

    A linkage group

    B gene.

    C chromosome.

    D cross.

5. Reproduction involving fusion of gametes produced by genetically related individuals is known as:

    A inbreeding.

    B outbreeding.

    C polyploidy.

    D interspecific hybridisation.

6. Sexual reproduction between members of different species is known as:

    A inbreeding.

    B outbreeding.

    C polyploidy.

    D interspecific hybridisation.

7. Reproduction involving fusion of gametes produced by genetically unrelated individuals is known as:

    A inbreeding.

    B outbreeding.

    C polyploidy.

    D interspecific hybridisation.

8. Transfer of human genes into bacteria e.g. to produce insulin is an example of:

    A infection.

    B vaccination.

    C hybridisation.

    D a transgenic technique.

Genetics

9. In an experiment using the χ^2 test, the critical value is 5.99 and $\chi^2 = 0.48$. The conclusion that can be drawn is:

A $\chi^2 = 0.48$ is below the critical value, therefore we cannot reject H_0.

B $\chi^2 = 0.48$ is above the critical value, therefore we cannot reject H_0.

C $\chi^2 = 0.48$ is below the critical value, therefore we should reject H_0.

D $\chi^2 = 0.48$ is below the critical value, therefore we must accept H_0.

10. A black Drosphila (fruit fly) with vestigial (small) wings was mated with an individual that showed wild type for both characteristics. The offspring were all wild type.

     a.   Which are the dominat alleles?

     b.   List the possible genotypes and phenotypes of the parents and offspring. Use the correct notation.

     c.   Use a Punnett square to predict the genotypes and phenotypes of the F_2 if the F_1 is allwed to interbreed.

     d.   Give the expected ratio's of the genotypes and phenotypes of the F_2.

11. a.   Write down Mendel's second law.

     b.   How does Mendel's second law apply to linked genes.

12. In Drosophila (fruit flies) the allele for grey colour is dominant over black. Straight wings are dominant over curly wings.

A heterozygous grey-straight winged fly was crossed with a black-curly winged fly. The offspring were as follows :

43   grey straight

12   grey curly

39   ebony curly

10   ebony straight

     a.   If these genes are not linked, which phenotypes would you expect in the F_1?

     b.   Give the expected numbers of the phenotypes.

     c.   Based on these results, would you expect the genes to be linked?

     d.   Can you find the map distance between them?

13. In rabbits, having coloured fur is dominant over producing no pigment (albino). Grey fur is dominant over black. An homozygous completely recessive albino rabbit is mated with a homozygous grey rabbit. The F_1 are allowed to interbreed.

     a.   What is the genotype and phenotype of the F_1.

     b.   Predict the genotypes of the F_2

c. What are the expected ratios?
d. Predict the phenotypes of the F$_2$
e. What are the expected ratios?

14. Describe an example of transgenic techniques in agriculture or horticulture.

Genetics

HUMAN REPRODUCTION

11

Chapter contents
- Production of gametes
- Fertilisation and pregnancy

Human Reproduction

11.1 PRODUCTION OF GAMETES

11.1.1
Draw the structure of the testis as seen using a light microscope.

© IBO 1996

The testis when studied under the light microscope (see diagram) is seen to consist of many seminiferous tubules. In between the **seminiferous tubules** (over 100 m per testis), you find interstitial cells and blood capillaries. The seminiferous tubules have an outer germ cell layer which is surrounded by the **basement membrane**. Development of spermatozoa takes places from the outside of the tubule and developing spermatozoa are nourished by **Sertoli cells**. Spermatozoa will eventually leave the tubule via the lumen.

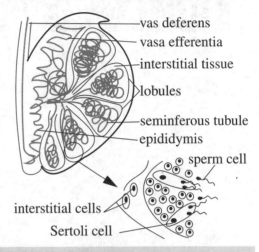

11.1.2
Describe the processes involved in spermatogenesis including mitosis, cell growth, the two divisions of meiosis and cell differentiation (cross reference 7.2, 7.3 and 10.1).

© IBO 1996

Spermatogenesis is the process of producing sperm cells. It involves mitosis, meiosis I and II, and cell differentiation. Mitosis produces the germ cell layer and the spermatogonia. Growth then produces primary spermatocytes which undergo meiosis I and II. The spermatids have the correct amount of genetic material (haploid) but need to differentiate into spermatozoa.

The process is illustrated by the diagram.

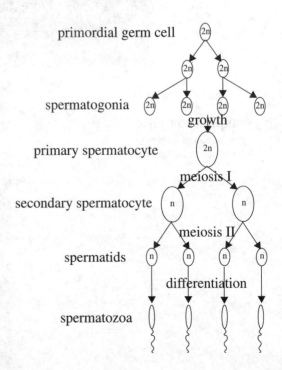

AHL

11.1.3
Outline the origin and the role of the hormones FSH, testosterone and LH in spermatogenesis.

© IBO 1996

Three hormones play a role in spermatogenesis: FSH, LH and testosterone. FSH and LH are produced in the pituitary gland. Testosterone is produced by the Leydig cells in the testis.

- **FSH** stimulates sperm production in the seminiferous tubules.
- **LH** stimulates the interstitial cells (Leydig cells) to produce testosterone.
- **Testosterone** stimulates the growth of male accessory sex organs, the development of secondary sexual characteristics and the increase of libido.

11.1.4
Draw the structure of the ovary as seen using a light microscope.

© IBO 1996

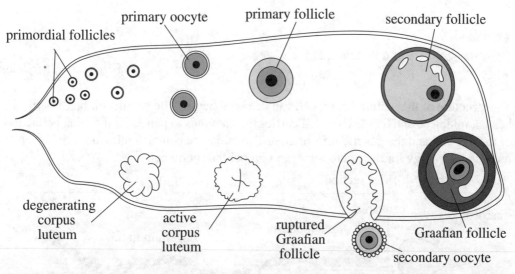

SCHEMATIC SECTION THROUGH A MAMMALIAN OVARY

The ovaries contain follicles in different stages of development, containing **developing oocytes**. Remember that the development until the stage of primary oocyte occurs before birth. The developing follicle is also known as a **Graafian follicle**. The developing oocytes are surrounded by a **zona pellucida**. During development, meiosis I occurs, followed by an unequal cell division. This results in a **secondary oocyte** and a **polar body** (sometimes also referred to as a secondary oocyte). **Ovulation** then takes place and meiosis II will occur after fertilisation.

11.1.5
Explain the processes involved in oogenesis including mitosis, cell growth, the two divisions of meiosis and the unequal division of cytoplasm and the degeneration of polar bodies (cross reference 7.2 and 10.1).

© IBO 1996

Human Reproduction

Oogenesis is the development of egg cells (ova). It includes mitosis, cell growth, meiosis I and II, and the unequal division of cytoplasm leading to the formation of polar bodies.

Mitosis produces the germ cell layer and the oogonia. Growth then produces primary oocytes which undergo meiosis I and II. Unequal divisions remove excess genetic material while allowing the maximum amount of cell material to stay with the future ovum.

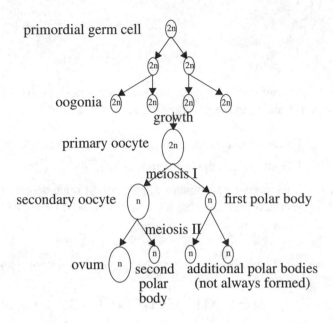

11.1.6
Draw the structure of a mature sperm and egg.

© IBO 1996

The structure of the mature sperm cell can bee seen below. The sperm cell has 3 parts: a head, a midpiece and a tail. The head carries the enzymes to penetrate the zona pellucida of the egg cell and the genetic information. The midpiece contains mitochondria which provide the energy for the tail to move and propel the sperm cell.

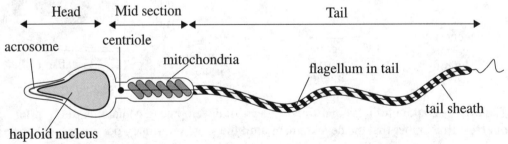

The structure of the egg cell can be seen in the diagram.

[A Graafian follicle has a cavity, follicles of other vertebrates do not have cavities and hence are called 'follicle'.]

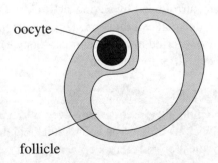

Biology

11.1.7
Outline the role of the epididymis, seminal vesicle and prostate gland in the production of semen.

© IBO 1996

The sperm cells produced in the seminiferous tubules travel to the head of the epididymis (via the vasa efferentia). Here they mature and become somewhat motile. Fluid from the Sertoli cells in the seminiferous tubules which carried the sperm cells to the epididymis is reabsorbed, concentrating the sperm.

During one ejaculation, approximately 3 cm^3 of semen is produced. Only 10 % of this are sperm cells. Most of the fluid in the semen is produced by the seminal vesicles. The fluid they produce contains fructose for energy and prostaglandins which cause contractions in the female reproductive system (helping the sperm to move towards the egg cell).

The fluid from the prostate is alkaline and helps to neutralise the normally acidic environment of the female reproductive tract. Normal pH is around 4 but presence of prostate fluid will make it around pH 6 which is the optimum pH for sperm motility.

Cowper's glands (or the bulbo-urethral glands) produce a clear fluid which will lubricate the penis and facilitate copulation (the female secretions play a more important role in this).

11.1.8.
Compare the processes of spermatogenesis and oogenesis including number of gametes, timing of the formation and release of gametes.

© IBO 1996

Comparing spermatogenesis and oogenesis, some similarities as well as some differences become obvious. Similarities are obvious in that both produce gametes by meiosis in gonads. Differences can be found in the following :

number of gametes produced in total:

spermatogenesis produces large numbers of gametes, oogenesis produces few.

number of gametes per cell:

spermatogenesis produces 4 spermcells from 1 primary spermatocyte, oogenesis produces 1 ovum from 1 primary oocyte.

time of formation:

males produce gametes continuously, from puberty until old age;

females produce primary oocytes before birth and then 1 ovum per month, from puberty until menopause

release of gametes:

males can release gametes at any time, females are on a monthly cycle.

Human Reproduction

11.2 FERTILISATION AND PREGNANCY

11.2.1
Describe the process of fertilisation including, the acrosome reaction, penetration of the egg membrane by a sperm, and the cortical reaction.

© IBO 1996

Fertilisation: the fusion of male and female gametes.

For male and female gametes to fuse in humans, the sperm cell needs to penetrate the egg cell. From the diagram in 11.1 you may remember that at the moment of ovulation, the secondary oocyte is surrounded by a zona pellucida and a corona radiata.

Zona pellucida: a **mucoprotein** (a complex of protein and polysaccharide) membrane surrounding the secondary oocyte of mammals. It is secreted by the ovarian follicle cells.

Corona radiate: layer of follicle cells surrounding the zona pellucida.

The acrosome of the sperm cell (see section 11.1) contains proteolytic enzymes. As the sperm touches the cells of the corona radiata, the membrane around the acrosome fuses with the membrane of the cells, releasing the proteolytic enzyme and digesting the cell. The head of the sperm cell can thus penetrate this layer.

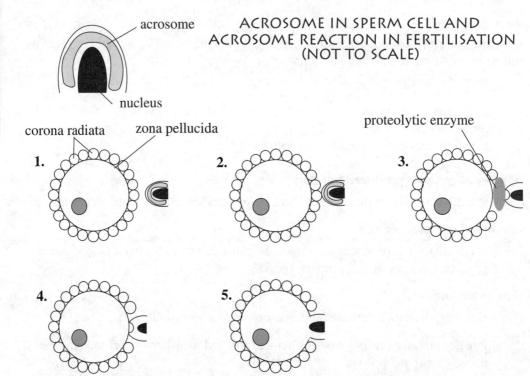

ACROSOME IN SPERM CELL AND ACROSOME REACTION IN FERTILISATION (NOT TO SCALE)

The sperm cell then reaches the zona pellucida (a thick jelly like layer surrounding the secondary oocyte). The zona pellucida has special receptors to bind the sperm cell.

Another acrosome reaction will digest a hole in the zona pellucida and the sperm will pass through.

The head of the sperm cell will fuse with the membrane of the secondary oocyte and special lysosome, the **cortical granules**, will release enzyme to thicken the zona pellucida so that it becomes a fertilisation membrane. This cannot be penetrated by other sperm cells. The ovum is therefore fertilised by only one sperm cell. The reaction of the cortical granules is called the **cortical reaction**.

11.2.2
Describe the role of human chorionic gonadotrophin (HCG) in early pregnancy and pregnancy testing.

© IBO 1996

Approximately 8 days after fertilisation, the blastocyst will embed itself into the endometrium. The outer cells, the trophoblastic cells, start to secrete human chorionic **gonadotropin** (HCG). This hormone sustains the corpus luteum which therefore will continue to produce progesterone, maintaining the endometrium. Gradually the placenta will start to produce progesterone and at approximately 10 weeks of pregnancy, the corpus luteum is no longer neccessary.

HCG is excreted via urine and can be detected with a pregnancy test. A specially manufactured stick is placed in urine. The urine will flow through the stick (capillary action) and reach an area of antibodies against HCG which have a pigment attached to them. HCG will attach to the antibodies and the complex will travel further, with the movement of fluid. A second group of antibodies is found higher up in the stick. These antibodies are fixed. The HCG-antibody complex will attach to these antibodies, showing as a blue line.

If no HCG is present, the first group of antibodies (with pigment) cannot bind to it. Therefore no HCG complex is formed. As a result the second group of (fixed) antibodies cannot bind to an HCG complex so the coloured pigment will not concentrate in one line.

11.2.3
Describe the structure and functions of the placenta including its hormonal role (oestrogen and progesterone) in the maintenance of pregnancy.

© IBO 1996

Progesterone is needed throughout pregnancy to sustain the endometrium and to inhibit release of FSH. During pregnancy, a woman should not ovulate because she cannot be pregnant with 2 babies of different age. Oestrogen also plays a role in inhibiting FSH. It also stimulates further growth of the endometrium.

Human Reproduction

EXERCISE

1. The function of Sertoli cells is to:

    A nourish developing spermatozoa.
    B produce testosterone.
    C produce ova.
    D produce spermatozoa.

2. Oogenesis is:

    A the production of oestrogen.
    B the production of spermatozoa.
    C the development of egg cells.
    D birth.

Questions 3 and 4 refer to the diagram of a sperm cell.

3. Point X is the:

    A tail sheath.
    B acrosome.
    C haploid nucleus.
    D centriole.

4. Point Y is the:

    A tail sheath.
    B acrosome.
    C haploid nucleus.
    D centriole.

5. The prostate gland produces:

    A a neutral fluid.
    B a slightly acidid fluid.
    C a slightly alkaline fluid.
    D a strongly alkaline fluid.

6. Which one of the following statements is not true?

    A spermatogenesis produces large numbers of gametes, oogenesis produces few.
    B spermatogenesis produces 4 spermcells from 1 primary spermatocyte.
    C oogenesis produces 1 ovum from 1 primary oocyte.
    D males produce gametes continuously, from birth until old age.

7. The fusion of male and female gametes is known as:

    A   pregnancy.
    B   fertilisation.
    C   menstruation.
    D   birth.

8. Pregnancy is often detected by testing the level of:

    A   human chorionic gonadotropin (HCG) in the urine.
    B   human chorionic gonadotropin (HCG) in the blood.
    C   testosterone in the urine.
    D   progesterone in the urine.

9. Progesterone is needed throughout pregnancy:

    A   to maintain correct blood pressure.
    B   to prevent the release of oestrogen.
    C   initiate ovulation.
    D   to sustain the endometrium.

10.
    a.   What is oogenesis?
    b.   What is the function of mitosis in oogenesis?
    c.   What is the function of meiosis in oogenesis?
    d.   Why does a spermatogonia yield 4 spermatozoa while an oogonia yields one ovum and 2 or 3 polar bodies?

11.
    a.   Which structure secretes HCG?
    b.   What is the function of HCG to the embryo?
    c.   Why is HCG only produced early in pregnancy?
    d.   How can HCG be used in pregnancy testing?

DEFENSE AGAINST INFECTIOUS DISEASE

12

Chapter contents
- Agents that cause infectious disease
- Types of defense

Defense against infectious disease

12.1 AGENTS THAT CAUSE INFECTIOUS DISEASE.

12.1.1
Define pathogen.

© IBO 1996

Pathogen: an organism causing disease.

12.1.2
State one example of an infectious disease caused by members of each of the following groups: viruses, bacteria, fungi, protozoa, flatworms and roundworms.

© IBO 1996

Many different organisms can cause disease. Below you find some examples (see also section 5.3.1)
Viruses: e.g. chicken pox, poliomyelitis.
Bacteria: e.g. tetanus, tuberculosis.
Fungi: e.g. athlete's foot (tinea), ringworm.
Protozoa: e.g. malaria, sleeping sickness.
Flatworms: e.g. bilharzia (caused by blood fluke).
Roundworms: e.g. elephantiasis (caused by threadworms living in and blocking lymphatic system).

12.1.3
List six methods by which disease-causing agents are transmitted and gain entry to the body.

© IBO 1996

Methods by which disease-causing agents are transmitted and gain entry to the body:

airborne droplets in the air	breathe in pass through lung epithelial
water-borne water contaminated by human faeces	drink pass through intestinal epithelium
food-borne food contaminated by human faeces or pass through intestinal epithelium	eat insects/other animal vectors
insect-borne insect vectors carrying pathogens	bite/sting injected straight into bloodstream
sexually transmitted pathogens transmitted during intercourse	intercourse/close proximity pass through skin
direct contact spread by skin contact/saliva	pass through skin/into digestive tract

Biology

12.1.4
Describe the cause, transmission and effects of one human bacterial disease.

© IBO 1996

Cholera is caused by the bacterium *Vibrio cholerae*. It is transmitted easily in areas where clear drinking water is not available. It is typical for places where a (natural) disaster has just occurred (floods, earthquake, war) which has interfered with the normal infra structure.

Cholera bacteria will leave the body of a patient with the faeces. If these faeces contaminate water, many other people drinking this water or washing their food with it can be infected. A few days later these people will suffer from severe diarrhoea, caused by the toxins produced by the bacteria. It is not unusual for these people to die of dehydration.

12.1.5
Explain the cause, transmission and social implications of AIDS.

© IBO 1996

Cause:
    HIV virus; origin of the virus is not confirmed.

The most common methods of transmission are:
    via blood (mother/child, transfusion, contaminated needles).

    via sexual intercourse (homosexual and heterosexual)

Social implications:
    ostracising of homosexuals ('homophobia').

    ostracising of HIV positive people.

    unease over bloodtransfusions.

    changes in sexual behaviour of homo- and heterosexuals including reductions in promiscuity and the increased use of condoms.

Defense against infectious disease

12.2 TYPES OF DEFENSE

12.2.1
Describe the process of clotting involving, thrombokinase, prothrombin, Ca^{2+} ions, fibrinogen, platelets and vitamin K.

© IBO 1996

Blood is too precious a fluid to allow it to escape in large amounts. As a result, blood clots when there is a cut. It is equally important that blood does not clot at other times and obstructs the bloodvessels. The finely regulated process of bloodclotting involves the following :

Blood from a cut will react with air and substances from damaged cells and platelets. Damaged cells will release thrombokinase (=thromboplastin) which, together with factor X and VII and Ca^{2+} will change prothrombin into thrombin. Thrombin will hydrolyse soluble fibrinogen into smaller insoluble fibrin molecules. These will form a network which captures erythrocytes and becomes a clot.

SCHEMATIC DIAGRAM OF THE PROCESS OF BLOOD CLOTTING

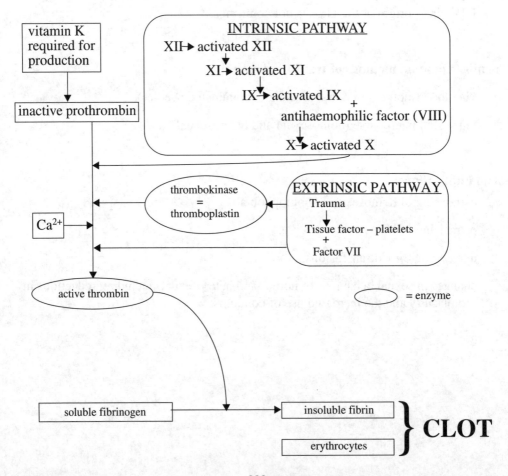

12.2.2
Outline the principle of challenge and response, clonal selection and memory cells as the basis of immunity.

© IBO 1996

The macrophages which are the first to encounter a pathogen, will ingest the pathogen (**phagocytosis**) but do so incompletely. Parts of the bacterial cell wall and cell membrane will be displayed on the outside of the cell membrane of the macrophage. It will then travel to the lymph node.

Inside the lymph node, the macrophage displaying the antigen will select a T-helper cell which has receptors on its membrane complementary to the antigen that the macrophage carries. As a result, these T-helper cells will divide by mitosis, forming a clone. The clone of Th cells will activate B-cells which have surface receptors complementary to the antigen. The B-cells will also form a clone. The B-cells will then differentiate into plasma cells and memory cells. The plasma cells will make large amounts of antibodies.

Clonal selection: the macrophage selecting which T-cells and B-cells have the required surface receptor.

Clonal expansion: the T-cells and B-cells forming clones by mitosis to produce the large numbers of cells required to deal with the infection.

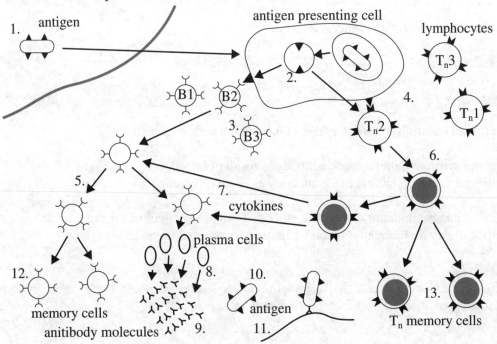

1-2 antigen presentation, 3-4 clonal selection, clonal expansion, 7 T_n secrete cytokines
8 plasma cells secrete antibodies, 9 antibody molecules in blood and lymph
10 antibody coats bacteria, 11 phagocytosis, 12-13 memory cells remain

Defense against infectious disease

12.2.3
Define active immunity.

© IBO 1996

Active immunity: immunity due to production of antibodies by the organism itself after the body's defence mechanisms have been stimulated by the invasion of foreign microorganisms. This means that people who have had some disease such as measles are very unlikely to be infected for a second time.

12.2.4
Define passive immunity.

© IBO 1996

Passive immunity: immunity due to acquisistion of antibodies from another organism in which active immunity has been stimulated including antibodies received via the placenta or in colostrum. Passive immunity gives protection from immunity immediately. This is unlike active immunity which takes some time to develop.

12.2.5
Define natural immunity.

© IBO 1996

Natural immunity: immunity due to infection.

12.2.6
Define artificial immunity.

© IBO 1996

Artificial immunity: immunity due to inoculation with a vaccine.

Natural active immunity: made antibody as result of infection
Artificial active immunity: made antibody as result of vaccination

Natural passive immunity: obtained antibody from placenta/colostrum
Artificial passive immunity: obtained antibody from other organism.

12.2.7
Explain the roles of B-cells, MHC proteins, helper T-cells, cytotoxic T-cells, memory cells and immunoglobulins in the antigen/antibody response.

© IBO 1996

As was stated above, the **B-cells** can be found in the lymph nodes. They will form a clone and then differentiate into plasma cells (and memory cells) producing one kind of antibody.

MHC (major histocompatability complex) proteins are membrane proteins found on macrophages. T cell receptors do not respond to antigens unless the antigens are associated with MHC proteins (on the macrophages).

The macrophage will present the antigen to the **T-helper cell**. The T-helper cell will secrete a substance which activates nearby B-cells. The B-cells will divide, form a clone and differentiate into plasma cells producing specific antibodies.

Cytotoxic T-cells are involved in the cell mediated response. In response to substances secreted by the T-helper cell, cytotoxic T-cells will kill pathogens and viruses which have invaded cells. A cell which contains viruses or other pathogens will be detected (these cells display special proteins on their cell surfaces) and destroyed.

B-cells can form memory cells. This means that the second response to exposure of the antigen is much faster and stronger than the first response. It has been argued that T-helper cells can also form memory cells but their existence has not yet been proven.

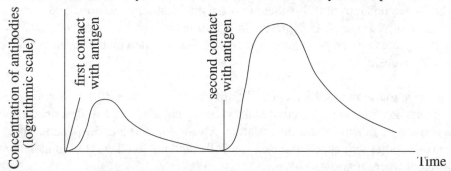

There are 5 different classes of antibodies:

Classes of antibodies	IgA	IgD	IgE	IgG	IgM
mass	light/int	light	light	light	heavy
abundance	common	rare	very rare	very common	not so common
#Ab binding sites	2/4		2	2	10
site of action	saliva, tears, mucous, milk	surface of B-cells	tissue, tissue fluid	blood, tissue fluid, can also cross placenta	blood, cannot cross placenta
functions	stops bacteria sticking to cells, stops colonies on mucous	role in B-cell activation	release histamines, allergies	macrophages, antitoxins, agglutinates	agglutinates, macrophages

Defense against infectious disease

12.2.8
Describe the production of monoclonal antibodies and one use in diagnosis and one use in treatment.

© IBO 1996

Monoclonal antibodies are obtained from single B-cell clones. You inject a mammal (e.g. a mouse) with the antigen (e.g. human red blood cells - type A). The mouse plasma cells will produce antibodies against the human red blood cell. You can extract the plasma cells and fuse them with B-cell tumour cells. The resulting hybridoma cells will grow in culture and produce identical antibodies.

If you add human blood to the antibodies and you see clotting, then you know that the blood contained type A cells.

This way you can diagnose pregnancy. Obtain monoclonal antibodies against the HCG (human chorionic gonadotropin). Fix them in place on a testing stick/strip. Add urine to the testing stick/strip. If the HCG is present in the urine (as it will be if the woman is pregnant), it will attach to the antibodies. The test has been so designed that this will give a colour showing a positive test.

It is also possible to use monoclonal antibodies in treatment of disease. An example is in cancer treatment. Again you first need to make monoclonal antibodies as described above. This time you attach an anti-cancer drug to the mAb and inject it into the patient. The monoclonal antibodies will attach to the cancer cells (and not to others) bringing the drug to the exact spot where it needs to be.

12.2.9
Explain the need for immunisation against the bacterial infections: diphtheria, whooping cough and tetanus, and against the viral infections: measles, polio and rubella.

© IBO 1996

As explained in section 12.2.7, the secondary response to an antigen is much faster and stronger than the first response. This fact is used in vaccination. By deliberately exposing someone to a weakened/dead/related pathogen, this person develops memory cells against the antigen. If the disease is very serious (bacterial diseases: diphtheria, whooping cough, tetanus; viral diseases: measles, polio, rubella) you may want to vaccinate against it. If the person then comes into contact with the pathogen, the body will have a much faster and stronger response and the person is unlikely to become ill.

12.2.10
Outline the process of immunisation.

© IBO 1996

As has been described above, immunisation involves a deliberate exposure (often injection) to the pathogen in order to produce memory cells. To avoid becoming ill as a result of this, the pathogen is killed, weakened or a related strain is used (cowpox for smallpox).

12.2.11
Discuss the benefits and danger of immunisation against bacterial and viral infection.

© IBO 1996

To prevent epidemics, it is important to develop 'herd immunity'. This means that almost everyone in a population is immune against a disease. This interrupts transmission so that the few who are suseptible do not come into contact with the pathogen.

Vaccination has certainly saved many lives. Small pox has been eradicated as a result of vaccination. However, some vaccines are not always safe and some people develop the disease as a result. People vaccinated with a live vaccine may pass the pathogens out in their faeces and infect others. Another possibility is that some people do not respond to a vaccine and are not immune even though they think they are. In the case of malnutrition, antibodies (proteins) are not always formed. Finally a virus like the common cold mutates regularly and renders previously formed antibodies useless.

EXERCISE

1. Influenza and the common cold are caused by:

    A   viruses.
    B   bacteria.
    C   fungi.
    D   protozoa.

2. Which one of the following is not an insect born disease?

    A   malaria.
    B   sleeping sickness.
    C   yellow fever.
    D   tetanus.

3. The small tropical town of Utopia has just been hit by a major earthquake. The arriving rescue and aid workers find that there has been extensive damage to buildings, the electricity and piped water supplies have been destroyed and the railway bridge in the town has collapsed. The major disease likely to result from this disaster and that the aid workers should be concerned about is:

    A   AIDS.
    B   ringworm.
    C   cholera.
    D   tetanus.

Defense against infectious disease

4. Which one of the following is not considered to be a means of the transmission of AIDS?

    A   sexual intercourse.
    B   insect bites.
    C   blood transfusion.
    D   contaminated needles.

5. Imelda had measles as a child. Today she is a doctor and has just treated a child with measles. Imelda is very unlikely to develop measles as a result of this contact because:

    A   she has active immunity.
    B   she has passive immunity.
    C   she has artificial immunity.
    D   she has been vaccinated.

6. Large numbers of B-cells can be found in the:

    A   blood.
    B   urine.
    C   heart.
    D   lymph nodes.

7. Chai has just received a second vaccination against hepatitis B. His body's reaction to this is most likely to be to:

    A   produce fewer antibodies than on the first vaccination.
    B   produce more antibodies than on the first vaccination.
    C   give Chai a mild attack of the disease.
    D   produce complete immunity to the disease.

8. The most effective way of preventing epidemics is:

    A   immunisation of most of the community.
    B   immunisation of the elderly.
    C   chlorination of the water supply.
    D   isolation of individuals who develop the disease.

9.   a.   What is the cause of AIDS?
       b.   How can AIDS be transmitted?
       c.   Name two ways in which many people have changed their behaviour since AIDS has become widespread in the early 1980s.

10. a. What is the role of B-cells in the antigen/antibody response?
b. What is the role of MHC proteins in the antigen/antibody response?
c. What is the role of helper T cells in the antigen/antibody response?
d. What is the role of cytotoxic T cells in the antigen/antibody response?
e. What is the role of immunoglobulins in the antigen/antibody response?

Defense against infectious disease

Biology

CLASSIFICATION AND DIVERSITY

13

Chapter contents
- Classification
- Diversity

Classification and Diversity

13.1 CLASSIFICATION

13.1.1
Describe the value of classifying organisms.

© IBO 1996

The value of classifying organisms:
- we need a classification scheme to make sense of the vast diverstiy of organisms all around us.
- when trying to study living things, it helps to be able to order them.
- classification tries to reflect evolutionary relationships.

13.1.2
Outline the binomial system of nomenclature.

© IBO 1996

The binomial system of nomenclature was invented by Carolus Linnaeus in the 18th century. It is still in use today. It is based on the idea that every species has a Latin name, made up of 2 parts.

The first part is the name of the genus, the second part specifies the species. The name should be printed in italics (or underlined when handwritten) and the first part (but not the second) is capitalised. Humans are *Homo sapiens*.

The photograph shows the black whaler shark *Carcharinus obscurus* (and dinner).

13.1.3
Discuss the definition of the term species.

© IBO 1996

A **species** is a group of organisms which could interbreed and produce fertile offspring.

13.1.4
Outline the features used to classify organisms into the kingdoms: Prokaryotae, Protoctista, Fungi, Plantae and Animalia.

© IBO 1996

The most common system of classification used today divides living organisms into 5 kingdoms: Prokaryotes, Protoctista, Fungi, Plantae and Animalia.

Prokaryotes:
> Unicellular organisms lacking distinct nuclei and other membrane bound organelles. DNA is mainly circular and is not organised in chromosomes. (see section 1.2)
>
> Examples are bacteria and cyanobacteria.

Protoctista:
> Unicellular and multicellular eukaryotic organisms. They may be auto- or heterotrophic, live in salt and fresh water.
>
> Examples are *Euglena* and *Paramecium*.

Fungi:
> Eukaryotic filamentous or unicellular. Filamentous fungi grow a mycelium from which mushrooms or toadstools grow. They are heterotrophic and they feed by absorption of nutrients. Their cells have cell walls containing chitin (as opposed to cellulose in plants).
>
> Examples are yeast and mushrooms.

Plantae:
> Eukaryotic, multicellular, photosynthetic organisms. Cell walls contain cellulose, most cells contain chlorophyll. Generally they are non-motile.
>
> Examples are mosses, ferns, flowering plants e.g. buttercup.

Animalia:
> Eukaryotic, multicellular, heterotrophic organisms. Often motile, feeding by ingestion.
>
> Examples are humans and jellyfish.

13.1.5
List the seven levels in the hierarchy of taxa: kingdom, phylum, class, order, family, genus and species using an example from each of two different kingdoms.
© IBO 1996

A **kingdom** is the largest group in the system of classification.

A kingdom consists of one or more phyla (sometimes called divisions for prokaryotes, protista and fungi), which is divided into one or more classes, which is divided into one or more orders, which is divided into one or more families, which is divided into one or more genera, which is divided into one or more species. (see below).

The classifications of *Homo sapiens* and *Taraxacum officinale*, the dandelion, are illustrated in the following diagram.

Classification and Diversity

KINGDOM	Prokaryotes	Protoctista	Fungi	Plantae	Animalia
PHYLUM/ DIVISION	Bacteria	Ciliophora	Ascomycetes	Tracheophyta	Chordata
CLASS				Angiospermae	Mammalia
ORDER				Asterales	Prima
FAMILY				Asteraceae	Hominidae
GENUS	*Escheria*	*Paramecium*	*Penicillium*	*Taraxacum*	*Homo*
SPECIES	*E. coli*	*Paramecium cordatum*		*Taraxacum officinale*	*Homo sapiens*
common name				dandelion	human

13.1.6
Design and/or apply a key for a group of up to eight organisms.

© IBO 1996

In Biology, a key is used to identify an organism. Imagine that you are taking someone to a farm who does not know the appearance of a cow, a horse, a chicken and a pig. You could give this person the following key:

1. Animal is taller than 1.5 meters .... go to 2
    Animal is smaller than 1.5 meters .... go to 3

2. Animal is black and white .... cow
    Animal is brown .... horse

3. Animal has feathers .... chicken
    Animal is pink with curly tail .... pig

You can make a key to identify items of any group of things. The important thing is that it works.

In Biology, keys are most commonly used to identify plants, insects and birds. These are often area specific, for example, the Plants of Northern Europe.

13.2 DIVERSITY

13.2.1
Outline the wide range of metabolic activity of prokaryotes including fermentation, photosynthesis and nitrogen fixation.

© IBO 1996

Prokaryotes can obtain energy in various ways. Some of them are photosynthetic, using light energy to produce large organic materials which they later break down, producing ATP.

Others are heterotrophic, breaking down large organic molecules obtained from other organisms. This can be done in the presence or absence of oxygen. The latter is called fermentation.

It is also possible to use the energy released from a reaction involving inorganic molecules to produce ATP. Bacteria involved in nitrogen fixation are an example of this (section G 5).

13.2.2
State that a wide range of organisms including algae and protozoa axe classified in the protoctista.

© IBO 1996

The kingdom of the Protoctista is the kingdom with the widest range of organisms. It has been said that it contains everything that does not belong elsewhere. Protoctista may be motile or not, autotroph or heterotroph, mostly unicellular but some multicellular, aquatic and terrestrial, with sexual and asexual reproduction.

Algae are plant like protoctista. They can be unicellular or multicellular (seaweed). They live in salt and fresh water and on land in moist environments. They are photosynthetic and contain chlorophyll a, and possibly other coloured pigments.

Protozoa are animal like protoctista. They can be single celled or colonial. They live in salt and fresh water, in soil and inside other organisms. They are heterotrophic and mostly motile.

13.2.3
Describe how fungi obtain nutrients using one parasite and one saprotroph as examples.

© IBO 1996

Fungi absorb nutrients they need from their surroundings. They are therefore heterothophs. Many fungi are saprotrophic, obtaining nutrients from the remains of dead plants and animals. Some are parasitic, obtaining nutrients from their host.

Fungi secrete digestive enzymes onto organic material around them and absorb the nutrients after digestion. By hydrolysis, large organic molecules are broken down until

Classification and Diversity

they are small enough to be absorbed.

Saprotrophic fungi play an important role in the breaking down of dead organisms and recycling of materials. The genus *Penicillum* (phylum Ascomycota, sometimes classified as phylum Deuteromycota) is a saprotrophic fungus. It secretes a substance which inhibits the growth of bacteria. This reduces interspecific competition but the substance has also made a great impact on modern medicine.

Parasitic fungi such as those causing athlete's foot belong to the phylum of Deuteromycota. They also secrete enzymes onto their source of food (in this case their host) and absorbe the small molecules resulting from the hydrolysis of large compounds.

13.2.4
Outline the wide range of diversity in the plant kingdom as exemplified by the structural differences between bryophytes, filicinophytes, coniferophytes and angiospermophytes.
© IBO 1996

The kingdom of the Plantae can be divided into several phyla, some of which are:

Phylum Bryophyta: mosses and liverworts.
Phylum Filicinophyta: ferns.
Phylum Coniferophyta: coniferous plants.
Phylum Angiospermatophyta: flowering plants.

The Bryophyta live in damp areas and still require water for reproduction. They are small simple plants. They do not have vascular tissue or true roots, leaves or stems. Absorption of water takes places over all surface areas. Mosses do not produce flowers. The male gamete is motile, swimming to the female gamete.

The moss plant we see is the haploid gametophyte. It forms haploid gametes which meet and a diploid zygote is formed. This grows out to form a diploid sporophyte which sits on top of the haploid gametophyte. The diploid sporophyte forms haploid spores by meiosis. The haploid spores can grow out to form the haploid gametophyte. The gametophyte is the dominant stage.

Bryophyta are homosporous, producing one kind of spore.

The Filicinophyta are vascular plants. This means that they have a specialised transport system: vascular bundles (see section 16.2). They have true roots, leaves and stems.

The fern plant we see is the diploid sporophyte. It forms haploid spores by meiosis. The haploid spores can grow out to form the haploid gametophyte, the prothallus. This is a small flat structure which produces haploid gametes. The gametes meet and form a diploid zygote which can grow out to form the diploid sporophyte.

Filicinophyta are heterosporous: this means that they form 2 kinds of spores. Microspores which form the male gametophyte and megaspores to form the female gametophyte.

Filicinophyta do not produce flowers. Male gametes are motile and swim to the female gamete.

The Coniferophyta are also vascular plants with true roots, leaves and stems. The sporophyte is the dominant generation. Gametes are not motile.

Coniferophyta are heterosporous, the microspores are called pollen grains, the megaspores the embryosac. The seeds are not enclosed in an ovary and the plant bears no fruit. Spores develop on cones.

The xylem contains only tracheids, vessels are missing; companion cells are absent from phloem (see section 16.2).

The Angiospermophyta are also vascular plants with true roots, leaves and stems. The sporophyte is the dominant generation. Gametes are not motile.

Angiospermophyta are heterosporous. Spores develop in flowers. Seeds are enclosed in an ovary which develops into a fruit after fertilisation. Xylem contains vessels, phloem contains companion cells (see section 16.2).

13.2.5
Outline the wide range of diversity in the animal kingdom as exemplified by movement in earthworm, swimming in a bony fish, flying in a bird, walking in arthropods.

© IBO 1996

There are large variations in behaviour in the animal kingdom. A good example is the different means of movement found in various animals.

- **earthworm**. The earthworm moves by compressing and expanding its body segments.

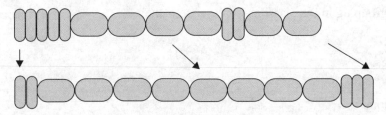

- **fish** generally move by moving their fins in a horizontal plane. This is in contrast to dolphins and whales who move their tail fins in a vertical direction. Other sea creatures use completely different methods from simply drifting on currents to the more exotic jet propulsion used by squid.

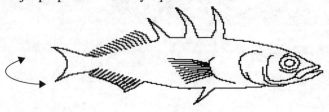

- **birds**, of course, generally fly by flapping their wings. However, not all birds can fly. The ostrich has to walk everwhere. There is also considerable variety in the means of flight used by other birds. Hummingbirds are small and can both fly and hover. Many sea birds can dive into the water to catch fish. Large birds such as the albatross can be more correctly described as gliding than flying. The generally take off by jumping from cliffs and are very adept at using upcurrents of air to reduce the energy required to keep them airborne.
- **Arthropod**s such as man move by walking. This is a complex process which has proved difficult to mimic using robots.

There are similarly large variations in other activities such as feeding and reproduction.

EXERCISE

1. Fire coral, *Millepora platyphylla* is of the genus:

    A   coral.

    B   *Millepora.*

    C   *platyphylla.*

    D   *Millepora platyphylla.*

2. Unicellular organisms lacking distinct nuclei and other membrane bound organelles are known as:

    A   Prokaryotes.

    B   Protoctista.

    C   Fungi.

    D   Plantae.

3. The largest grouping in the classification system is the:

    A   genus.

    B   species.

    C   class.

    D   kingdom.

4. The organisms that fix nitrogen in the roots of some plants are:

    A   Prokaryotes.

    B   Protoctista.

    C   Fungi.

    D   Plantae.

5. Parasitic fungi obtain nutrients from:

   A the air.
   B their surroundings.
   C their host.
   D the earth.

6. Mosses and liverworts are members of the Phylum Bryophyta. They absorb water:

   A through their roots.
   B through their leaves.
   C through their stems.
   D through their whole surface.

7. Ferns reproduce by producing:

   A seeds.
   B spores.
   C cones.
   D pods.

8. a. Design a key for the following imaginary insects.

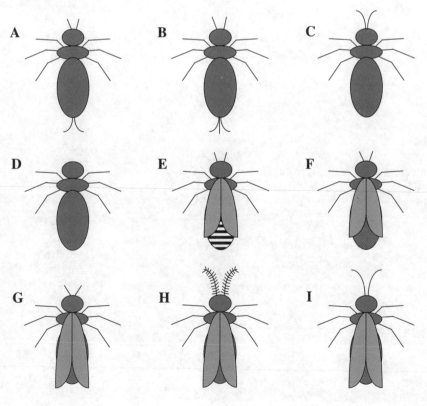

Classification and Diversity

9.. In the plant kingdom, we find Bryophyta, Filicinophyta, Coniferophyta and Angiospermatophyta.

Which of these groups have/are

a. motile gametes.

b. real roots, stems and leaves.

c. flowers.

d. fruits.

e. vascular bundles.

NERVES, MUSCLES AND MOVEMENT

14

Chapter contents
- Nerves
- Muscles and Movement
-

Nerves, muscles and movement

14.1 NERVES

14.1.1
Outline the general organisation of the human nervous system including CNS (brain and spinal cord) and PNS (nerves).

© IBO 1996

The human nervous system can be divided into several sections.

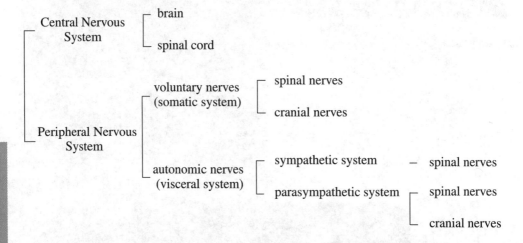

14.1.2
Draw the structure of a motor neuron.

© IBO 1996

A **motor neuron** is a nerve cell which transmits impulses from the brain to a muscle or gland. It structure has been drawn below.

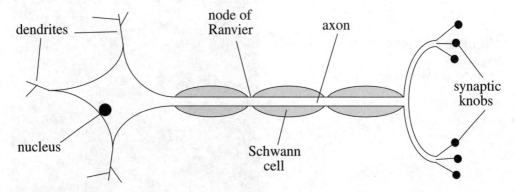

The cell body contains the following cell organelles:

nucleus - rER and sER - Golgi bodies - ribosomes - lysosomes - mitochondria.
The mitochondria are also found in the axon.

14.1.3
Define resting potential.

© IBO 1996

Resting potential: an electrical impulse across a cell membrane when not propagating an impulse.

14.1.4
Define action potential.

© IBO 1996

Action potential: the localised reversal and then restoration of electrical potential between the inside and outside of a neuron as the impulse passes along it.

14.1.5
Explain how a nerve impulse passes along a non-myelinated neuron (axon) including the role of Na^+ ions, K^+ ions, ion channels, active transport and changes in membrane polarisation.

© IBO 1996

The nature of nerve impulses is discussed in this section.

At rest there is a potential difference between the outside of the membrane of an axon and the inside. This resting potential varies a little in different species and different parts of the nervous system but the usual (average) value used is -70 mV. This means that the outside is positive compared to the inside.

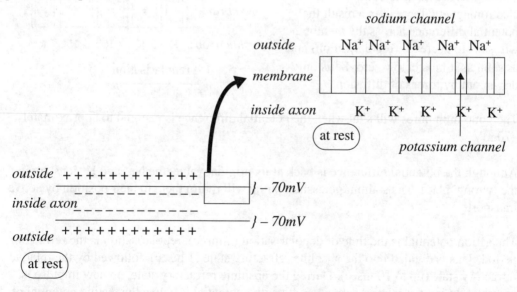

It has been found that the concentration of Na^+ is higher outside while K^+ concentration is higher inside the axon. Since both have a charge of +1, this does not create a potential difference. The distribution of Cl^- and negatively charged organic ions are responsible for the resting potential. The situation is maintained by the properties of the selectively permeable membrane.

Information travels down a neuron as an action potential. An action potential is generated by a stimulus of a receptor or from an action potential from another neuron.

In an action potential first the sodium pores suddenly open. Due to the difference in concentration of sodium outside and inside, the sodium diffuses in. Also the electromagnetic forces will cause Na^+ to go from a positively charged environment into a negatively charged environment. This influx of positive ions reduces the potential difference and is called depolarisation.

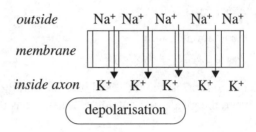

As soon as the potential difference is above zero, the Na^+ is now only driven by diffusion forces. The inside of the axon is now more positive than the outside and the movement of Na^+ is into the more positively charged area. When the value has reached +40mV, the sodium pores shut and the potassium pores open.

The K^+ moves through the potassium channels out of the axon. The forces behind this are diffusion and electromagnetic forces. As a result the potential difference across the membrane will start to decrease (repolarisation) and as soon as it falls below zero, K^+ is only driven out by forces of diffusion.

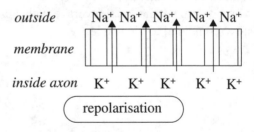

The potassium pores will shut when the potential difference is restored to approximately −70 mV.

Although the potential difference is back at its original value, the Na^+ and K^+ ions are in the 'wrong' place. The sodium/potassium pump will return them to their position by active transport.

The action potential is the time of depolarisation (1 msec); repolarisation is the refractory period. This is divided into the absolute refractory state (1 msec), followed by the relative refractory state (up to 10 msec). During the absolute refractory state, no new impulse is possible, during the relative refractory state, the potential is below the resting potential of −70mV and a stronger stimulus is required to generate an action potential.

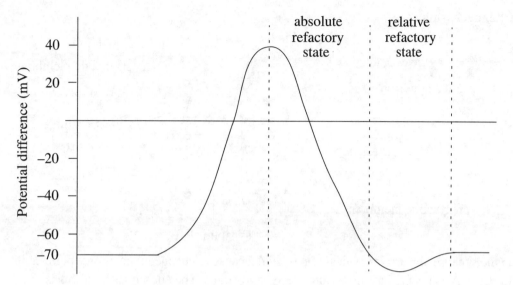

The refractory period is followed by the action of the Na/K pump which return the ions to their original sides of the membrane. A neuron can generate thousands of action potential before the change of concentrations of Na and K affects the cell.

Threshold potential.
An action potential is not generated in every case. A **threshold potential** needs to be reached (which means a depolarisation to –40 to –50mV) or the impulse fades out; it is called an 'all or nothing response'.

So, for a certain neuron, all action potentials are identical. The strength of the stimulus is conveyed by the frequency of the action potentials. Using the ear as an example: sounds of a different frequency (pitch) will stimulate different neurons and end up in slightly different places in the brain. Sounds of the same frequency but different loudness will have a different frequency of action potentials in the same neuron.

14.1.6

Explain the principles of synaptic transmission as exemplified by the neuromuscular junction including Ca^{2+} influx and release, diffusion and binding of neurotransmitter, polarisation of the post-synaptic membrane, and subsequent removal of neurotransmitter. Communication between neurons or between neurons and glands or muscles takes place in synapses.

© IBO 1996

Electrical synapses.
Two neurons may have membranes pressed close together with minute pores through them (gap junctions). An impulse can travel from one membrane to another causing an action potential in the second neuron. Electrical synapses are faster than chemical ones. They occur in invertebrates' giant axons and are related to escape responses. Vertebrate fish have electrical synapses to activate the tail flip which is used in quick starts for escape or in catching prey.

Nerves, muscles and movement

Chemical synapses.

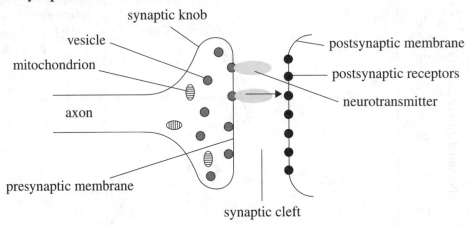

In the synapse, the arrival of an action potential causes a change in membrane permeability for Ca^{2+}. As a result, Ca^{2+} flows into the synaptic knob. The consequence of this is exocytosis of transmitter substance in vesicles. The neurotransmitter then diffuses across the synaptic cleft (20 nm) and attaches to receptors in the post synaptic membrane. The receptor sites change their configuration and open the Na^+ channels (in an excitatory synapse) which causes an action potential in the neuron. In an inhibitory synapse, the configuration change in the receptors opens the K^+ and Cl^- channels; K^+ moves out and Cl^- moves in, increasing polarisation of the neuron and increasing the distance from the threshold value.

After the post synaptic membrane has been affected, enzymes break down the neurotransmitter. (Acetyl)cholinesterase (found on the post synaptic membrane) changes acetylcholine into choline and ethanoic acid. These diffuse back to the synaptic knob, are absorbed and recycled into acetylcholine.

Acetylcholine is a common neurotransmitter found in synapses all through the nervous system. **Noradrenaline** is found in synapses of the sympathetic nervous system; **dopamine** and **serotin** are two examples of neurotransmitters found in the brain.

14.2 MUSCLES AND MOVEMENT

14.2.1
List the functions of the human skeleton.

© IBO 1996

The functions of the human skeleton are:
- **support:** supports the weight of the body against gravity; organs are attached to and suspended from the skeleton.
- **protection:** e.g. skull to protect brain, ribcage to protect the heart and lungs.
- **locomotion:** bones are rigid material to which muscles are attached. The skeleton provides a leverage system.

14.2.2
Describe the roles of nerves, muscles and bones in producing movement or locomotion.

© IBO 1996

When an animal moves, the signals pass along the nerves to the muscle, causing it to contract. The muscles are conneccted to the bones by tendons. The contraction causes the bones to move. In the case shown in the diagram, the leverage of the bones causes the end of the bone (A) to move much more than the amount by which the muscle has contracted. The movement is usually reversed by the contraction of a muscle on the opposite side of the bone. An example of an opposed pair of muscles is the bicep at the front of the upper arm and the tricep at the back. The bicep bends the arm and the tricep straightens it.

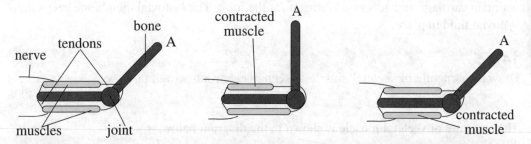

14.2.3
Draw a diagram of the human elbow joint including cartilage, synovial fluid, tendons, ligaments, named bones and named antagonistic muscles.

© IBO 1996

A diagram of human elbow joint can be found below. The opposed (antagonistic muscles are the bicep and tricep.

The joint is held together by ligaments.

Nerves, muscles and movement

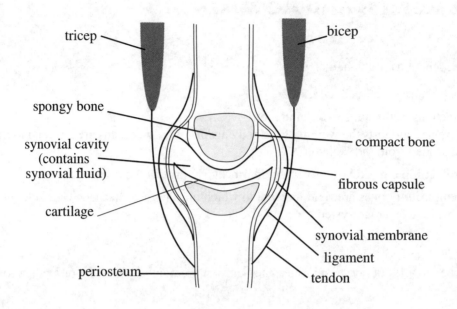

14.2.4
Outline the functions of the above named structures of the human elbow joint.
© IBO 1996

At a joint, two bones can move relative to each other. The end of the bones is made of spongy bone which is light and strong. A cartilage covering helps in smooth movement as well as absorbing shocks. Synovial fluid contains the required food and oxygen to maintain cartilage and acts as a lubricant for the joint. The synovial membrane keeps the synovial fluid in place.

14.2.5
Draw the structure of skeletal muscle as seen in electronmicrographs.
© IBO 1996

The structure of skeletal muscle is shown in the diagram below.

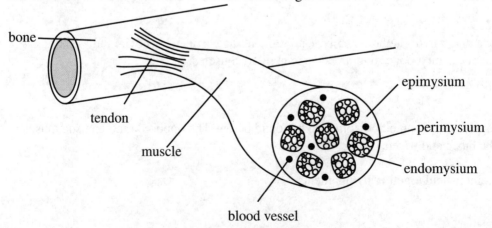

The individual muscle fibres have a structure illustrated in the diagram.

Muscles are groups of cells working together. Each muscle cell originally was many cells which fused and the resulting cell has many nuclei.

Muscles cells can be more than 1 cm long. Groups of cells (also called muscle fibres) are arranged together and have connective tissue around them. Muscles have tendons (connective tissue) at each end, attaching them to the bones.

Inside a muscle cell, you find many thin myofibrils. These thin fibres cause the typical striated (striped) pattern of skeletal muscles. Myofibrils contain two types of myofilaments: **myosin** and **actin**. They are made of a protein like substance.

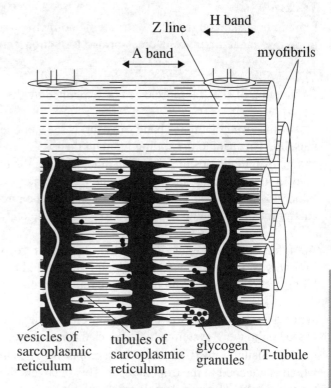

The 'unit' within a myofibril is the **sarcomere**. It contains a light section, a dark section, an intermediate section, a dark section and a light section. The thin actin filaments are attached to the Z line. This is the light section, the I band. They partly overlap with the thick myosin filaments which appears as a dark section, the A band. In the middle between the 2 Z lines, you find only myosin, showing as a gray section, the H band.

Across the fibres, you find the T system. These are tubules touching the sarcolemma and associated with vesicles which are part of the sarcoplasmic reticulum. A T tubule with a pair of vesicles is called a triad.

The vesicles are of great importance because they regulate the movement of calcium ions (Ca^{2+}) to/from the sarcoplasm. Since the Ca^{2+} concentration determines the activity of ATPase (which hydrolyses ATP, releasing its energy), this essentially determines the activity of the muscle.

Sarcolemma: membrane surrounding muscle cell/fibre.
Sarcomere: section of the myofibril between 2 Z lines.
Sarcoplasm: cytoplasm in a myofibril.
Sarcoplasmic reticulum: internal membranes within sarcoplasm.

Nerves, muscles and movement

14.2.6
Explain how skeletal muscle contracts including the roles of Ca^{2+} ions, troponin, tropomyosin, actin, myosin, cross-bridge formation, movement and breakage, and ATP.

© IBO 1996

Contraction of skeletal muscle can be explained by the sliding filament theory.

It was discoved that the A band (myosin) is the same length in contracted and relaxed muscles. This lead to the sliding filament theory. Essentially this theory says that the actin and myosin filament slide over each other to make the muscle shorter. Little 'hooks' on the myosin filaments attach to the actin and pull them closer. Then they release and repeat futher down the actin. This can be referred to as the ratchet mechamism. ATP provides the energy for this by being hydrolysed to ADP by the enzyme ATPase.

Actin filaments contain actin as well as 2 proteins: **tropomysosin** and **troponin**. Tropomyosin forms two strands which wind around the actin filament, covering the binding site for the myosin hooks. The muscle cannot contract now.

When a nerve impulse arrives at the muscle, the depolarisation of the motor end plate is passed on to the T system which causes the vesicles of the sacroplasmic reticulum to release calcium ions (Ca^{2+}) into the sacroplasm. The calcium ions attach to the troponin which is attached to the tropomyosin. This uncovers the binding sites on the actin for the myosin hooks. The muscle will now contract.

When no more nerve impulses arrive, calcium ions are moved back into the vesicles of the sacroplasmic reticulum by active transport. The binding sites on the actin will then be covered again and the muscle will relax.

EXERCISE

Questions 1 & 2 refer to the diagram of a motor neuron.

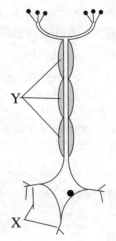

1. The parts labelled X are:

    A   axons.
    B   Schwann cells.
    C   nodes of Ranvier.
    (D) dendrites.

2. The parts labelled Y are:

    A   axons.
    (B) Schwann cells.
    C   nodes of Ranvier.
    D   dendrites.

3. The localised reversal and then restoration of electrical potential between the inside and outside of a neuron as the impulse passes along it is known as the:

    A   resting potential.
    (B) action potential.
    C   nerve potential.
    D   electrical potential.

4. The ions principally involved in transmitting nerve impulses are:

    A Na^+ & Cl^-.
    B K^+ & Cl^-.
    C Fe^{2+} & K^+.
    (D) Na^+ & K^+.

5. The diagram shows a chemical synapse. The synaptic cleft is labelled:

    A   A.
    B   B.
    (C) C.
    D   D.

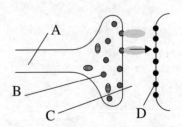

Nerves, muscles and movement

6. The movement of the knee joint is achieved by:

    A   a single muscle in the upper leg.

    B   flexing of the knee joint.

    C   a pair of opposed muscles in the upper leg,

    D   a single muscle in the back of the lower leg.

7. The digram shows a joint. The synovial fluid is contained in the part labelled:

    A   A.

    B   B.

    C   C.

    D   D.

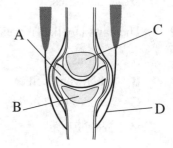

8. Muscles are joined to bones by:

    A   ligaments.

    B   tendons.

    C   cartilage.

    D   fibres.

9. Cytoplasm in a myofibril is known as:

    A   Sarcolemma.

    B   Sarcomere.

    C   Sarcoplasm.

    D   Sarcoplasmic reticulum.

10. a. Which ions are responsible for the -70 mV resting potential across the membrane of e.g. an axon.
b. What is the first thing that happens in an action potential?
c. What effect does it have on the potential across the membrane?
d. Which forces drive the movement of sodium ions at any stage of this process?
e. How is the original potential restored?
f. How is the original distribution of ions restored?

11. Explain the role of the following in muscle contractions?
a. Ca^{2+} ions
b. troponin and tropomyosin
c. actin and myosin
d. ATP

Nerves, muscles and movement

EXCRETION

15

Chapter contents
- Excretion
- The human kidney

15.1 EXCRETION

15.1.1
Outline the need for excretion in all living organisms.

© IBO 1996

All organisms produce waste in the process of metabolism. If this waste is allowed to accumulate, it will cause a problem for the organism. **Excretion** is the process of removal of these wastes.

15.1.2
State that excretory products in plants include oxygen, and in animals include carbon dioxide and nitrogenous compounds.

© IBO 1996

Obviously, animals secrete nitrogenous waste in their urine. However, carbon dioxide (excreted from the lungs) is also a waste product in animals. Oxygen (produced in photosynthesis, secreted via stomata) is a waste products in plants (during the day).

15.1.3
Discuss the relationship between the different nitrogenous waste products and habitat in mammals, birds, amphibians and fish.

© IBO 1996

Nitrogenous waste is produced when amino acids are broken down, for example, to be used for energy. The resulting product is ammonia. Ammonia is a highly toxic substance because it is very basic.

Animals living in fresh water (freshwater fish like trout, amphibian larvae like tadpoles) will take in a surplus of water due to osmosis. They will therefore need to get rid of large amounts of water. This can conveniently be used to dilute the ammonia and flush it out of the organism.

Marine animals do not take up large amounts of water so they need to convert ammonia to a less toxic substance. Urea [$(NH_2)_2 CO$] and trimethylamine oxide [$(CH_3)_3 NO$] are found in the urine of marine organisms.

Adult amphibians need to conserve water and will convert ammonia into urea. Birds cannot carry much water and excrete nitrogenous waste in the form of insoluble uric acid (the white part in bird droppings).

Mammals excrete urea which is produced in the liver in the ornithine cycle. The extent to which they are able to concentrate their urine depends on the structure of the kidney (length of Henle's loop, section 15.2.5). This determines the potential habitat of the organism.

15.2 THE HUMAN KIDNEY

15.2.1
Draw the structure of the kidney including cortex, medulla, pelvis, ureter and renal blood vessels.

© IBO 1996

Humans have two kidneys, situated low in the abdominal cavity, near the back. Each kidney has a renal artery leading to it and a renal vein and a ureter leading away from it. The renal vein takes the 'clean' blood away from the kidney while the ureter leads the urine to the bladder.

SCHEMATIC DIAGRAM OF THE HUMAN URINARY SYSTEM.

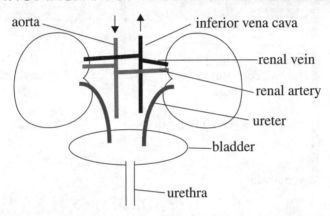

When you look into a kidney, you can see the cortex on the outside and the medulla on the inside. In the centre you see the renal pelvis.

SCHEMATIC DIAGRAM OF THE HUMAN KIDNEY.

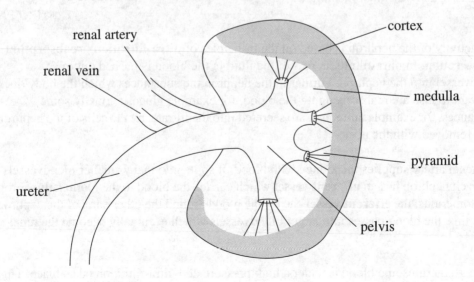

Excretion

15.2.2
Draw the structure of a glomerulus and associated nephron.

© IBO 1996

The functional unit in the kidney is the **nephron**. The diagram below is a schematic representation of a nephron.

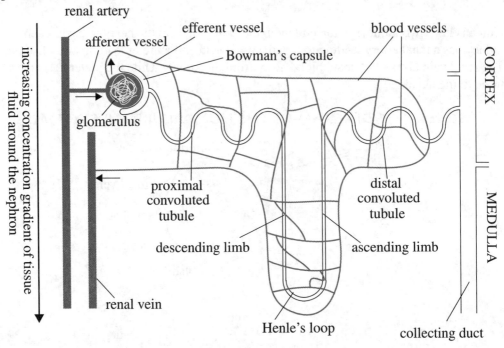

15.2.3
Explain the process of ultrafiltration including blood pressure, fenestrated blood capillaries, and basement membrane.

© IBO 1996

The activity of the nephron is based on the principles of **ultra-filtration**, **reabsorption** and **secretion**. In ultra-filtration, part of the fluid in the blood is pushed out of a bloodvessel into the nephron. Further in the nephron the substances which the body does not want to lose are reabsorbed into the blood, for example glucose. Finally some substances, for example ammonia, are secreted into the filtrate, by the cells of the nephron, to be removed with the urine.

The renal artery supplies the kidney with blood. It splits into many smaller bloodvessels and each nephron has an **afferent vessel** which carries the blood to the **glomerulus**. From the glomerulus, the **efferent vessel** carries the blood around the other parts of the nephron. After this, the blood passes into larger bloodvessels which eventually become the renal vein.

In the glomerulus, the blood is under a high pressure and ultra-filtration takes place. This means that some of the liquid and dissolved particles are pushed out of the bloodvessel; the cells and the larger molecules (e.g. proteins) are too big to pass through and will not be found in Bowman's capsule. All the blood in the body passes through the kidney every 5 minutes. Approximately 15 - 20% of the fluid in the blood will pass into Bowman's capsule which is about 200 litres a day!

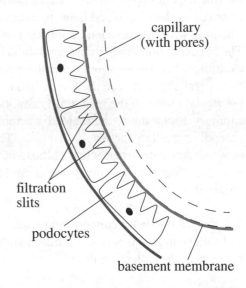

The filtrate needs to pass a 'barrier' made of 3 different layers: the **wall of the glomerulus**, the **basement membrane** of the glomerulus and the inner wall of **Bowman's capsule**. The wall of the glomerulus contains small pores. It is said to be **'fenestrated'** and allows blood plasma to pass through. The basement membrane is a protein membrane outside the cells; it contains no pores and serves as a filter during the ultrafiltration process (it acts as a dialysis membrane) and stops the blood cells and large proteins. The cells of the inner wall of Bowman's capsule are called **podocytes**; they have many extentions which fold around the blood vessels and a network of filtration slits that hold back the blood cells.

15.2.4
Explain the process of ultrafiltration including blood pressure, fenestrated blood capillaries, and basement membrane.

© IBO 1996

The first part of the nephron is Bowman's capsule. This is where the filtrate from the glomerules enters the nephron. Bowman's capsule joins onto the **proximal convoluted tubule**. The fluid in the proximal convoluted tubule is similar to plasma and contains glucose, amino acids, vitamins, hormones, urea, salt ions and, of course, water. Most of the reabsorption in the nephron occurs here: all the glucose, amino acids, vitamins, hormones and most of the sodium chloride and water are reabsorbed into the bloodvessels (peritubular capillaries). Osmosis drives the reabsorbtion of water as it follows the active transport of glucose and Na^+. Cl^- passively follows the actively transported Na^+. All these substances need to move across the wall of the proximal convoluted tubule. To facilitate this, the cells lining the lumen of the proximal convoluted tubule have a brush border: a row of microvilli (finger like extensions of the cell) which greatly increase the available surface area. Mitochondria are also prominent in these cells, providing the energy for active transport.

15.2.5
Explain the production of hypertonic urine including the roles of the loop of Henle, medulla, collecting duct, ADH and water potential gradients (cross reference 16.2.2).

© IBO 1996

Excretion

In the **descending limb** of **Henle's loop** water leaves the nephron by osmosis due to the increasing concentration (of salt). This water immediately passes into the blood capillaries and is removed from the area. Some salt diffuses into the filtrate. The **ascending limb** is impermeable to water and salt is lost from the filtrate by active transport. (The amount of salt actively transported from the ascending limb is greater than the amount which diffuses into the descending limb.) The salt remains near Henle's loop (it is not immediately removed by the blood) and helps to maintain a concentration gradient in the medulla. The fluid which leaves Henle's loop is less concentrated than the tissue fluid around it. The concentration gradient in the medulla is maintained by the **vasa recta countercurrent exchange**. The **vasa recta** are the blood vessels running along Henle's loop. There is no direct exchange between the filtrate and the blood but substances pass through the interstitial region of the medulla. The blood entering the medulla will, in the descending capillary, lose water to the region by osmosis and absorb salt and urea by diffusion. In the ascending capillary the reverse happens. The advantage is that the blood leaving the area is in a constant state, irrespective of the osmotic concentrations of the blood entering the medulla. Since the movements are caused by osmosis and diffusion, there is no energy required.

The wall of the **distal convoluted tubule** is permeable (permeability regulated by ADH) and water can pass from the ultrafiltrate into the bloodvessels to be carried away. The same happens in the **collecting duct**.

ADH increases the permeability of the walls of the distal convoluted tubule and the collecting duct. ADH is released from the posterior lobe of the pituitary gland when there is a lack of water. The dilute filtrate coming from Henle's loop can then lose water (by osmosis) in the distal convoluted tubule and again in the collecting duct. The water is reabsorbed by the blood. When ADH is absent and the walls are impermeable, water is not removed from the filtrate in the distal convoluted tubule and the collecting duct and ends up in the bladder as dilute urine.

15.2.6
Compare the composition of blood in the renal artery, and renal vein, and glomerular flitrate and urine.

© IBO 1996

Blood in the renal artery is rich in oxygen and contains more urea, more salt and possibly more water than the set value.

Glomerular filtrate is similar to blood plasma without large proteins. The selection process that has occurred in producing glomerular filtrate was based on size of the molecules only, not on any other criteria.

Urine contains less water, less salt, no glucose, no proteins or amino acids but a lot more urea (more concentrated) than glomerular filtrate.

Blood in the renal vein contains carbon dioxide but the correct amounts of water and salts and very little urea.

15.2.7
Outline the structure and action of kidney dialysis machines.

© IBO 1996

If both kidneys fail to function properly, the person becomes dependent on a **kidney dialysis machine**. The principle behind this is simple. Lead the patient's blood from a vein through a machine which cleans it and return it to the body. Cleaning the blood is done by running it through **partially permeable tubes** (e.g. visking tubing). On the other side of this partially permeable membrane a dialysis fluid is found. The dialysis fluid contains the same solute concentration as blood plasma, but no urea. Urea will diffuse through the membrane into the dialysis fluid, removing it from the body. Water and solutes can be added to the returning blood as needed.

The problems arise when trying to apply this simple principle. The surface area required for effective dialysis is very large and the process is slow. This means that the patient will have to spend at least 4 hours in dialysis, 3 times a week. This obviously has a great impact on school or work.

EXERCISE

1. An organism is found to excrete oxygen. It is most likely that the organism is:

    A   a vertebrate.
    B   a bacterium.
    C   a fungus.
    D   a plant.

2. Bird droppings (guano) is a useful fertilizer because it contains the element:

    A   oxygen.
    B   hydrogen.
    C   nitrogen.
    D   iron.

3. Waste products are removed from the kidney by the:

    A   ureter
    B   urethra
    C   inferior vena cava
    D   aorta

Excretion

Questions 4 & 5 refer to the diagram of a human kidney.

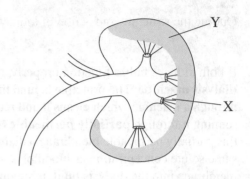

4. The point labelled X is:

   A   cortex.
   B   medulla.
   C   pyramid.
   D   ureter.

5. The point labelled Y is:

   A   cortex.
   B   medulla.
   C   pyramid.
   D   ureter.

6. The functional unit in the kidney is the:

   A   aorta.
   B   ureter.
   C   cortex.
   D   nephron.

7. In the descending limb of Henle's loop water leaves the nephron by:

   A   diffusion.
   B   osmosis.
   C   evaporation.
   D   condensation.

8. Kidney dialysis machines remove waste products from:

   A   the urine using a partially permeable membrane.
   B   the blood using a partially permeable membrane.
   C   the blood by replacing it with fresh blood.
   D   the bladder directly.

9. a.   Name 4 nitrogenous waste products.
   b.   For each of these products, give an example of an organism or group of organisms) that uses it.

c. What are the demands that each of these products places on the organisms environment/lifestyle?

10. Complete the table below.

	proximal convoluted tubule	descending branch of Henle's loop	ascending branch of Henle's loop	distal convoluted tubule	collecting duct
glucose	*	-------------	-------------	-------------	-------------
amino acids		-------------	-------------	-------------	-------------
Na+		in diffusion		-------------	-------------
Cl-	out - almost all passive	in diffusion	out diffusion and active transport	-------------	-------------
water					
urea		in diffusion	in diffusion	-------------	some out ** ADH dependent diffusion
foreign substance	out - all active transport	-------------	-------------	-------------	-------------

* except in diabetes patients.

** ADH increases permeability of the wall of the collecting duct to water and urea.

Excretion

PLANT SCIENCE

16

Chapter contents
- Dicotyledenous plant structure
- Transport in Angiosperms
- Germination
- Plants and people

Plant science

16.1 DICOTYLEDENOUS PLANT STRUCTURE

16.1.1
Draw a diagram to show the external parts of a named dicotyledonous plant including root, stem, leaf, buds, sepal, petal, stamen and carpel.

© IBO 1996

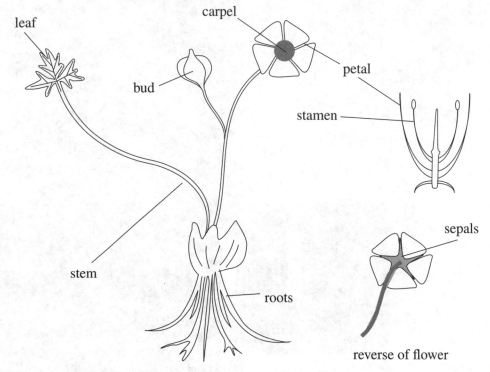

This part of the course is best completed as a practical using a local plant.

16.1.2
Draw plan diagrams to show the distribution of tissues in the stem, root and leaf of a generalised dicotyledonous plant.

© IBO 1996

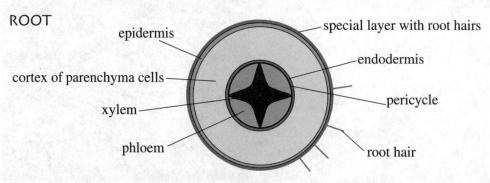

276

Biology

STEM

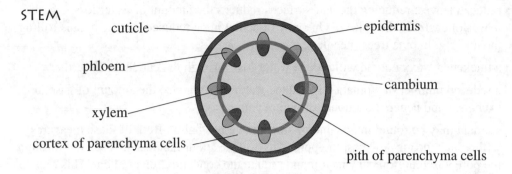

LEAF

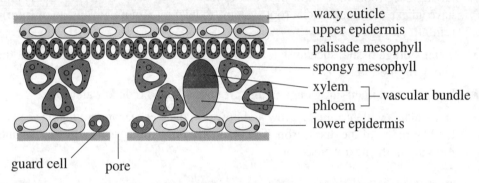

16.1.3
Explain the relationship between the distribution of tissues in the leaf and their functions.
© IBO 1996

The waxy cuticle is found on the top and bottom surface of the leaf. It is (almost) impermeable to water and serves to reduce water loss. The upper and lower epidermis cover the leaf. They form a barrier against infection. The mesophyll is divided into two layers: the palisade mesophyll is the primary layer for photosynthesis; the spongy mesophyll also performs photosynthesis, but since light intensity is less as well a having a lower number of chloroplasts (compare with the palisade layer), it is less important in photosynthesis. Its spongy structure with air spaces allows rapid diffusion of gases. The pore allows gaseous exchange with the environment while the guard cells can close the pore when too much water is lost.

Chloroplasts are found in the mesophyll and, in some species, in the guard cells.

16.1.4
Outline four structural adaptations of xerophytes.
© IBO 1996

Xerophytes are plants that are able to live in dry environments. They can have any of the following adaptations so that they can obtain the maximum amount of water and not lose it.

Plant science

- reduced leaves: reducing the leaf surfaces reduces the amount of water loss, e.g. the leaves of cacti are the spikes. One way of achieving a reduced surface area is rolling the leaf e.g. in pine trees (needles).
- a thickened waxy cuticle will reduce water loss through the cuticle even further.
- a reduced number of stomata will reduce water loss but also the amount of gaseous exchange and hence the amount of photosynthesis.
- stomata may be found in pits and/or surrounded by 'hairs'. Both of these measures will reduce the air flow past the pore. Any water vapour that has diffused out through the pore will therefore stay near it and reduce the concentration gradient. This reduces further diffusion of water out of the plant.
- having deep roots may allow the plant to reach water deep in the soil. Some plants have an extensive superficial root systems to take maximum advantage of (rare) rainfall.
- water storage tissue, such as is found in the stems of cacti, will help the plant through a long dry period.
- being a small plant, growing near the ground will also reduce water loss.
- some plants will germinate, grow and flower in the wet season, producing seeds before the beginning of the dry season. The plant will die but the seeds will survive and start growing in the next wet season.

16.1.5
Outline two structural adaptations of hydrophytes.

© IBO 1996

Hydrophytes are waterplants. They too have special adaptations to survive in their environment.

- airspaces will allow the plant to float on the surface so that it can obtain the most light.
- as the water provides the upwards force supporting a plant, a typical hydrophyte has little strengthening tissue.
- being surrounded by water, the roots mainly serve for anchorage and are reduced in size.

Water lilies at the artist Monet's garden at Giverny, France

16.2 TRANSPORT IN ANGIOSPERMS

16.2.1
Define water potential.

© IBO 1996

Water potential is a measure of the tendency of water to move between regions. In practice it is the force acting on water molecules in solution when separated from pure water by a membrane permeable to water only (i.e. partly permeable).

16.2.2
State that water moves down a water potential gradient (cross reference 15.2.5).

© IBO 1996

Water will move down a water potential gradient. It will go from an area of higher water potential (= higher concentration of water = lower concentration of solutes) to an area of lower water potential (= lower concentration of water = higher concentration of solutes).

16.2.4
Explain how the root system provides a large surface area for mineral ion and water uptake by means of branching, root hairs and cortex cell walls.

© IBO 1996

Plants take up water (and minerals) via their roots. As we have seen in other systems, materials need a certain amount of time to cross a barrier. Therefore, to allow an adequate uptake of these molecules, a large surface area is needed. In, for example, the small intestine, villi and microvilla greatly increase the surface area. Roots are branched and have root hairs to further increase their surface area. The structure of the cortex is such that it also facilitates water uptake (see 16.2.5).

16.2.5
Explain the process of water uptake in roots by osmosis.

© IBO 1996

Roots take up water from the soil by osmosis. The principle behind this is that the water in the soil contains a lower concentration of solutes (i.e. a higher water potential) than the cytoplasm, so water moves in. By transporting water from the cells to the xylem and up into other areas of the plant, the concentration gradient is maintained and the root can continue to take up water.

Water is mainly absorbed by the root hairs (see section 16.1.2) in the process of osmosis. To get the water from the root hair to the xylem, 3 paths are possible.

- **the apoplast pathway:** the water does not enter the cell. It travels through the cell walls until it reaches the endodermis. The cells of the endodermis have a Casparian strip around them which is impermeable to water. To pass the endodermis, the water

Plant science

will have to follow the symplast pathway.

- **the symplast pathway:** the water enters the cytoplasm but not the vacuole. It will pass from cell to cell via plasmodesmata (tiny connections of cytoplasm between one cell and the next (see also section 9.2.7).
- **the vacuolar pathway:** the water will enter the cell and move into the vacuole. From here it will travel through the cytoplasm and cell wall to the vacuole of the next cell.

Water commonly travels mostly via the apoplast pathway. Minerals tend to follow the symplast pathway (see section 16.2.3).

Again, the water potential of the soil will be higher than that of a root hair, which will be higher than that of a cell in the epidermis, etc.

The **Casparian strips** in the endodermis are thought to be a protective measure. By forcing the water to go through the cell (rather than past it, through its cell wall) the plant has a measure of control over the substances that enter.

From the endodermis, water can be actively secreted into the xylem and/or pulled up by transpiration forces (see section 16.2.7).

16.2.3
Describe the process of mineral ions uptake into roots by active transport.

© IBO 1996

Since it has been found that the mineral concentration inside the root may exceed that of a specific mineral in the soil water, it can be concluded that minerals are taken up by active transport. This is supported by the fact that mineral uptake requires energy. In experiments, mineral uptake has been brought to a halt by depriving the roots of oxygen or adding a substance that blocks cellular respiration.

16.2.6
State that terrestrial plants support themselves by means of cell turgor and xylem.

© IBO 1996

Plants do not have a skeleton to keep them upright. Trees and shrubs have woody stems that support them but herbaceous plants depend mainly on **turgor** for their support. As the vacuole takes up water, the cell swells up. This will stretch the cell wall until it has stretched to the limit. The vacuole will still have a lower water potential than the fluid surrounding the cell. It will continue to draw in water. However, the force of the cell wall will force water out at the same rate. The result is like a water filled balloon in a cardboard box. It is quite firm and a number of these cells on top of each other will not need external support.

Of course, the xylem vessels have some supporting tissue which will assist in the process of keeping the plant upright but this by itself is insufficient as you know if you have ever seen a wilting plant.

16.2.7
Define transpiration.

© IBO 1996

Transpiration: the loss of water vapour from the leaves and stems of plants.

16.2.8
Explain how water is carried by the transpiration stream including an outline structure of xylem vessels, transpiration pull, cohesion and evaporation.

© IBO 1996

The specialisation of different parts of the plant means that it is necessary to move substances from one place to the next. Water and minerals are absorbed by the roots and need to be moved to other parts while sugar is produced by the leaves and therefore also requires distribution.

Water and minerals are transported by the xylem tissue while the products of photosynthesis (mainly sucrose) are transported by the phloem tissue. Xylem tissue is dead while phloem is living tissue.

There are two kinds of conducting elements found in xylem tissue. Tracheids are narrow cells which are arranged in columns. They overlap at their tapered ends allowing them to have a support function. The overlapping ends have pits for water to move rapidly from one cell to the next. Tracheids are less efficient than xylem vessels and all vascular plants have them.

Xylem vessels are larger. They are also made of columns of cells. When the cells die, the wall in between them disappear partly or completely. Xylem vessels are wider in diameter than tracheids. This makes the xylem vessel more efficient in water transport. The walls on the side are reinforced with lignin so that xylem also plays a role in supporting the stem. Xylem vessels are generally only found in angiosperms.

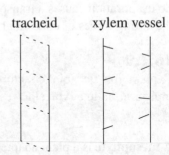

tracheid   xylem vessel

The structure of xylem vessels somewhat resemble a drainpipe. Yet water does not run up a drainpipe without assistance. Several forces cooperate in getting the water from the roots of a plant up to the leaves, which may be 20 metres higher.

A tree has many leaves exposed to direct sunlight. The leaves are thin and large, so that it can catch a lot of light and perform a lot of photosynthesis. It also creates a lot of surface area from which water can (and will) evaporate. On a hot, dry day, a fair sized tree can evaporate more than 1000 litres of water.

Inside the leaf, the water will evaporate from the spongy mesophyll cells into the airspaces (see section 16.1.2). From there, it will diffuse out through the pores of the stomata. New water molecules will then evaporate from the cells to replaces the ones lost by diffusion.

Plant science

As the water evaporates from the mesophyll cells, their water potential will decrease. As a result, water from the vascular bundle will move into the mesophyll cells. The xylem vessels of the vascular bundle are filled with water. The water molecules have strong cohesion forces and moving some water molecules out of the vessel will mean that others will want to move up to. It is like pulling a cord. So all the water molecules in the vessel will move up a bit and at the other end, in the roots, some water will move from the soil into the plant. This is called **transpiration pull**.

Yet transpiration pull would not work if water molecules did not have strong forces of cohesion.

Cohesion: the attraction between molecules of a liquid resulting from intermolecular forces. Water molecules are not straight as you can see in the diagram. See also section 2.1.6.

Oxygen has two pairs of electrons that are not covalently bonded ('lone' pairs of electrons, see '*Chemistry for use with IB*' p 97). They make a concentrated negative charge, pushing the two pairs of bonding electrons towards each other. All this creates a non symmetric molecule where the centre of the positive forces does not coincide with the centre of the negative forces. The molecule is therefore a dipole and the positive side of one water molecule strongly attracts the negative side of the neighbouring water molecules. This causes strong forces of cohesion.

So evaporation causes a transpiration pull which will pull water into the roots because of the strong forces of cohesion between water molucules.

16.2.9
Explain how the abiotic factors light, temperature, wind and humidity affect the rate of transpiration in a typical terrestrial mesophytic plant.

© IBO 1996

A **mesophyte** is a plant adapted to conditions of average water supply, i.e. not a xerophyte or a hydrophyte.

A typical terrestrial mesophytic plant could be an oak tree or a dandelion. The rate of transpiration depends on several factors:
- temperature: the rate of evaporation is doubled for every 10°C increase in temperature.
- humidity: evaporation is much higher in dry air than in air which is already (partly) saturated with water vapour.
- wind: air currents will take water vapour away from the leaf. This will maintain the concentration gradient and as a result water will continue to diffuse out of the leaf.
- light: plants generally open their stomata in the light to allow diffusion of carbon dioxide into the leaf and hence allow photosynthesis to take place. However, this will also greatly increase evaporation.

16.2.10
Define translocation.

© IBO 1996

Translocation: the movement of substances from one part of a plant to another in the phloem.

16.2.11
Outline the role of phloem in active translocation of various biochemicals.

© IBO 1996

Translocation occurs through the phloem. The xylem is made up of different element (tracheids and vessels) and so is the phloem. The elements making up the phloem in Angiosperms are sieve tube members and companion cells.

Sieve tubes are stacks of **sieve tube member cells**. The walls between them have pores and are called **sieve plates** which greatly facilitate transport. Sieve tube members are living cells (unlike vessels and tracheids in xylem). However, many of the cell organelles of a sieve tube member will disappear as it matures. Next to the sieve tube member, you will usually find a special **companion cell**, which will assist the sieve tube member in those metabolic processes it can no longer do itself, especially to generate energy required by the sieve tube member.

Sieve tubes carry products of photosynthesis. The fluid in them tastes sweet. The content of the sieve tubes from maple trees can be collected and boiled for some time to make maple syrup.

Sieve tubes can carry material in either direction. For example, in summer the maple tree will transport sugars from the leaves down to the roots. In spring, the sieve tubes will carry the sugar from the roots (where it was stored) to the branches to allow new leaves to grow.

The rate of transport in the sieve tubes may vary. It has been found to be as high as 1 m/s. This is higher than could be achieved by diffusion alone.

16.2.12
Describe an example of food storage in a plant.

© IBO 1996

An example of food storage in plants is the food stored in the roots. For example, carrots store a large amount of food in the cortex of the root. This may be used to allow the growth of stem and leaves after the winter.

Another example of food storage in plants is the food stored in the seeds. Seeds need a certain amount of stored food to allow them to grow a stem and a few leaves. As soon as these are big enough, photosynthesis will provide the growing plant with food.

Plant science

16.2.13
Outline the gas exchange pathways in the root and leaf of a typical terrestrial mesophytic plant.

© IBO 1996

Gas exchange in the leaf of a typical terrestrial mesophytic plant, such as an oak tree or a dandelion, takes place mainly through the stomata. See section 16.1.2 for a diagram of the leaf.

During the day, under normal conditions, the stomata are open. Carbon dioxide will diffuse in while oxygen and water vapour will diffuse out. As carbon dioxide is used during photosynthesis, it diffuses into the mesophyll cell from the intercellular spaces. The reverse is true for oxygen, produced by photosynthesis. Please remember that cellular respiration goes on in any living cell, including a plant cell in the light. However, the amount of photosynthesis far exceeds the amount of cellular respiration. The net result will be uptake of carbon dioxide and excretion of oxygen.

During the night, stomata will mainly be closed so any gas exchange will be greatly reduced. Cellular respiration will continue to proceed but photosynthesis has stopped. The net result will be uptake of oxygen and excretion of carbon dioxide (and water).

The roots also have gaseous exchange. The cells in the roots do not have photosynthesis (no light) but they do require cellular respiration. So oxygen will diffuse from the airspaces in the soil into the roots.

16.3 GERMINATION

16.3.1
Draw the external and internal structure of a named dicotyledonous seed.

© IBO 1996

Flowering plant (angiosperms) are divided into two main groups, the monocotyledons and the dicotyledons.

Monocotyledons only have one seed leaf. In the mature plant, the leaves tend to be long and narrow with parallel veins. Examples are lilies, grasses, palmtrees.

The picture shows the leaves of a palm tree.

Dicotyledons have two seed leaves. In the mature plant, the leaves are broader and shorter and form a net like pattern. Examples are the oak tree and the buttercup.

The picture shows the leaves of a passion fruit vine.

The diagram shows the structure of a bean seed.

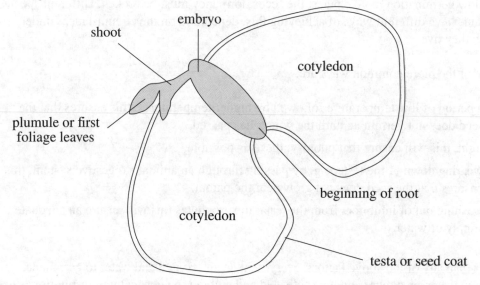

The function of the cotyledon is to store large amounts of reserve food which the embryo will use when it starts to grow. The function of the testa is to protect the seed from, for example, fungi.

Plant science

16.3.2
Describe the metabolic events of germination in a typical starchy seed.

© IBO 1996

A seed has very few metabolic processes going on. The seed grew in the parent plant and became dormant when it left the parent. Germination is the resumption of growth or development from a seed.

The first step in the process of germination is the absorption of water. As the water content of the seed is very low, a large amount of water, sometimes as much as the mass of the seed itself, needs to be absorbed to allow metabolic processes to start. The presence of water will activate hydrolytic enzymes such as amylase.

Amylase breaks down the stored starch into maltose. The maltose will be moved to the embryo and it will be used in cellular respiration to provide energy but also to make cellulose for cell walls of new cells.

The stored proteins and lipids will also be hydrolysed (broken down by adding water, see section 2.2). The amino acids thus produced will be used to make new proteins which the growing embryo can use in cell membranes or as enzymes. The fatty acids and glycerol from the stored lipids are used in cell membranes (phospholipids) and for energy.

16.3.3
Explain the conditions needed for the germination of a typical seed.

© IBO 1996

To allow germination to take place, the seeds dormancy must be broken. Different species of plants have different ways of achieving this, depending on the circumstances under which they live.

Some of the more common ways are:

- a period of low temperature followed by higher temperatures; this ensures that the seed does not germinate until the winter has passed.
- light; this will ensure that photosynthesis is possible.
- wearing down of the testa, e.g. by passing through an animal's digestive system; this ensures that the seed does not grow near the parent.
- washing out of inhibitors from the testa; in xerophytes this will ensure an adequate supply of water.

Other than any of the above factors, every seed needs oxygen and water to germinate. Without water, enzymes cannot be activated and without oxygen, cellular respiration is not possible.

16.4 PLANTS AND PEOPLE

16.4.1
Outline the importance of plants to people in terms of food, fuel, clothing, building materials and aesthetic value.

© IBO 1996

Plants are important to people in many ways.

They serve the following functions:
- food: you only need to think of your own dinner tonight.
- fuel: people have burned not only wood but also peat.
- clothing: cotton and linen are two common examples.
- building materials: again wood is used but also palm leaves for covering a roof.
- aesthetic value: most cities have several parks and people will often go for a walk in the park.

Plant science

16.4.2
State one example of a plant in each of the five categories above.

© IBO 1996

An example of each of the categories above:
- **food:** rice
- **fuel:** wood or peat
- **clothing:** cotton
- **building materials:** oak as supporting beams
- **aesthetic value:** palm trees and grass

16.4.3
Describe the cultivation of a plant of economic importance.

© IBO 1996

Maize is cultivate as animal fodder in large parts of Europe. Maize is a fast growing C4 plant with a small and simple root system. It grows best on well fertilised ground and weeds only get a chance to grow when the plants are still small. This means that herbicides are routinely used only once.

Rice is cultivated in (submerged) rice paddies. The conditions in these paddies are such that they promote the growth of the rice plants and discourage others. Cultivating rice is often still done by hand and weeds are removed this way.

16.4.4
Discuss two techniques used in cultivation which have led to improvements in yield (cross reference 10.6).

© IBO 1996

As long as agriculture exists, people have selected seeds to plant the following year. The seeds were often from the most desirable plants, i.e. those with the biggest yield or sweetest fruits. The assumption was that these characteristics would be passed on to the next generation. This is the simplest form of selective breeding.

Hybridisation can be used to bring in, for example, higher resistance to disease or insects. Most of our cultivated plants have a high yield and we find them tasty. The main problem is that they are not very resistant to disease or insects and therefore require large amounts of pesticides. Hybridisation with another, often less cultivated but closely related species, can be tried in an attempt to combine the desirable characteristics of both. Hybrids may have an odd number of

chromosomes and, as a result, may not be able to produce fertile offspring so the new generation may have to be created from the two different species.

Genetic engineering is used in an attempt to transfer the NIF genes (the nitrogen fixation genes) into non-leguminous plants to reduce the need for expensive fertilisers. This way the yield may be increased since the plant no longer depends on (expensive) fertiliser being provided.

For additional examples see section 10.6.

EXERCISE

1. The diagram shows a cross section of a root. The xylem is labelled:

   A   A
   B   B
   (C)   C
   D   D

   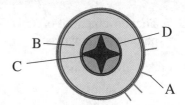

2. The diagram shows a section through a flower. The part labelled X is:

   (A)   a stamen.
   B   a sepal.
   C   a petal.
   D   a bud.

   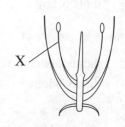

3. Plants that are able to live in dry environments are called:

   A   Hydrophytes.
   B   Algae.
   (C)   Xerophytes.
   D   Mesophytes.

4. Roots take up water from the soil by:

   A   evaporation.
   B   suction.
   C   diffusion.
   (D)   osmosis.

Plant science

5. The loss of water vapour from the leaves and stems of plants is known as:

- **(A)** transpiration.
- **B** diffusion.
- **C** dessication.
- **D** osmosis.

6. Water has a high surface tension and comparatively high boiling point because its molecules are:

- **A** covalent.
- **(B)** polar.
- **C** ionic.
- **D** large.

7. The movement of organic substances from one part of a plant to another in the phloem is known as:

- **A** diffusion.
- **B** osmosis.
- **C** evaporation.
- **(D)** translocation.

8. Palms and grasses are examples of:

- **A** monocotyledons.
- **B** dicotyledons.
- **C** conifers.
- **D** hydrophytes.

9. Which one of the following plants are capable of fixing nitrogen in their roots:

- **A** hydrophytes.
- **B** legumes.
- **C** dicotyledons.
- **D** algae,

10. a. Draw an annotated diagram of the distribution of tissues in the leaf.
  b. List the functions of each tissue.
  c. For each tissue, explain how the structure of this tissue is suited to its function.

11. a. Explain the process of water uptake in the roots.
  b. Explain how water is moved from the roots to other parts of the plant which are above the roots.

12. a. List the conditions that are always required for germination.

b. Explain why these conditions are necessary.

c. List 2 other conditions that may be necessary in some plants and explain how they help to ensure survival of the seedling.

13. a. Name two techniques in plant cultivation that have improved the yield.

b. Briefly outline the principals behind them.

OPTION A: DIET AND HUMAN NUTRITION

17

Chapter contents
- Diet
- Biochemistry of nutrition
- Diet and health

Option A: Diet and human nutrition

A.1 DIET

A.1.1
State that diet is the total food taken in by an individual.

© IBO 1996

Diet: the total food taken in by an individual.

A.1.2
Define nutrient.

© IBO 1996

Nutrient: a substance needed in the diet of an organism. Examples include proteins, vitamins and minerals.

A.1.3
List the constituents of a diet including carbohydrate, protein, lipid, minerals, vitamins, water and fibre.

© IBO 1996

A diet consists of macronutrients, micronutrients, fibre and water.

- The **macronutrients** are carbohydrates, proteins and lipids; these substances are needed in bulk.
- **Micronutrients** are vitamins and minerals of which only small quantities are required.
- **fibre** and **water** are not 'digested' but are necessary for a healthy diet.

A.1.4
Explain the functions of the constituents of the diet listed above.

© IBO 1996

The main functions of the constituents of diet are:

- **carbohydrates** - starches and sugars that are sources of energy and also building blocks for larger molecules.
- **protein** - used mainly for growth and for the repair of damaged tissue.
- **lipids** - fats and oils that are an energy source that the body can store for long term usage.
- **minerals** - inorganic elements and componds such as iron that are needed in the synthesis of important componds such as haemoglobin.
- **vitamins** - organic compounds required in small quantities for good health.
- **water** - about 65% of the human body is water. If a person does not ingest sufficient quantities of water they become dehydrated and lose energy. Water deprivation results in death in days wheras starvation does not usually cause death for months.

- **fibre** consists of a mixture of indigestible compounds mainly derived from plant cell walls.

A.1.5
List several examples of common foods which supply some of the various constituents of the diet.

© IBO 1996

Most foods contain many of the constituents of a good diet. Some examples of foods that are rich in the various components are listed below:

- **carbohydrates** are present in a variety of foods. A typical breakfast cereal is about 50% carbohydrate and vegetables about 20% carbohydrate.
- **proteins** are also present in many foods, though meat and fish are good sources.
- **lipids** - fats (which are solid at room termperature) are present in fatty meat, milk and many processed foods. Oils (liquid at room temperature) can be obtained from sources such as olive oil, fish etc.
- **minerals**

    Calcium (required for bones and teeth) obtained from, for example, milk.

    Phosphorus (required for the synthesis of many important compounds such as DNA, RNA and adenosine triphosphate. Some sea foods can supply phosphorus. Milk is also a good source.

    Potassium is important in the conduction of nerve impulses. Vegetables are a good source of potassium.

    Sodium is required for similar reasons to potassium. It is present in a variety of foods because it is artifically added in the form of sodium chloride (table salt). Excess consumption of sodium is associated with high blood pressure. Chlorine is also acquired from these sources.

    Sulphur is needed in the synthesis of proteins. Lean meat, fish and milk are good sources.

    Nitrogen is needed in the synthesis of proteins. Lean meat, fish and milk are good sources.

    Magnesium is needed in bones and teeth and is present in many foods, particularly vegetables.

Trace elements (required in very small amounts) include:

    Manganese is needed in bone growth and is present in many foods, particularly vegetables.

    Iron is essential in the synthesis of the oxygen carrying compound haemoglobin. Red meat and spinach are good sources of iron.

Cobalt, necessary for the synthesis of vitamin B_{12} is obtained from liver and red meat.

Fluorine improves the durabilty of teeth and is now often added to public water supplies and toothpaste. These measures have greatly reduced the incidence of dental caries.

Iodine. The absence of this element causes goitre (a throid disease). It is present in seafoods and is often added to table salt.

- **vitamins**

    A (retinol) obtained from many sources including fish oil, milk and spinach.

    D (caciferol) from fish oil and dairy products. This vitamin is unusual in that, under the action of sunlight, it can be synthesised in the skin.

    E (tocopherol) obtained from green vegetables, liver and and wheat germ.

    K (phylloquinone) from fresh green vegetables.

    B_1 (thiamine) from wholemeal flour, liver, kidney and yeast extract.

    B_2 (riboflavin) from yeast extract and dairy products.

    B_3 (nicotinic acid) from meat, wholemeal flour and yeast extract.

    B_5 (pantothenic acid) and B_6 (pyroxidine) is present in a variety of foods.

    B_{12} (cyanocobalamin) from meat and dairy foods.

    M (folic acid) from fish and green vegetables.

    H (biotin) from yeast, liver and kidney.

    C (ascorbic acid) from fruits (particularly citrus) and green vegetables.

- **water** - obviously from drinking water, but also from many foods. A variety of 'prepared drinks' such as commercial soft drinks and alcoholic drinks contain diuretics. Drinking too much of these can actually reduce the water content of the body.

- **fibre** - necessary for the efficient functioning of the digestive tract is particularly present in fruit, vegetables and cereals.

A.1.6
Describe a balanced diet as an equilibrium between food intake and energy expenditure and in terms of meeting bodily needs for growth, replacement and healthy functioning.
© IBO 1996

Balanced diet: an equilibrium between food intake and energy expenditure. A balanced diet should meet bodily needs for growth, replacement and healthy functioning. Note that an unbalanced diet not only results from the under consumption of important parts of a balanced diet. Over consumption of energy producing foods such as fats and carbohydrates can lead to **obesity** (substantial overweight).

Several health agencies issue advice as to a balanced diet.

The search criteria *balanced+diet* recently yielded almost 20000 matches on an internet search engine. Many of these give dietary information.

Students are not required to memorise the **recommended daily allowances** (RDA's) of the various components of a balanced diet.

Some examples are:

- **energy**, which is usually measured in kiloJoules. The amount of energy required to raise the temperature of 1 gm of water 1°C is called a calorie. 1 kCal (1000 calories) is called a kilo calorie. The unit kJ (kiloJoule) is related to the calorie. There are about 4.2 joules to the calorie (and 4.2 Kj per kCal).

The recommended daily intake of energy depends upon a variety of factors such as age and lifestyle. Young people generally need more energy than old people and active people more than those with sedentary lifestyles. Males generally use more energy than females.

An approximate figure for the daily energy intake for a teenage female is 9000kJ and for 20 year old male is 11000 kJ.

These figures should not be taken as exact. Athletes need much more energy than, for example, computer operators.

- **vitamins**. The RDA for these depends very much on the vitamin. About 40mg of vitamin C are required daily for an adult whereas the requirement for vitamin B_1 is about 1mg.
- **protein** requirements also vary with about 40-50 gms per day being a typical recommendation.

A.1.7
Evaluate common packaged food items by the interpretation of dietary information printed on them.

© IBO 1996

Food labels vary from country to country and this part of the course is best covered by students using the labels on local products.

These may give a variety of information. Examples are: a percentage of daily requirements, a straight percentage, grams per serving, grams per hundred grams etc.

Option A: Diet and human nutrition

Some example of the information on a variety of labels are given below:

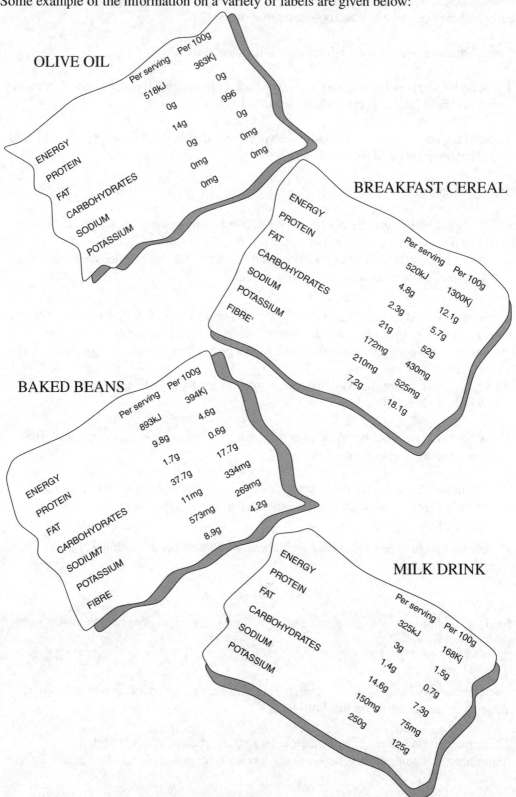

OLIVE OIL

	Per serving	Per 100g
ENERGY	518kJ	363Kj
PROTEIN	0g	0g
FAT	14g	996
CARBOHYDRATES	0g	0g
SODIUM	0mg	0mg
POTASSIUM	0mg	0mg

BREAKFAST CEREAL

	Per serving	Per 100g
ENERGY	520kJ	1300Kj
PROTEIN	4.8g	12.1g
FAT	2.3g	5.7g
CARBOHYDRATES	21g	52g
SODIUM	172mg	430mg
POTASSIUM	210mg	525mg
FIBRE	7.2g	18.1g

BAKED BEANS

	Per serving	Per 100g
ENERGY	893kJ	394Kj
PROTEIN	9.8g	4.6g
FAT	1.7g	0.6g
CARBOHYDRATES	37.7g	17.7g
SODIUM7	11mg	334mg
POTASSIUM	573mg	269mg
FIBRE	8.9g	4.2g

MILK DRINK

	Per serving	Per 100g
ENERGY	325kJ	168Kj
PROTEIN	3g	1.5g
FAT	1.4g	0.7g
CARBOHYDRATES	14.6g	7.3g
SODIUM	150mg	75mg
POTASSIUM	250g	125g

A.1.8
Calculate, compare and evaluate the nutritional content of foods and diets.

© IBO 1996

It is a very bad idea to rely on one food source alone as no single source of food provides all the requirements of a balanced diet in the correct proportions. People who, for reasons of poverty, are forced to eat only one food are usually malnourished.

An example is rice, which is a staple food in many countries.

A typical analysis of rice is: (for each 100g)

Energy	Protein	Fat	Carbohydrates	Fibre	Vitamin C
651kJ	2.8g	0.4g	34.6g	0.5g	0.0mg

Approximate RDAs are:

Energy	Protein	Fat	Carbohydrates	Fibre	Vitamin C
10000	55	90	170	12	40

If a person were to eat the right amount of rice to give them sufficient energy, they would need to eat: $\frac{10000}{651} \approx 15$ lots of 100g or about 1.5 kg of the food. If a person were to do this then they would have the following intakes of some components of a balanced diet.

	Energy	Protein	Fat	Carbohydrates	Fibre	Vitamin C
Surplus	−235	−13	−84	349	−5	−40

These figures indicate that, if a person were to eat 1.5 kg of rice per day they would get their correct energy requirement and more than the carbohydrate requirement but would be seriously deficient in protein, fat, fibre and vitamin C (as well as other requirements not considered).

Option A: Diet and human nutrition

A.2 BIOCHEMISTRY OF NUTRITION

A.2.1
List two sources for each of monosaccharide, disaccharide and polysaccharides in the diet (cross reference 2.2.8 and 2.2.9).

© IBO 1996

- monosaccharides include fructose and glucose, both present in honey. Monosaccharides also occur in many fruit juices.
- disaccharides include sucrose (refined sugar from sugar cane) and lactose or milk sugar (in milk).
- polysaccharides such as starch and cellulose (both present in many plants).

A.2.2
Outline the fate of the products of ingested carbohydrates including storage (as glycogen or lipid) and cell respiration.

© IBO 1996

Carbohydrates are first broken down in the small intestine:

$$\text{Starch (a polysaccharide)} \xrightarrow{\text{Pancreatic amylase}} \text{Maltose (a disaccharide)}$$

$$\text{Disaccharides} \xrightarrow{\text{Maltase (etc.)}} \text{Monosaccharides}$$

Once the carbohydrate has been broken down to glucose, the epithelial cells of the villi will absorb the molecule (facilitated diffusion) and pass it to the capillaries. The blood will carry the glucose via the hepatic portal vein to the liver. The liver regulates the blood glucose levels with the aid of the hormones insulin and glucagon, produced by the islets of Langerhans in the pancreas.

The glucose may be used for energy (cellular respiration) in any body cell or may be stored as glycogen in the liver or muscles. For long term storage, glucose would be changed into lipids and stored in fat cells. Further information is in Appendix A.17.1.

A.2.3
List three sources of lipids in the diet.

© IBO 1996

Lipids (fats and oils) are are present in both meat and fish. There are also many vegetable oils such as sunflower oil, olive oil etc.

A.2.4
Describe the fate of the products of ingested lipids including storage, growth of membranes and cell respiration.

© IBO 1996

Fat globules $\xrightarrow{\text{Bile salts}}$ Fat droplets

Fat droplets $\xrightarrow{\text{Lipase}}$ Fatty acids and glycerol

Once the lipid has been broken down to fatty acids and glycerol, the epithelial cells of the villi will absorb the molecules and reassemble them into lipids. They will then be packaged by the Golgi apparatus and leave the cell via exocytosis. The lipids then enter the lacteal, which is part of the lymphatic system. The lipids are transported by the lymph and will drain into the subclavian artery.

Lipids can become part of the cell membrane of any cell (phospholipids) or they can be stored in fat cells. The fatty acids can be broken down into segments of 2 Carbon atoms and enter the process of cellular respiration as acetyl CoA.

A.2.5
Discuss the variation in energy requirements (in kJ or MJ) depending on age, gender, activity and condition.

© IBO 1996

Children up to approximately 12 years need less food than adults since their bodies are smaller. From 12 to 15-18 years, young people need to supply their growing bodies with more food than they will need later. Old people are usually less active and need less food. Generally the higher the level of activity, the more food the person needs. Also in general, men need more food than women. Pregnant and lactating women need up to 25% more food than they would at other times.

A.2.6
List four sources of protein in the diet.

© IBO 1996

Meat, fish, dairy products and beans are all good sources of proteins.

A.2.7
Outline the fate of the products of ingested proteins including protein synthesis and deamination.

© IBO 1996

Polypeptides $\xrightarrow[\text{Chymotrypsin}]{\text{Trypsin}}$ Smaller polypeptides

Small polypeptides and dipeptides $\xrightarrow[\substack{\text{Aminopeptidase}\\\text{Carboxypeptidase}\\\text{Dipeptidase}}]{}$ Amino acids

Option A: Diet and human nutrition

Once the protein has been broken down to amino acids, the epithelial cells of the villi will absorb the molecules by active transport and pass it to the capillaries. The blood will carry the amino acids via the hepatic portal vein to the liver. The liver regulates the amino acids by allowing some to stay in the blood, assembling some into proteins, changing some into other amino acids (**trans-amination**) and de-aminate some to use them for energy.

A.2.8
State that essential amino acids are those which must be ingested and cannot be synthesised.

© IBO 1996

Essential amino acids cannot be produced by trans-amination. They have to be obtained from food. Different proteins contain different amino acids. So a person who obtains her protein from only one or a few sources, may not take in all essential amino acids. Vegans or vegetarians have to plan their meals carefully so that they vary their sources of protein.

A.2.9
Describe the functions of calcium, iron, iodine and potassium.

© IBO 1996

There are many essential components in a balanced diet. Calcium, iron, iodine and potassium are four of these. Their main uses and the consequences of deficiency are summarised in the following table:

Element	Main use in the body	Main results of deficiency
calcium	synthesis of bones and teeth	osteoporosis (brittle bones) and weak teeth.
iron	synthesis of haemoglobin, the compound that carries oxygen in the red blood cells	anaemia which results in loss of energy and general weakness
iodine	synthesis of the hormone throxine	goitre (swelling of the thyroid gland)
potassium	needed to maintain nerve impulse conduction.	Muscular weakness and, eventually, paralysis

A.2.10
Outline the functions of the following vitamins: retinol, cyanocobalamin, ascorbic acid, calciferol and tocopherol.

© IBO 1996

The following table lists the main functions of retinol, cyanocobalamin, acorbic acid, calciferol and tocopherol. See also appendices 17.2 & 17.3.

Vitamin	Main use in the body	Main results of deficiency
retinol (A)	Controls growth. It is also important in the synthesis of the visual pigment rhodopsin.	Dry skin. Also poor vision leading, in extreme cases, to blindness.
cyanocobalamin (B_{12})	RNA synthesis.	Pernicious anaemia.
ascorbic acid (C)	Collagen synthesis. Essential for the production of skin and connective tissue.	Scurvy. The skin becomes weak, wounds fail to heal, gums crack and teeth fall out. Anaemia and death due to heart failure.
calciferol (D)	Controls the absorption of calcium and hence bone formation.	Rickets: poor formation of bones, usually in children. Osteomalacia: an adult condition associated with pain and frequent bone fractures.
tocopherol (E)	Thought to be connected with the reproductive system and the maintenance of red blood cells.	Anaemia.

A.2.11

Discuss the importance of fibre in the diet.

© IBO 1996

Fibre is a mixture of indigestible compounds. Its function is mainly to facilitate the movement of food through the digestive tract.

There is still some debate about the exact function of fibre in the diet though it seems that a high fibre diet can reduce the incidence of a number of the 'diseases of civilisation' such as heart disease, high blood pressure, obesity and various cancers. It is certainly the case that cultures in which the consumption of fats is low and of fibre is high, have lower incidences of these diseases.

It is thought that a high throughput of fibre helps the body expel excess fats that may have been eaten by physically attaching the fat droplets to the fibre before they are excreted. This helps to reduce obesity and heart disease. It is possible that a similar thing happens to carcinogenic (cancer causing) toxins that may be produced during digestion and which could be responsible for cancers of the digestive tract.

Option A: Diet and human nutrition

A.3 DIET AND HEALTH

A.3.1
Explain the significance of saturated and unsaturated lipids in relation to a healthy diet.
© IBO 1996

Since lipids contain a high amount of energy per gram, a diet rich in lipids is likely to contain more energy than the body uses. This leads to fat being deposited in cells under the skin (**subcutaneous fat**). A person who has large subcutaneous fat deposits is said to be obese.

The body mass index indicates whether someone is underweight, acceptable, overweight or obese.

$$\text{Body mass index} = \frac{\text{body mass (kg)}}{(\text{height (m)})^2}$$

Having calculated the body mass index, an individual can now be (approximately) classified:

BMI	category
< 20	underweight
20 - 24	acceptable
25 - 30	overweight
> 30	obese

It is most important to recognise that people vary. It would be a mistake to conclude that you or other people are the wrong weight either on the basis of this calculation or fashion!

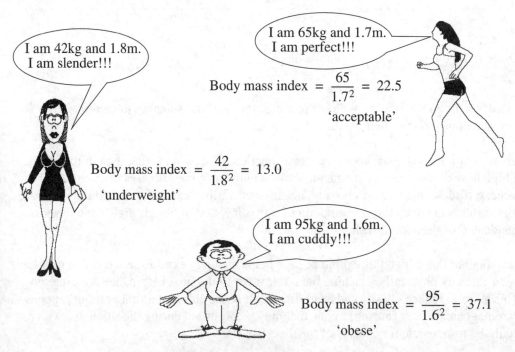

I am 42kg and 1.8m. I am slender!!!

I am 65kg and 1.7m. I am perfect!!!

$$\text{Body mass index} = \frac{65}{1.7^2} = 22.5$$

'acceptable'

$$\text{Body mass index} = \frac{42}{1.8^2} = 13.0$$

'underweight'

I am 95kg and 1.6m. I am cuddly!!!

$$\text{Body mass index} = \frac{95}{1.6^2} = 37.1$$

'obese'

Being obese is a health hazard in several ways. It puts more strain on all the body systems to deal with the extra amount of weight. High blood pressure and high levels of cholesterol often go together with obesity. It also increases the risk of developing diabetes, arthritis, cancer (of colon/rectum) and stroke.

CHD is **coronary heart disease**. The other cardiovascular disease is stroke. CHD involves the arteries which supply the heart with blood. A build up of fatty material on the inside of the coronary arteries will reduce the blood supply to the cells that make the heart muscle. This will cause a lack of oxygen in these cells which might cause them to work less well or even to die.

A.3.2
Explain the significance of saturated and unsaturated lipids in relation to a healthy diet.
© IBO 1996

There is a direct correlation between the level of cholesterol in the blood and the risk of CHD. The higher the cholesterol level, the higher the chances of developing CHD.

A correlation has also been found between a high intake of saturated (animal) fat and salt and high levels of cholesterol. Recent research has not always supported this and scientists are divided on this issue.

So a diet rich in saturated fat is likely to lead to high cholesterol levels, and high cholesterol levels are likely to lead to CHD.

A.3.3
Outline how the body synthesises and uses cholesterol.
© IBO 1996

Cholesterol can be made from lipids by the smooth endoplasmic reticulum (sER). This cell organelle also changes the cholesterol into steroid hormones.

Structure of cholesterol

sER is particularly abundant in tissue that produces steroid hormones such as the interstitial tissue of the testis (see section 5.6 and 11.1).

A cell membrane containing a lot of cholesterol will be less fluid than a membrane containing little cholesterol.

A.3.4
Discuss the effects of additional dietary cholesterol.
© IBO 1996

Eating more lipids will often increase the amount of cholesterol that the cells produce and hence the blood cholesterol level. Excess can be controlled by diet and, in extreme cases, drugs.

A.3.5
Describe the main features of diets that demonstrate the various degrees of vegetarianism (vegans, lacto-vegetarians, pesco-vegetarians and ovolacto-vegetarians).

© IBO 1996

Vegans do not use any animal products. Their diet consists of only plant material.
Lacto-vegetarians will eat plant products and dairy products.
Pesco-vegetarians will eat plant products and fish.
Ovolacto-vegetarians will eat plant products, eggs and dairy products.

A.3.6
Explain how each of the above diets can achieve a balance.

© IBO 1996

A balanced diet can be reached in each of the above cases by being aware of the nutrient value of the different foods.

Essential amino acids (easily found in meat) are also found in beans and peas. Essential fatty acids are also found in rice.

Vegans tend to eat vegetables, beans or peas and wheat/rice with every meal.

A.3.7
Discuss the possibility of a deficiency in calcium, iron, calciferol and cyanocobalamin in the above diets.

© IBO 1996

Calcium is found in dairy products. If you do not consume these, you have to find a source of calcium in other foods. Calcium needs to be consumed in combination with calciferol to ensure efficient absorbtion. Iron can be found in beef but also in some vegetable products.

A.3.8
Define malnutrition.

© IBO 1996

Malnutrition: the result of feeding on a diet that is not balanced.

A.3.9
Suggest how malnutrition can be caused by any (or a combination) of the following conditions: social, economic, cultural and environmental.

© IBO 1996

Malnutrition can be caused by (a combination of) the following factors : social, economic, cultural and environmental.

- **Social:** Anorexia nervosa can be brought on by excessive dieting in an attempt to fit society's ideal image.
- **Economic:** some people cannot afford to buy sufficient food.
- **Cultural:** a few years ago many people became vegetarians because it was the thing to be. 'Fast food' is extremely popular but should not be the only source of nutrients.
- **Environmental:** environmental disasters, such as drought and flood, can lead to lack of (proper) food.

A.3.10
Discuss one example of global malnutrition using published data.

© IBO 1996

An example of global malnutrition is the diet common to those in the western world. The diet contains too much fat and salt and too little fibre found in vegetables and fruits. A typical meal of a fast food chain will confirm this.

An internet search using the word 'malnutrition' (May 1999) produced nearly 40000 matches. Some of these (giving current information) were:

>http://www.idrc.ca/mi/org_cfni.htm
>http://www.vegan-straight-edge.org.uk/foodproc.htm
>http://www.idrc.ca/mi/org_brea.htm
>http://www.who.int/inf-fs/en/fact119.html
>http://www.dmaonline.org/legb10.html

A.3.11
Discuss the nutritional issues of one of the following: anaemia, vitamin deficiency, osteoporosis.

© IBO 1996

Osteoporosis: a disease caused by insufficient calcium in the bones. The bones become brittle and break easily. Osteoporosis can be caused by an insufficient intake of calcium or vitamin D (which is needed for efficient calcium absorption). Oestrogens and progesterones also play a role in the movement of calcium in the body. After menopause, females produces less of these hormones and are more at risk of developing osteoporosis. It is possible to use hormone replacement therapy, giving the hormones as tablets when the natural production declines. This is not suitable for all women because it can have undesirable side effects.

Some of the consequences of vitamin deficiency were discussed in earlier sections.

A.3.12
State that chemical additives can act as preservatives, antioxidants, colouring, flavouring, stabilisers and acid-regulators.

© IBO 1996

Option A: Diet and human nutrition

Chemical additives can act as preservatives, antioxidants, colouring, flavouring, stabilisers and acid regulators. The inclusion of these in processed foods is generally documented in the nutritional information on the label.

A.3.13
Outline that some additives may have deleterious effects and the importance of consumer protection.

© IBO 1996

Some **additives** may have a deleterious (= harmful) effect on some people. Most people do not react at all to these additives, otherwise they would never have been permitted. However, some people are allergic to some of them, children may have very clear behaviour responses to some food colourants. It is therefore vital that each product has a complete list of all ingredients clearly printed on the outside of the wrapping.

A.3.14
Explain the importance of hygienic methods of food handling and preparation.

© IBO 1996

Bacteria can divide every 20 minutes. So, starting with 1 bacteria at time 0, after 20 minutes you have 2, after 40 minutes you have 4 and after 6 hours you have more than 16 million.

Most bacteria are killed when the food is heated. When raw chicken is placed on a plate, then cooked and then returned to the same plate, you are simply allowing the bacteria left behind to contaminate the cooked meat. It is therefore important to be very aware of what has touched the raw meat. It must be washed before it touches other food.

Food must of course not be contaminated with harmful bacteria before packaging. The package must keep out all bacteria and an expiry date should be indicated. Food that can spoil (go off) must be refrigerated or frozen since this slows down the bacterial growth.

APPENDIX A.17.1: DIGESTION

Place	pH	Gland	Secretion	Enzyme	Substrate	Product
mouth	7	salivary glands	saliva	salivary amylase	starch	maltose
stomach chief cells parietal cells	2	gastric glands hydrochloric acid	gastric juice	pepsin	protein	polypeptides
small intestine	8	pancreas	pancreatic juice	pancreatic amylase trypsin lipase	starch protein lipid	maltose polypeptides glycerol + fatty acids
		intestinal glands	intestinal juice	maltase (sucrase, lactase) peptidases	maltose sucrose lactose polypeptides	glucose glucose + fructose glucose + galactose amino acids
liver			bile	bile salts (not enzymes)	emulsification of fats	

Option A: Diet and human nutrition

APPENDIX 17.2: MACRONUTRIENTS

Name	Other name	Found in	Function	Deficiency disease
Carbohydrates monosaccharides disaccharides polysaccharides		potatoes, sugar apple, orange sugar beet, cane potatoes, rice	energy	starvation/ marasmus
proteins		beef, chicken, fish, soya beans	growth, repair, enzymes	kwashiorkor
fat		butter, cheese, peanuts	energy, insulation	
fibre	roughage	bread, cereals, fresh fruit, vegetables	stimulates peristalsis of the gut	constipation
water		drinks and food	replace urine and sweat	dehydration

APPENDIX 17.3: MICRONUTRIENTS

Name	Other name	Found in	Function	Deficiency disease
Retinol*	vitamin A*	liver, carrots, codliver oil	rod cells in the eyes	nightblindness
Cyanocobalamin	vitamin B_{12}	beef, liver, milk	making red blood cells	anaemia
Ascorbic acid	vitamin C	fruit, vegetables	keeps cells together	scurvy
Calciferol*	vitamin D*	milk, codliver oil, made in skin	hardening teeth and bones	rickets, osteoporosis
Tocopherol	vitamin E	green leaves, vegetable oil	protects r.b.c, retinol from oxidation	anaemia
Calcium	Ca	milk, cheese	hardening teeth & bones	rickets
Potassium	K	meat, milk, fruit	acid/base balance, muscle & nerve activity	muscular weakness, death
Iodine	I	seafood, salt	making thyroxine	goitre
Iron	Fe	organ meat, cocoa	part of haemoglobin	anaemia

* fat soluble vitamins, harmful in large amounts.

Option A: Diet and human nutrition

EXERCISE

1. a. List the constituents of a complete diet.
   b. For each constituent, list 2 examples of food that are good sources of this constituent.
   c. Describe the function of each constituent.

2. a. Pregnant women 'eat for two'. Does that mean double the amount the woman would normally eat? Explain.
   b. Who requires more food: a 15 year old boy or a 65 year old woman?
   c. Describe a situation in which your answer to 'b' would not be correct.

3. a. Outline the diet of a vegan, a lacto-vegetarian, a pesco-vegetarian and a ovolacto-vegetarian.
   b. What is osteoporosis?
   c. Which groups are especially susceptible to osteoporosis?

OPTION B: PHYSIOLOGY OF EXERCISE

18

Chapter contents
- The skeleton, joints and muscles
- Coordination of muscle activity
- Muscles and energy
- Fitness and training
- Injuries

Option B: Physiology of exercise

B.1 THE SKELETON, JOINTS AND MUSCLES

B.1.1
Compare the activity of fast (twitch) and slow (tonic) muscle fibres in terms of speed and stamina.

© IBO 1996

The skeleton is sub-divided into axial and appendicular parts. The 80 **axial bones** form the axis of the body. They form the skull, backbone and ribcage. The 126 **appendicar bones** are found in the arms, legs, shoulders and hips.

B.1.2
Explain the structure of a long bone including the hollow shaft and spongy head, in relation to strength and shock absorption.

© IBO 1996

The diagram below shows a schematic diagram of a long bone.

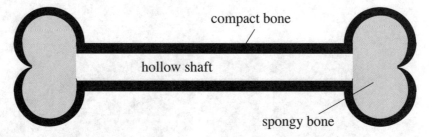

Bone is a tissue. Specifically, bone is a connective tissue which means that it has cells as well as an extracellular matrix (material between the cells). The extracellular matrix of bone has two components : calcium phosphate crystals and collagen (a protein). The crystals make the bone rigid, the collagen give it some flexibility.

As you can see in the diagram, the ends of the bone are made of spongy bone. Spongy bone is made of many little bone plates around spaces, looking like a sponge. The bone plates are arranged so that the bone is the strongest in the direction of the most usual force it experiences. If the direction of the force changes, the bone plates will eventually rearrange so that the spongy bone is strongest in the direction of the new, most common force. The rest of the bone is made of a hollow pipe. The 'pipe' is made of compact bone which is very strong. The centre is filled with marrow and some spongy bone.

B.1.3
Draw a diagram of the human elbow joint including cartilage, synovial fluid, tendons, ligaments, named bones and named antagonistic muscles.

© IBO 1996

A diagram of the human elbow joint can be found below.

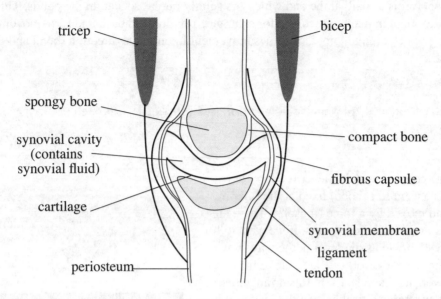

The muscles in the upper arm will move the lower arm. The flexor muscle of the upper arm (humerus) is called the biceps. It is attached to the shoulder and the lower arm (ulna). Contracting this muscle (shortening it) will pull the lower arm towards the shoulder, i.e. the arm bends. The extensor muscle is the triceps. It is found on the other side of the upper arm. It is attached to the shoulder and runs via the elbow to the lower arm (radius). Contracting this muscle will straighten the lower arm. These muscles work as **antagonists**: one doing the opposite of the other. Body builders showing their muscles will contract both biceps and triceps at the same time. The lower arm will not move but the muscles become very visible.

B.1.4
Outline the functions of the of the above named structure of the human elbow joint.

© IBO 1996

At a joint, two bones can move relative to each other. The end of the bones is made of spongy bone which is light and strong. A cartilage covering helps in smooth movement as well as in absorbing shocks. Synovial fluid contains the required food and oxygen to maintain cartilage and acts as a lubricant for the joint. The synovial membrane keeps the synovial fluid in place. Tendons are strong cords of connective tissue, attaching a muscle to a bone. Ligaments are bands of strong connective tissue keeping the parts of a joint together (bone to bone attachments).

B.1.5
Describe the movements at the hip joint and the knee joint.

© IBO 1996

Several different types of joints are recognised by the type of movement they allow. Synovial joints have the most freedom of movement. These are, for example, the elbow, hip and knee joints. The hip joint is a typical 'ball and socket' joint. The bone of the upper

Option B: Physiology of exercise

leg (femur) forms a 'ball' at the end which fits snugly into a 'socket' in the pelvis. This allows movement in many directions, for example, rotation and back and forth movement. The knee joint is a hinge joint and allows movement in only one direction (bend and straighten).

B.1.6
Outline the principles of levers including effort, fulcrum, load, with reference to the elbow joint.

© IBO 1996

Fulcrum: point on which a lever is supported. A lever is a rigid rod (bone) fixed at the fulcrum (joint) and moved by a force (muscle). There are two ways in which we use levers. If we want to lift a heavy weight we us the lever in the way shown. This magnifies the effort (we are able to lift heavier weights than usual). In this arrangement, the effort must move further than the load will be lifted.

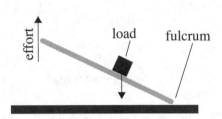

The other arrangement is to exchange the positions of the load and effort. In this case, the fulcrum would need to be a hinge. In this arrangement, we would need to exert a larger effort than the load. However, the load will move a larger distance than the effort. This is the way in which the body generally uses muscles and joints.

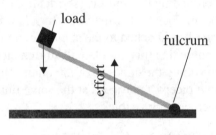

This diagram is of the arm of an athlete doing a 'bicep curl' with a 20 kg weight. In approximate terms, the forearm is about 30 cm long and the bicep is attached to the forearm about 6cm from the elbow. The load is 5 times further from the fulcrum than the effort (muscle). Thus the bicep must exert 5 times force of the the load (100 kg). However, if the bicep contracts 1cm the load (weight) will move 5 cm.

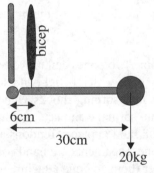

Leverage used in this way means that our muscles have to be a lot stronger than we perhaps realise that they are. However, leverage magnifies the amount that the muscles actually move.

B.1.7
Outline the structure of skeletal muscle in terms of muscle fibres, myofibrils, actin, myosin filaments.

© IBO 1996

Biology

Muscles are groups of cells working together. Each muscle fibre was originally many cells which fused. The resulting cell has many nuclei.

Muscles fibres can be more than 1 cm long. Groups of muscle fibres are arranged together and have connective tissue around them. Muscles have tendons (connective tissue) at each end, attaching them to the bones.

Inside a muscle cell, you find many thin myofibrils. These thin fibres cause the typical striated (striped) pattern of skeletal muscles. Myofibrils contain two types of myofilaments: **myosin** and **actin**. They are made of a protein like substance.

The structure of a skeletal muscle is shown in the diagram below.

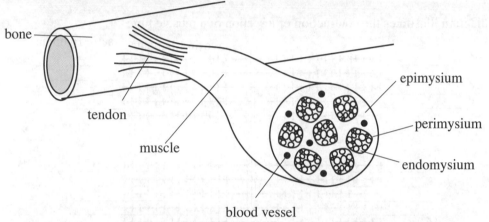

The individual muscle fibres have a structure illustrated in the diagram.

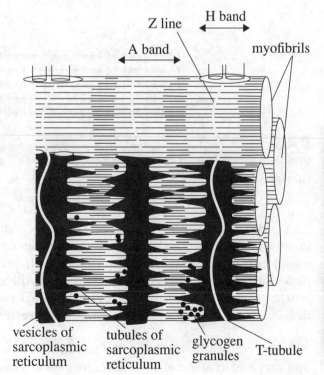

Option B: Physiology of exercise

B.1.8
Explain how skeletal muscle contracts by the sliding action of actin and myosin filaments with ATP as an energy source.

© IBO 1996

The basis of muscle action is that the muscle becomes shorter. This means that a contracting muscle is shorter than a relaxed muscle. The shortening of a muscle is brought about by the two filaments named earlier. The actin and myosin filaments of the muscle can slide relative to each other. So, in the relaxed muscle, they only overlap partly. In a contracted muscle, the actin an myosin filaments have slid towards each other so that they overlap almost completely. This makes the muscle shorter. The relative movement of the actin and myosin filaments is an active process and requires ATP.

The diagram illustrates the contraction of a section of a muscle fibre.

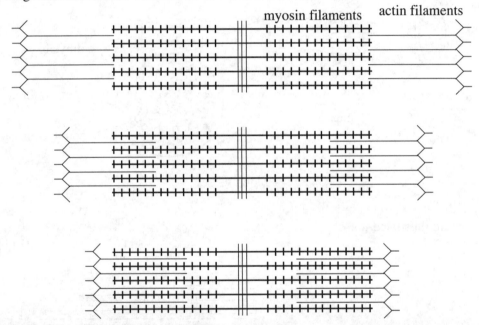

B.1.9
Compare the activity of fast (twitch) and slow (tonic) muscle fibres in terms of speed and stamina.

© IBO 1996

All humans seem to have two types of muscle fibres: fast (twitch) muscle fibres and slow (tonic) muscle fibres. The fast fibres have a greater oxygen need than normal, have little myoglobin but provide maximum work for a short time. Slow fibres have a very good blood supply and a lot of myoglobin. Fast fibres will rely of anaerobic respiration to provide them with energy. Slow fibres can have high rates of aerobic respiration because of their good blood supply and can therefore sustain activity for long periods of time.

Although everyone has both types of muscle fibres, their relative amounts can vary. Sprinters tend to have many fast fibres, marathon runners have many slow fibres.

Biology

B.2 COORDINATION OF MUSCLE ACTIVITY

B.2.1
Outline the general organisation of the human nervous system including CNS (brain and spinal cord) and PNS (nerves).

© IBO 1996

The human nervous system can be divided into several sections.

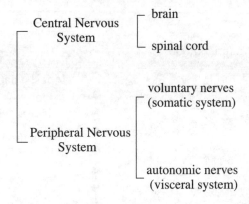

B.2.2
Draw the structure of a sensory and a motor neuron.

© IBO 1996

A schematic diagram of a **motor neuron**:

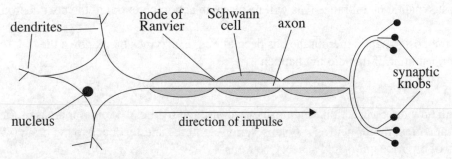

A schematic diagram of a **sensory neuron**:

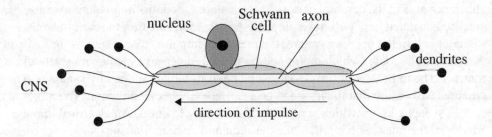

Option B: Physiology of exercise

B.2.3
Outline synaptic transmission including arrival of an electric impulse, release, diffusion and destruction of a neurotransmitter substance and the subsequent propagation of another electrical impulse.

© IBO 1996

Communication between neurons or between neurons and glands or muscles takes place in synapses. There are electrical and chemical synapses. Only chemical synapses will be discussed here.

Chemical synapse.

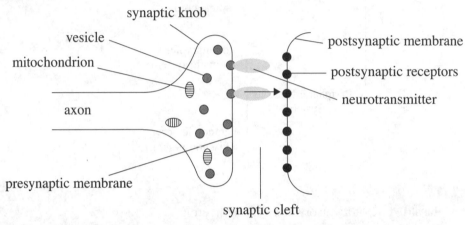

In the **synapse**, the arrival of an action potential causes a series of events which will lead to the release of transmitter substance into the synaptic cleft (via exocytosis). The neurotransmitter then diffuses across the synaptic cleft (20 nm) and attaches to receptors in the post synaptic membrane. This will result in an action potential in the next neuron.

After the post synaptic membrane has been affected, enzymes break down the neurotransmitter. If this did not happen

B.2.4
Explain how the contraction of a muscle is controlled by means of motor areas of the cerebral cortex, motor neurons, synapses, muscle fibres and feedback to the brain by means of proprioceptors and sensory neurons.

© IBO 1996

The motor area in the cerebral cortex is responsible for controlling voluntary muscles. The impulse generated here will travel down a specific motor neuron to reach the muscle that will contract. In the nervous system, the message (impulse) that is sent, is the same in every case. The route (which motor neuron is used) determines which muscle will contract. The impulse will travel via several neurons which interact by synapses. The synapses will ensure that the message gets carried only in one direction. The motor neuron ends in synaptic knobs which are located on the muscle fibres. An electrical impulse arriving at the muscle fibre will make the actin and myosin filaments slide over each other (using ATP), shortening the muscle.

There are several feedback mechanisms which tell the brain what is happening. Proprioceptors, receptors in muscles, joints and tendons, will inform the brain on the body's movement and position. The proprioceptors will pass the information to a sensory neuron which will pass it to the brain.

B.2.5
Explain the role of inhibitory neurons in coordinating the activity of antagonistic muscles at a joint.

© IBO 1996

A limb can be moved in two directions. In section B.1, the biceps and triceps are discussed as antagonistic muscles. Contracting the biceps will bend the arm, contracting the triceps will extend it. Contracting both at the same time does not move the arm; it just requires energy to keep both muscles contracted. So, normally, when the brain sends a message to the biceps to contract, it will also send a message to the triceps to relax. This means that the biceps can contract easily, stretching the triceps without too much resistance.

The nerve that will tell the biceps to contract is called an **excitatory neuron**. The nerve that will tell the triceps to relax is called an **inhibitory neuron**. Inhibitory neurons will make other (excitatory) neurons less sensitive to impulses received from another excitatory neuron.

If an impulse from an excitatory neuron arrives at a synapse, the next excitatory neuron will normally pass on the impulse. However, if, just before the impulse of the excitatory neuron arrives, an impulse from an inhibitory neuron arrives on the neuron, it may not send the impulse. The inhibitory neuron has made the excitatory neuron less sensitive to an impulse from another excitatory neuron.

So the message to contract the biceps will be sent via excitatory neurons. At the same time, an impulse will be sent via an inhibitory neuron to the triceps. Any message to contract the triceps that might accidently be sent at the same time will be ignored so that the triceps will remain relaxed as the biceps contracts.

Option B: Physiology of exercise

B.3 MUSCLES AND ENERGY

B.3.1
State that the energy is released by respiration in the form of ATP.

© IBO 1996

Cellular respiration releases energy from organic molecules. This energy is captured in ATP. See section 4.2 and 9.1.

B.3.2
Explain that ATP supplies in muscles provide enough energy for only the first few seconds of exercise, that anaerobic respiration can supply energy for up to 2 minutes of high intensity exercise and aerobic respiration can supply energy indefinitely for low intensity activity.

© IBO 1996

The muscles store a small supply of ATP. The amount of ATP that can be stored is limited; at the same time the amount of ATP needed for muscle contraction is large. As a result, the stored ATP provides the energy for the muscle contraction for only a few seconds of exercise.

To continue the exercise, muscles can switch to anaerobic respiration (see section 9.1.8). Anaerobic respiration is the breakdown of glucose in the absence of oxygen. This is less desirable because anaerobic respiration produces less energy per molucule of glucose than aerobic respiration. It also produces lactate in muscles which may cause muscle fatigue. It may be necessary to use anaerobic respiration because the body cannot provide the muscles with sufficient amounts of oxygen to release the required amounts of ATP in aerobic respiration. Anaerobic respiration can supply muscles with ATP for approximately 2 minutes of intense exercise such as a sprint.

To supply energy to muscles for a much longer time, for example, in a marathon, the muscles have to use aerobic respiration. The level of intensity of the exercise that can be sustained is much lower than that described above (sprinters run faster than marathon runners). Training improves heart and lungs so that the muscles can be better provided with oxygen (and organic molecules).

B.3.3
Explain how 2-oxopropanoate (lactate) produced in anaerobic respiration is passed to the liver and creates an oxygen debt.

© IBO 1996

The lactate produced in the muscles during anaerobic respiration needs to be broken down or converted into other organic molecules. The circulatory system carries the lactate from the muscles to the liver. The lactate metabolism in the liver requires oxygen. This is one of the causes of the temporary continuation of increased breathing after an exercise has stopped. The production of lactate has allowed the muscles to continue their work anaerobically by producing lactate. This has created the 'oxygen debt' as the oxygen is required later to metabolise lactate.

B.3.4
Explain how the oxygen debt is repaid.

© IBO 1996

When a person stops exercising, his or her rate of breathing will not immediately go back to resting levels. Neither will the cardiac frequency immediately return to resting levels. This can be verified in a simple practical.

The reason for the additional time of increased respiratory and cardiac frequency is the oxygen debt. The haemoglobin of the blood and the myoglobin in the muscles have given up their oxygen during the exercise. These molecules need to be replenished with oxygen. The lactic acid produced during anaerobic respiration needs to be metabolised which requires oxygen. As these processes are underway, less oxygen is needed and hence the respiratory (and cardiac) frequency decrease.

B.3.5
Outline the role of myoglobin in muscles.

© IBO 1996

Myoglobin is a protein similar to haemoglobin. It is found in the muscles and it can bind oxygen. Myoglobin has a high affinity for oxygen and will not release its oxygen until the oxygen level in the muscle is low, i.e. during exercise. It is therefore a valuable, if somewhat limited, store of oxygen. Slow muscle fibres have a particularly high myoglobin content. Here it also plays a role in obtaining oxygen and delivering it to the muscle fibres. As fast muscle fibres tend to use anaerobic respiration, they have lower levels of myoglobin.

B.3.6
Explain the role of adrenaline in increasing supplies of oxygen and glucose to muscles.

© IBO 1996

Adrenalin (also called epinephrine) is a hormone produced by the adrenal glands. These are located just above the kidney. Adrenalin is released when a person is excited or afraid. Adrenalin increases cardiac frequency (heart rate) and blood pressure, and it dilates the bronchioles. The combination of these improves the supply of oxygen to the muscles. Adrenalin also stimulates the conversion of glycogen into glucose. All these factors together allow the muscles to have as much aerobic respiration as possible, releasing the maximum amount of ATP for contraction.

B.3.7
Explain the causes of muscle fatigue in terms of lactate (2-oxopropanoate) accumulation and depletion of carbohydrate supplies in muscles.

© IBO 1996

Option B: Physiology of exercise

Muscle fatigue is a term used to describe the decrease in the muscle's ability to generate a force due to extensive use of the muscle. Several factors are thought to play a role in muscle fatigue:
- build-up of lactic acid produced by anaerobic respiration.
- depletion of the amounts of muscle glycogen available.

Trained athletes derive more energy from the aerobic breakdown of fatty acids.

B.4. FITNESS AND TRAINING

B.4.1
Define fitness.

© IBO 1996

Fitness is the physical condition of the body which suits it to the particular exercise which it performs. There are two generally recognised types of fitness:
- **health related fitness** that reflects and individual's ability to resist infection and also the absence of conditions such as obesity and high blood pressure, strength of the muscles and skeleton etc.
- **performance related fitness** which is more related to the fitness needed to complete specific tasks such as climbing a mountain, sprinting 100 metres or winning a tennis match.

All fitness can be improved by training.

B.4.2
Describe the principles of training including specificity, progressive overload, frequency, intensity, duration, regularity, measurability, and quality not quantity.

© IBO 1996

Training is any process that improves an individual's fitness. However, the type of training that is undertaken determines the overall fitness effect. If a person wants to become a competitive gymnast, their program is likely to include activities that will improve strength and flexibility. Endurance is of less importance than it would be to a marathon runner who is much less interested in strength and who would almost certainly not want to build up a lot of muscle in their upper body. This requirement that fitness programs are tailored to a particular desired outcome is known as **specificity**.

The whole purpose of a training program is to improve performance in a particular task. If a person wants to be able to lift heavier weights, they need to develop strength by lifting weights that are a bit heavier than they can lift comfortably. This is known as **overload**, which does not imply that it is a good idea to lift weights that are too heavy as the most likely result of this is an injury. It is important that overload training is progressive. If an individual can comfortably lift 30kg, they should train by lifting a bit more than this

(35kg). As strength improves, this should be increased. Overload in running is achieved by running a bit further or faster than is comfortable in every session. This is known as **progression** and the whole process as **progressive overload**. It is a common mistake for people on training programs is to attempt to make too rapid changes in their fitness levels by running too far too soon and suffering physical pain or injury. A good session should finish with a feeling of well-being, not exhaustion or pain.

Effective training programs must be **regular** if they are to maintain or improve fitness. The **frequency** of training depends on how much improvement an individual wants to make. It is generally a good idea to mix the program. If a person decides that they can expend one hour every second day they are unlikely to reach the levels of fitness required of elite athletes. However, this amount of training could be expected to make a substantial difference to the person's general health and their ability to, for example, enjoy a holiday in which they might suddenly want to walk to the top of a hill to watch a sunset. Such a program should include a variety of activities to improve strength (weight training), endurance (running, cycling) and flexibility (stretching). A frequency of 'every other day' is thought to be the minimum frequency that will result in a good level of fitness.

It is a common mistake to think that training is only effective if it is 'flat out' or 'hurts'. This is not now considered to be true. The **intensity** of a program can be measured using a number of methods, but heart rate is commonly used as it is a convenient and easily measured indicator of the stress a body is under. It can be measured by feeling the pulse in the wrist, neck, temple etc. Also, there are a variety of electronic devices on the market which detect the electrical signals from the heart and give a digital readout of heart rate. The heart rate at which a particular individual should train depends on a variety of factors. As a rough guide, the maximum heart rate advisable can be estimated by subtracting one's age from 220. Thus a 40 year old has a maximum heart rate of $220 - 40 = 180$ beats per minute. Regular exercise at 60% (108 beats per minute) of this will produce general health benefits. Higher levels are required to produce the fitness desirable for competitive events. Heart rate is an example of a **measurable** indicator of intensity.

As well as getting the right intensity of training, it is also important to consider its **duration**. Again, there are definitive rules for this and for most people, the duration of training is determined by the amount of time that they are prepared to spend on it. This is not a simple as saying that, if we are training for a two hour long distance race we should always train for three hours. Endurance athletes frequently follow programs that mix periods of extended running with periods of recovery. In the period before a race the athletes scale back the intensity and duration of their training.

As a general principle, training should be matched to an individual's fitness objectives. It is the quality more than the quantity that counts. In considering a training program it is important to have goals in mind. Am I training to improve my general health or do I want to become a professional soccer player?

Finally, it can be a very good idea to have a medical check before embarking on a fitness program. Whilst a training program generally produces large benefits, some individuals have medical conditions that can make exercise dangerous.

Option B: Physiology of exercise

B.4.3
Discuss flexibility, agility, speed and stamina as measures of fitness.

© IBO 1996

The American College of Sports Medicine lists four different kinds of fitness :
- **cardiorespiratory endurance.**
- **body composition** (proportion of fat).
- **flexibility.**
- **strength.**

If total fitness is a combination of these four kinds, we can expect a 'fit' person to be more flexible, agile, faster and have more stamina compared to the 'unfit' person.

Many kinds of exercise will improve both cardiorespiratory fitness and muscle strength. If the body through exercise becomes better at taking up oxygen and moving it to the muscles, then the stamina of that person will improve. If muscles become stronger then, for example, speed will improve. As exercise improves the flexibility of muscles and joints, the person becomes more flexible and agile.

B.4.4
Explain how training affects the cardiovascular system, the lungs and the muscles (cross reference B.3).

© IBO 1996

Training improves cardiorespiratory fitness. This means that the following changes occur in the body:
- heart will increase its stroke volume; i.e. per stroke, more blood will be pumped through. As a result, resting cardiac frequency will decrease.
- as muscle tissue builds up, strength improves. Muscle fibres become larger and stronger with certain types of exercise such as weight lifting.

Strength is the ability to exert maximum force during a single effort.

B.5 INJURIES

B.5.1
Explain the need for warm-up and cool-down routines.

© IBO 1996

Before starting a serious training session, you should do a series of warm up exercises. Stretching is a good way to warm up. Stretching muscles will make them (and the tendons) stronger and more flexible and it will increase the range of movements. It will improve the blood circulation. Muscles and tendons that are not stretched are tenser and more likely to tear.

You can supplement the stretching by doing some other warming up exercises. The exercise usually are those movements that will be required later but less intense. For Example, brisk walking/jogging when running is required later.

Similarly, it is important to cool down after strenuous exercise. This again can be done by stretching. This allows muscles to ease back into a resting state.

B.5.2
Describe injuries to the muscles and joints including sprains, torn muscles, torn ligaments, dislocation and intervertebral disk damage.

© IBO 1996

Some of the most common injuries are those to muscles and joints.

A **sprain** is the result of a twisted joint. Ligaments are torn and blood vessels are damaged. A sprain can be relatively minor in which case it will heal with some rest. A very serious sprain may have torn loose a tendon and may require intensive medical treatment. Sprains are a common result of exercise without warming up; alternatively they are often caused by accidents.

Torn muscles: severe damage to the muscle tissue.

Torn ligaments may occur in a severe sprain or dislocation. The ligaments that keeps the bones of the joint together may be torn off during a twisting of the joint. Torn ligaments (when complete separated from the bone) have to be surgically repaired. It is not easy to reattach a ligment to a bone.

Dislocation occurs when the 'head' of the joints no longer rests in the socket. This is more common in, for example, the shoulder than the hipjoint. The hip joint is deeper (and less movable) which reduces this risk. A dislocation is very painful because it stretches/damages the tissues around the joint. The longer the duration of the dislocation, the more extensive, and possibly, lasting the damage. Returning a dislocated joint to its proper position should not be attempted without proper knowledge of the procedure.

Intervertebral disks are the cartilage disks between the vertebrae. They allow the movement of the vertebrae relative to one another and they cushion some of the shocks. **Damage to these disks** is likely to cause pressure on the nerves exiting from the spinal cord and is often painful.

B.5.3
Explain the use of rest, ice, compression and elevation in the treatment of soft tissue injuries.

© IBO 1996

Immediate treatment for minor sports injuries is called **RICE**. This stands for **Rest**, **Ice**, **Compression** and **Elevation**. RICE will in many cases immediately reduce the pain of the injury and in many cases reduce its severity.

Option B: Physiology of exercise

It makes sense to rest the part of the body that is injured. Applying ice will cool down the area. It deadens pain and reduces bleeding and swelling. Be careful not to put the ice directly on the skin; it should be packed in a cloth (towel). Compression (e.g. an elastic bandage) will reduce swelling. As swollen tissue tends to hurt (for some time), this is a longer term benefit. Be careful to check the bandage regularly because some swelling may occur despite it. The bandage could then become too tight and restrict circulation. Elevation is another way of reducing swelling as the fluid drainage from injured area is assisted by gravity.

EXERCISE

1. a. Name the two filaments found in muscle.
    b. Explain how these filaments are involved in muscle contraction.

2. a. What are propioceptors?
    b. Are they involved in initiating a muscle contraction?
    c. Outline how a muscle contraction is initiated and passed on to the muscle.

3. a. What is the role of adrenalin?
    b. Explain how it achieves this.

4. Outline what parameters you could measure to establish that a training programme is have an effect.

5. a. What does the acronym RICE stand for in the treatment of tissue injuries.
    b. What is the reason for doing each of the things listed above?

OPTION C: CELLS AND ENERGY

19

Chapter contents
- Membranes
- Proteins
- Enzymes
- Photosynthesis
- Cell respiration

Option C: Cells and energy

C.1 MEMBRANES

C.1.1
Explain the dynamic relationship between the nucleus, rough endoplasmic reticulum (rER), Golgi apparatus and cell surface membrane.

© IBO 1996

As can be seen from the fluid mosaic model, membranes are not static structures. Not only can the molecules in them move around but also the amound of membrane can change. Since the structure of the cell surface membrane is essentially the same as that of the nuclear envelope, the rER and the Golgi apparatus, it is possible to exchange membrane sections between them.

If you remember section 1.3, the function of the **Golgi apparatus** is to prepare substances for exocytosis. This involves wrapping it in a bit of membrane from the Golgi apparatus. This membrane then joins the cell surface membrane in the process of exocytosis.

Many of the substances which the cell 'exports' are proteins and hence the following organelles are involved:
- **nucleus:** chromosomes contain genes coding for proteins, mRNA is made by transcription.
- **rough endoplasmic reticulum (rER):** contains the ribosomes which make protein intended for export by translation.

The protein then goes into the lumen of the rER, is surrounded by membrane and leaves through the cell surface membrane by exocytosis.

C.1.2
Describe the ways in which vesicles are used to transport materials within a cell and to the cell surface.

© IBO 1996

Vesicles can be used to transport materials within the cell and to/from the cell surface membrane. See section 1.4.

C.1.3
Describe the membrane proteins and their positions within membranes.

© IBO 1996

As said in section 1.4.1, proteins can be extrinsic or intrinsic to the cell membrane. This depends on their polar character: a hydrophilic (polar) protein will be embedded between phophate 'heads' or be found just outside the membrane; proteins which have hydrophilic and hydrophobic regions will arrange themselves in the membrane so as to have hydrophilic meet hydrophilic and hydrophobic meet hydrophobic.

Proteins can serve as 'anchors'. The shape of the cell depends on the organisation of its cytoskeleton. This network of microfilaments interconnects in various ways and is also connected to certain proteins in the cell membrane. How the protein is attached (to a bundle of microfilaments or to a planar network) also decides the shape of that particular segment of membrane.

C.1.4
Outline the functions of membrane proteins as antibody recognition sites, hormone binding sites, catalysts for biochemical reactions and sites for electron carriers.

© IBO 1996

Many of the functions of proteins in membranes have already been mentioned in the above sections.

They include:

- **antibody recognition sites:** section 1.4.1 - glycoproteins.
- **hormone binding sites:** section 1.4.1 - glycoproteins.
- **catalysts** for biochemical reactions and
- **sites of electron carriers:** 1.4.1 - refer to sections 4.2, C.4/9.2 and C.5/9.1.

C.2 PROTEINS

C.2.1 & C.2.2
Explain the four levels of structure of proteins, indicating their significance.
Outline the difference between fibrous and globular proteins, with reference to two examples of each type.

© IBO 1996

Four 'levels' are distinguished in the structure of a protein :

- **primary structure:** the sequence of the amino acids in the chain. The linear sequence of amino acids with peptide linkages affects all the subsequent levels of structure since these are the consequence of interactions between the R group. Each amino acid is characterised by its R group. Polar R groups will interact with other polar R groups further down the chain and the same goes for non-polar R groups.
- **seconday structure:** the coils of the chain, e.g. α helix and β pleated sheath. α helix structures are found in hair, wool, horn, feathers; β pleated sheath is found in silk. Hydrogen bonds are responsible for the secondary structure. Fibrous proteins like collagen and keratin are in helix or pleated-sheet form caused by a regular repeated sequence of amino acids. They are structural proteins.
- **tertiary structure:** the way the helix chain is folded caused by the interactions of the R groups. **Hydrophobic groups** cluster together on the inside (away from the water) as do hydrophilic groups which are found on the outside (near the water). Some amino acids have a sulfur molecule in their R group. Two of these molecules may come together and the sulfur atoms will form a covalent bond :the **disulfide bridge**.

Option C: Cells and energy

Hydrogen bonds are also involved in the formation of the tertiary structure. Bonds between an ion serving as a cofactor (see section C.3/8.6) and the R group of a certain amino acid may also be responsible for the folds in the polypeptide. Proteins which have a globular (folded) shape are **globular proteins** e.g. **haemoglobin**. **Microtubules** are globular proteins which are structural. Enzymes are globular proteins. The folding of the polypeptide creates the 'active site', i.e. the location where the substrate binds to the enzyme so that the reaction can take place. (see section C.3/8.6).

- **quaternary structure:** many proteins (esp. large globular proteins) are made of more than one polypeptide chain. Together with the greater variety in amino acids, this causes a greater range of biological activity. The different polypeptide chains are kept together by hydrogen bonds, attraction between positive and negative charges, hydrophobic forces and disulfide bridges or any combination of the above. Cofactors may also assist in the quaternary structure.

C.2.3
Explain the significance of polar and non-polar amino acids (cross reference C.1.3, 1.4.1 and 1.4.2).

© IBO 1996

Of the 20 amino acids commonly used to build proteins, 8 have non-polar (hydrophic) R groups. The others have polar R groups and are soluble in water. The non-polar amino acids in the polypeptide chain will cluster together in the centre of the molecule and contribute to the tertiary structure of the molecule. In general, the more non-polar amino acids a protein contains, the less soluble it is in water. In membranes, proteins are found in between phospholipids. The phospholipid layer is polar on the outside and non-polar in the centre. The protein will often arrange itself so that it exposes hydrophilic portions to the outside of the membrane and hydrophobic sections to the centre. (see section 1.4)

C.2.4
State six function of proteins, giving a named example of each.

© IBO 1996

Functions of proteins:
- **enzymes:** all enzymes are (globular) proteins, e.g. amylase which catalyses the reaction: strarch → maltose.
- **hormones:** some hormones are proteins, others are steroids. An example of a protein hormone is insulin.
- **antibodies:** antibodies or immunoglobulins are globular proteins assisting in the defense against foreign particles.
- **structural proteins:** collagen is a fibrous, structural protein which build tendons and is an important part of your skin.
- proteins are part of the **cell membrane**, playing a role in the passage of substances into and out of the cell.

- **haemoglobin** is a protein which easily binds to oxygen due to the haem group attached to it.

C.3 ENZYMES

C.3.1
State that metabolic pathways consist of chains and cycles of enzyme catalysed reactions.
© IBO 1996

Very few, if any, chemical changes in a cell result from a single reaction. **Metabolic pathways** are chains and cycles of enzyme catalysed reactions.

The reactions will rarely occur spontaneously at reasonable speed at room temperature. Therefore enzymes are used to speed up a reaction. You may safely assume that biological reactions only occur in the presence of enzymes.

C.3.3
Explain that enzymes lower the activation energy of the chemical reactions that they catalyse.
© IBO 1996

How do enzymes work?

The presence of an enzyme will speed up the reaction because the active site will facilitate the chemical change. This happens by a means of lowering the activation energy. Every reaction requires a certain amount of **activation energy**. If two molecules are going to react with each other, they need to collide with a certain speed. The higher the activation energy, the higher the speed required. At a low temperature, only few molecules will have this speed which means that the rate of reaction is low.

The active site of the enzyme assists in the chemical reaction by lowering the required activation energy. This means that more molecules are able to react and the rate of reaction will increase.

The diagram shows what happens to the energy level during an exothermic reaction.

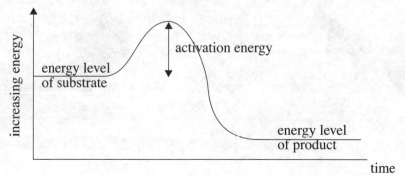

Option C: Cells and energy

Now if we added enzyme to this reaction, we could plot the following diagram:

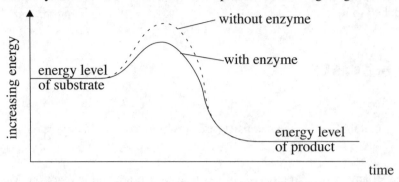

As you can see, the enzyme has reduced the required activation energy.

C.3.2
Describe the "induced fit" model.

© IBO 1996

As we discussed earlier (2.3.3) a lock and key model exists to explain the specificity of enzymes. The lock and key model was first suggested by Emil Fischer in 1894. Soon after, it appeared that certain enzymes can catalyse several (similar) reactions. The **induced fit model** suggests the following: The active site may not be as rigid as orginally was thought. Its shape will adapt somewhat to allow several slightly different substrates to fit. The active site will interact with the substrate and adapt to make the perfect fit. It is like a glove which will fit on several hands but not on a foot.

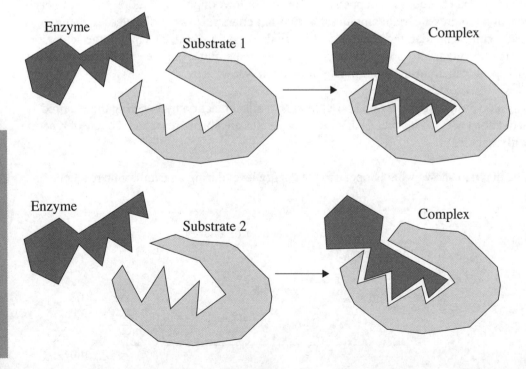

Biology

C.3.4
Explain the difference between competitive and non-competitive inhibition with reference to one example of each type.

© IBO 1996

A number of molecules exist which can reduce the rate of an enzyme controlled reaction. These molecules are called **inhibitors**. There are two kinds of inhibitors:

- **competitive inhibitors:** the inhibiting molecule is so similar to the substrate molecule that it binds to the active site of the enzyme and prevents the substrate from binding. Adding more substrate will reduce the effect of the inhibitor.

example:
Prontosil (an antibiotic) which inhibits synthesis of folic acid (vit B, which acts as a coenzyme) in bacteria. The drug will bind to the enzyme which makes folic acid. The folic acid will no longer be made and the bacterial cell dies. The animal cells are not damaged since they do not make folic acid but absorb it from food. The animal cells therefore lack the enzyme and the drug has no effect.

- **non-competitive inhibitors:** the inhibiting molecule binds to the enzyme in a place which is NOT its active site. As a result, the shape of the active site of the enzyme changes and the substrate molecule will no longer fit. Adding more substrate will have no effect on the reaction rate.

example:
Cyanide (CN^-) will attach itself to the -SH groups in an enzyme. It thereby destroys the disulfide bridges (-S-S-)and changes the tertiary structure of the enzyme. The shape of the active site is changed and cellular respiration will be disturbed. This means that energy is no longer released. If this happens in a large number of cells, the organism dies.

C.3.5
Explain the role of allostery with respect to feedback inhibition and the control of metabolic pathways.

© IBO 1996

A special kind of non-competitive inhibition is **allostery**. Allosteric enzymes are made of two or more polypeptide chains. The activity of allosteric enzymes is regulated by compounds which are not their substrates and which bind to the enzyme at a specific site well away from the active site. They cause a reversible change in the structure of the active site. The compounds are called allosteric effectors and are divided into two categories: allosteric activators (which speed up a reaction) and allosteric inhibitors (which slow down a reaction).

End products of a metabolic pathway can act as allosteric inhibitors. An example is found in glycolysis (part of cellular respiration).

Option C: Cells and energy

The following reactions take place:

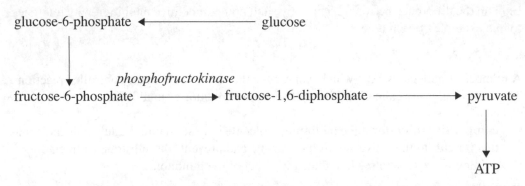

Phosphofructokinase catalyses the reaction of fructose-6-phophate to fructose-1,6-diphosphate. The chain of reactions will continue and eventually form ATP. If ATP is already present, it will bind to the phosphofructokinase and change the shape of the active site, decreasing the activity. However, when ATP is not present, phosphofructokinase is in its active form and ATP will be produced. So ATP is the allosteric inhibitor of phosphofructokinase. This is an example of negative feedback.

C.4 PHOTOSYNTHESIS

C.4.1
Draw the structure of a chloroplast as seen in electronmicrographs.

© IBO 1996

Photosynthesis occurs in the chloroplasts. These cell organelles found in the cells of green plants are 2 - 10 mm in diameter and ovoid in shape when found in higher plants (in green algea their shape varies).

STRUCTURE OF A CHLOROPLAST FROM EM

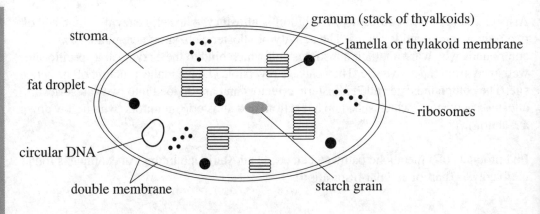

C.4.2
State that photosynthesis consists of light-dependent and light-independent reactions.
© IBO 1996

As you saw in section 4.2.6 and 4.2.7, photosynthesis is NOT a simple one step reaction. It consists of a series of reactions which can be grouped into a **light dependent stage** and a **light independent stage**. The light dependent stage will only take place in the light, the light independent stage can occur at any time, if provided with the required materials. Outside the laboratory, these materials (ATP and NADPH) come from the light dependent stage.

Some texts will still use the terms 'light stage' and 'dark stage'. These are incorrect since they imply that light is required for one stage and darkness for the other. So please do not use them.

C.4.3 & C.4.4
Explain the light-dependent reactions including the photoactivation of Photosystem II, photolysis of water, electron transport, cyclic and non-cyclic photophosphorylation, photoactivation of Photosystem I and reduction of $NADP^+$.
© IBO 1996

The diagram in section 4.2.6 gave you some idea of what is happening in the two stages. Now we will look at the light dependent stage in some more detail.

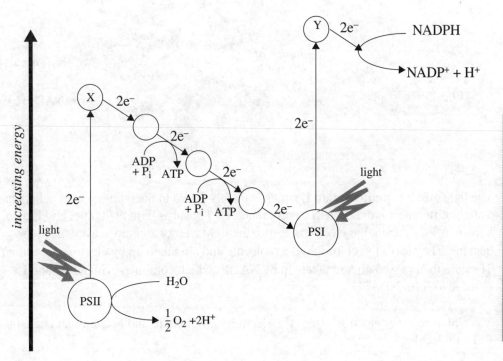

Option C: Cells and energy

Non-cyclic photophosphorylation

The light hits the pigments of **photosystem II (PS II)**, which are mainly found in the grana. The pigments involved are mainly **chlorophyll a** and they absorb light at 680 nm and are sometimes called P_{680}. Absorbing this light energy excites some electrons which as a result leave their normal position and move away from the nucleus. They are taken up by an electron acceptor X, resulting in a chlorophyll a molecule with a positive charge. The electrons are then passed through a number of electron carriers in the membrane via oxidation-reduction reactions (see Section 9.1.4/C.5.4) and will end up at PS I.

The presence of **Chl a⁺** will induce the **lysis of water** so that oxygen, H⁺ and electrons are released. P_{680}^+ is the strongest biological oxidant known.

- The electrons are taken up by Chl a⁺ (which returns to Chl a).
- The oxygen is released as a waste product.
- The H⁺ are pumped to the inside of the **thylakoids** (the lumen), they accumulate until the diffusion and electromagnetic forces are enough to drive them through proton channels in the ATP synthetase, driving the **chemiosmotic reaction** ADP + P_i → ATP (See Section C.5.4).

CHEMIOSMOTIC MECHANISM OF PHOTOPHOSPHORYLATION

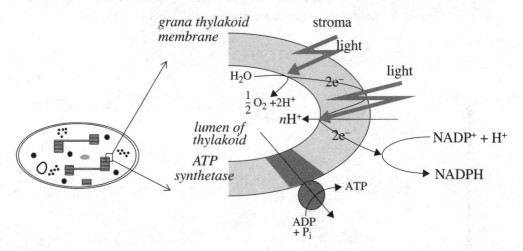

The light also hits **photosystem I**, which is mainly found in the intergranal lamella. Due to a slight difference in the protein environment, PS I absorbs light at 700 nm and is also known as P_{700}. Again, the electrons absorb the light energy and move away from the nucleus. They leave the chlorophyll a molecule and are taken up by electron acceptor Y. They are then passed on and taken up by NADP⁺ which combines with an H⁺ and is reduced to form NDAPH.

The Chl a⁺ receives electrons from the electron carrier chain and becomes an uncharged Chl a molecule.

In cyclic photophoshorylation, the electrons from PS I go to electron acceptor Y but instead of being used to produce NADPH, they go through the membrane via several electron carriers (redox reactions) and are returned to PS I. PS II is not involved. This process is cyclic, as its name suggests. It does not produce NADPH but it does produce ATP.

C.4.5
Explain the light-independent reactions including the roles of ribulose bisphosphate (RuBP) carboxylase, reduction of glycerate 3-phosphate (GP) to triose phosphate (TP or GALP), NADPH + H^+, ATP, regeneration of RuBP and synthesis of carbohydrate and other products.

© IBO 1996

The light independent stage also has some detail which was not included in 4.2.6. Below you find a diagram of the Calvin cycle.

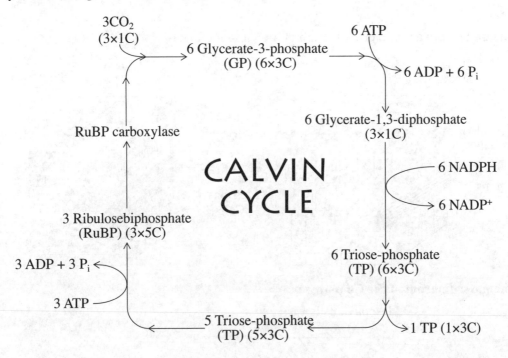

The Calvin cycle takes place in the stroma of the chloroplast. ATP provides the energy and NADPH provides the reducing power needed for biosynthesis using carbon dioxide. RuBP is the carbon dioxide acceptor and (catalysed by RuBP carboxylase) will take up CO_2, forming GP. GP will be reduced to TP but this conversion needs energy from ATP and reducing power from NADPH. TP can be converted to glucose, sucrose, starch, fatty acids and amino acids and other products. Of course, TP is also converted into RuBP to keep the cycle going. This process requires energy from ATP.

Option C: Cells and energy

C.4.6
Outline the differences in carbon dioxide fixation between C_3, C_4 and CAM plants, noting their adaptive significance.

© IBO 1996

The pathway of Carbon fixation described in section C.4.5 is called the C_3 pathway because GP (a three carbon compound) is the first recognisable compound after fixation of CO_2. Two alternatives to this exist : a C4 pathway and a CAM pathway.

The C_4 pathway uses PEP (propenoate-2-phosphate or phospho-enol-pyruvate) to accept CO_2 which will produce malate, a four carbon compound. The enzyme involved is PEP carboxylase. This enzyme has a higher affinity for CO_2 than RuBP carboxylase so that CO_2 can be fixed at lower concentrations. The malate will be moved into the bundle sheath cell where it is returned to pyruvate and CO_2. The CO_2 is then fixed in the normal way by RuBP.

Schematic diagram of a cross section of a C_3 and a C_4 leaf.

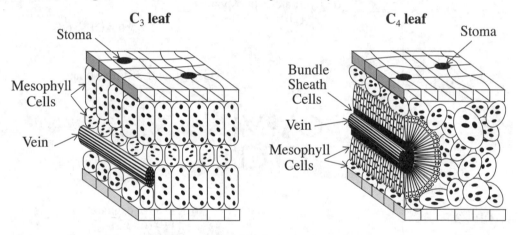

Schematic diagram of the C4 pathway.

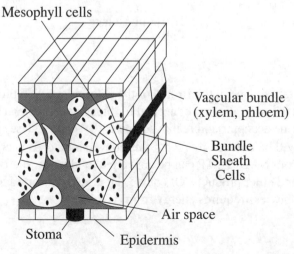

SCHEMATIC DIAGRAM OF THE HATCH-SLACK PATHWAY.

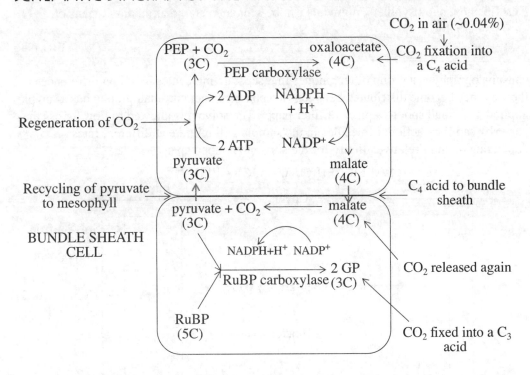

It also prevents **photorespiration**. Photorespiration is a reaction where RuBP carboxylase attaches oxygen to RuBP instead of fixing carbon dioxide. The resulting product is of no use to the plant. In effect, photorespiration causes a loss of organic molecules (the RuBP) to the plant. Using PEP carboxylase instead of RuBP carboxylase prevents photorespiration in C_4 plants.

CAM (Crassulaean Acid Metabolism) is an adaptation where some plants living in dry areas keep their stomata closed during the day to conseve water. They open them at night and fix the CO_2 using PEP carboxylase to form malate which is stored in a large vacuole in the mesophyll cells. During the day, the stomata are closed, PEP carboxylase is temporarily deactivated and malate is decarboxylated releasing the CO_2 for synthesis. These plants are particularly adapted to high light, high temperature and drought.

So in C_4 plants, the capturing and the fixing of CO_2 are separated in space (mesophyll and bundle sheath cells) while in CAM plants they are separated in time (night and day).

C.4.7
State one crop plant example for each of the following: a C_3, C_4 and CAM plant.

© IBO 1996

Examples of C_3 plants are: rice, wheat, potatoes.
Examples of C_4 plants are: sugar cane, maize (sweet corn).
Examples of CAM plants are: pineapple, prickly pear, vanilla orchid.

C.4.8
Describe how photosynthetic pigments can be separated and identified by means of chromatography.

© IBO 1996

Chromatography is a group of techniques that separate molecules based on differences in the way they become distributed between two phases. Paper chromatography has a sample applied as a small spot to a piece of filter paper. The solvent is allowed to flow across the paper by capillary action. Molecules in the sample will migrate at different rates depending on their relative affinity for the solvent and the paper.

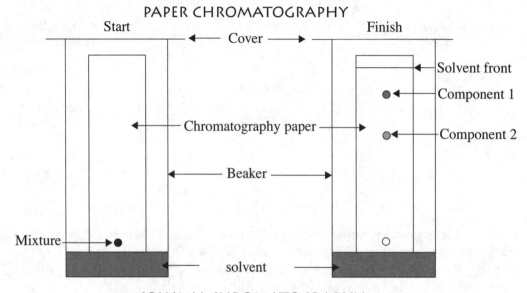

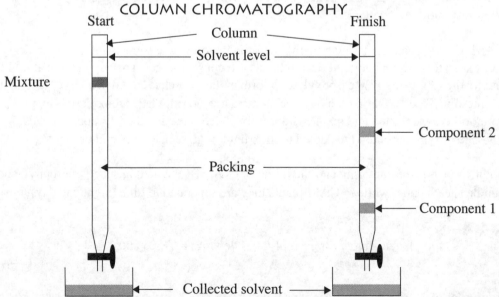

When a sample of photosynthetic pigments are subjected to this technique, you will see the different colours separating.

C.4.9 & C.4.10
Draw the action spectrum of photosynthesis.
Explain the relationship between the action spectrum and the absorption spectra of photosynthetic pigments.

© IBO 1996

The **action spectrum** of photosynthesis is a diagram which tells you how much photosynthesis goes on at any wavelength of light. The action spectrum depends largely upon how much light is absorbed. As you can see in the diagram below, it is closely related to the absorption spectra of several pigments.

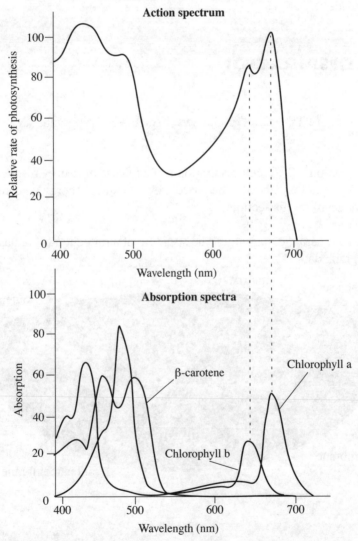

C.4.11
Explain the concept of limiting factors with reference to light intensity, temperature and concentration of carbon dioxide.

© IBO 1996

Option C: Cells and energy

Limiting factors have already been discussed in section 4.2.9.

The factor the furthest away from its optimum value will limit the amount of photosyntesis. This is then the **limiting factor**. If you improve this factor, the rate of photosynthesis will increase until another factor becomes the limiting factor. If you plot a graph of the amount of photosynthesis versus light intensity, the graph will go up until light is no longer the limiting factor. Then, the amount of photosynthesis will remain constant.

- Limiting factors for photosynthesis are: light intensity, temperature, concentration of carbon dioxide.

C.5 CELL RESPIRATION

C.5.5
Draw the structure of a mitochondrion as seen in electronmicrographs.

© IBO 1996

As you saw in Section 4.2.10, cell respiration is the release of energy from organic molecules made in photosynthesis. The process by which this energy is released will be studied in more detail in this section.

Below you find a schematic diagram of the structure of a mitochondrion, based on electronmicrograph data.

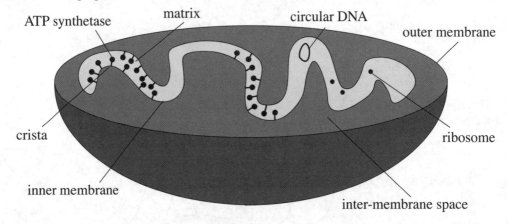

C.5.1
Outline that oxidation involves the loss of electrons from an element whereas reduction involves gain in electrons, and that oxidation frequently involves gaining oxygen or losing hydrogen; whereas reduction frequently involves loss of oxygen or gain in hydrogen.

© IBO 1996

Biology

In cell respiration, as in photosynthesis (see Section 9.2/C.4), reactions often involve the movement of electrons. This kind of reaction is called a **redox reaction**. In these reduction-oxidation reactions, one compound loses some electrons and the other compound gains them.

- **OIL RIG: O**xidation **I**s **L**oss (of electrons), **R**eduction **I**s **G**ain (of electrons).

The process of oxidation often involves gaining oxygen (hence its name) or losing hydrogen while reduction often involves the loss of oxygen or gain in hydrogen.

A substance which has been reduced, now has the power to reduce others (and become oxidised in the process); e.g. NADH and NADPH.

C.5.2
Outline what is achieved by the process of glycolysis including phosphorylation, lysis, oxidation and ATP formation.

© IBO 1996

Glycolysis takes place in the cytoplasm and produces 2 pryruvate molecules from every glucose in the following reaction:

$$\text{Glucose} + 2\text{ADP} + 2\text{P}_i + 2\text{NAD}^+ \rightarrow 2\text{Pyruvate} + 2\text{ATP} + 2\text{NADH} + 2\text{H}^+ + 2\text{H}_2\text{O}$$

<u>NO OXYGEN IS NEEDED IN THIS STEP OF THE REACTION.</u>

The structural formula of pyruvate is:

$$\begin{array}{c} \text{O}=\text{C}-\text{O}^- \\ | \\ \text{C}=\text{O} \\ | \\ \text{CH}_3 \end{array}$$

The steps that take place in **glycolysis** are shown in the diagram at the end of this section. You do not need to memorise them but should understand the major steps.

What happens is that in the cytoplasm one 6-C sugar is converted into two 3-C compounds (pyruvate) with a net gain of 2 ATP + 2 NADH + H^+.

To achieve this it is necessary to change glucose into fructose-1,6-diphosphate which is then split into two 3-C compounds (**lysis**). This process requires energy. Subsequently, the 3-C compounds are oxidised into pyruvate. During this process, energy is released (ATP is formed) and NAD is reduced into NADH.

Phosphorylation: the process of making ATP in vivo.

Option C: Cells and energy

The following steps take place in glycolysis:
(you do not need to memorise this but you must remember the principles).

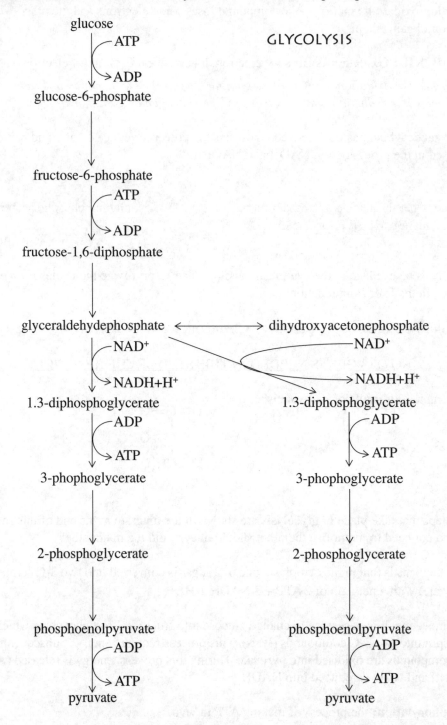

C.5.3

Outline aerobic respiration including oxidative decarboxylation of 2- oxopropanoate (pyruvate), Krebs cycle, NADH + H⁺ and electron transport chain.

© IBO 1996

If oxygen is present, pyruvate is transported to the mitochondrial matrix and the reactions continue in the following way:

Pyruvate + CoA + NAD^+ → Acetyl CoA + CO_2 + NADH + H^+ or
$CH_3.CO.COOH$ + CoA-S-H + NAD^+ → CO_2 + NADH + H^+ + CH_3CO-S-CoA

The reaction is known as the **link reaction** because it forms the link between glycolysis and the Krebs cycle. This process is known as oxidative decarboxylation of pyruvate (2-oxopropanoate).

Krebs cycle

The Krebs cycle occurs in the matrix of the mitochondria and produces CO_2, NADH + H^+, $FADH_2$ and ATP.

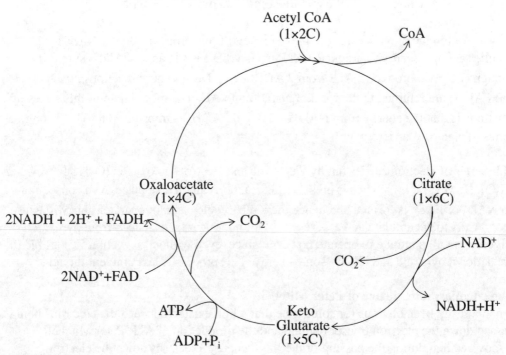

As you can see, one turn of the Krebs cycle yields:
- 2 × CO_2
- 3 × NADH + H^+
- 1 × $FADH_2$
- 1 × ATP

Option C: Cells and energy

Remember that one of the key roles of the co-enzymes NAD⁺ and FAD is to collect and remove H⁺.

The last step of aerobic respiration is the **Electron Transport Chain** (ETC). The ETC passes two hydrogens (and two electrons) from NADH or $FADH_2$ from one electron carrier to another (found in the inner membrane of the mitochondrion) by a series of redox reactions. The final acceptor is oxygen (producing water). This process produces 3 ATP for every 2 hydrogens (and 2 electrons) from NADH + H⁺ and 2 ATP for every 2 hydrogens (and 2 electrons) from $FADH_2$

C.5.4
Describe oxidative phosphorylation in terms of chemiosmosis including, proton pumps, a proton gradient and ATP synthetase (cross reference C.4.4).

© IBO 1996

Electron Transport Chain and Oxidative Phosphorylation
Oxidative phosphorylation occurs on the inner membrane of the mitochondria. As the electrons are passed down the electron transport chain, protons are being pumped across that membrane. The resulting proton gradient drives the production of ATP from (ADP and P_i) by ATP synthetase. This is the **chemiosmotic theory** of of Peter Mitchell.

As was said in the previous section, the net result of this process is that 1 NADH + H⁺ supplies enough energy to produce 3 ATP from 3 ADP + 3 P_i and 1 $FADH_2$ supplies enough energy to produce 2 ATP from 2 ADP + 2 P_i. During these reactions NADH + H⁺ and $FADH_2$ are returned to the form of NAD⁺ and FAD. The mechanism of this series of reactions is that the energy from NADH + H⁺ and $FADH_2$ is transferred to ATP through a series of electron carriers.

This series of electron carries finally yields H+ and electrons to oxygen (O_2) to form water (H_2O). However if no oxygen is present, this reaction cannot take place. As a consequence, no NAD⁺ or FAD⁺ is formed and hence the Kreb's cycle cannnot operate. This will cause acetyl CoA to accumulate and as a result it will no longer be produced from pyruvate. Glycolysis will continue to operate however, since, even without oxygen, it is possible to break down pyruvate and release some energy. This process is less efficient though.

The chemiosmotic theory of Peter Mitchell.
It had already been obvious for some time that a link existed between the electrons being passed down the electron transport chain and the production of ATP. Peter Mitchell discovered that during the passing of the 'high energy' electrons down the electron transport chain, protons are being pumped across the inner mitochondrial membrane.

There is a build up of H⁺ ions in the intermembrane space. The diffusion and electromagnetic forces will drive H⁺ through the ATP synthetase molecule. As the H⁺ ions go through the ATP synthetase molecule, the potential energy they possess will be used to drive ATP synthesis.

THE ELECTRON TRANSPORT CHAIN IN CELLULAR RESPIRATION

(Energy Level diagram showing the cascade: NADH → NAD, with successive ox/red pairs: FMN, Q, cyt b, cyt c_1, cyt c, cyt a, cyt a_3, ending with $2H^+ + \tfrac{1}{2}O_2 \rightarrow H_2O$)

FMN: flavin mono nucleotide
Q: coenzyme Q (ubiquinone)
cyt: cytochrome
ox: oxidised
red: reduced

C.5.6
Explain the relationship between the structure of the mitochondrion and its function.
© IBO 1996

Keeping in mind all of the above information, it is useful to return to the structure of the mitochondrion.

The outer membrane is a regular membrane, separating the mitochondrion from the cytoplasm. Its structure is based on the fluid mosaic model. It is impermeable to H^+ ions. The intermembrane space has a higher concentration of H^+ ions because of the electron transport chain. It pH is lower.

The inner membrane is folded into cristae to provide maximum space for the electron carriers and ATP synthetase. It is impermeable to H^+ ions. Its structure is based on the fluid mosaic model with the electron carriers and the ATP synthetase embedded among the phospholipid molecules. The ATP synthetase molecules can be seen on the cristae. The matrix contains the enzymes which enable the Krebs cycle to proceed.

Glycolysis takes place in the cytoplasm. Pyruvate is transported to the matrix of the mitochondrion and decarboxylated to acetyl CoA which enters the Krebs cycle. The resulting $NADH + H^+$ and $FADH_2$ give their electrons to the electron carriers in the inner membrane. The electrons move through the membrane as they are passed from one electron carrier to another in a series of redox reactions. During this process, H^+ ions are pumped from the matrix into the intermembrane space, creating a potential difference.

Option C: Cells and energy

Electromagnetic and diffusion forces drive the H$^+$ ions back to the matrix through the ATP synthetase which uses the energy released to combine ADP and P$_i$ into ATP, which is released into the matrix.

C.5.7
Describe the central role of ethanoyl (acetyl) CoA in carbohydrate and fat metabolism.

© IBO 1996

In the above sections, you have seen the key role in the carbohydrate metabolism, played by acetyl CoA, essentially linking the Krebs cycle to glycolysis. Acetyl CoA plays a similar key role in the metabolism of fatty acids. Since fats contain more energy per gram than carbohydrates or proteins (see Section 2.2.7), an efficient system must exist to break down fatty acids in cellular respiration. The long chains of fatty acids are oxidised, effectively breaking off sections of 2 Carbon molecules. These are changed into acetylCoA and enter the Krebs cycle.

C.5.8
Outline fermentation to 2-hydroxypropanate (lactate) and to ethanol, and the circumstances in which they occur in cells.

© IBO 1996

Anaerobic Respiration
When no oxygen is available, cells are capable of anaerobic respiration. Although anaerobic respiration releases some energy, the yield is much lower than that of aerobic respiration because a lot of the energy of the glucose remains 'locked up' in the end product (lactic acid or ethanol).

aerobic respiration:
$$C_6H_{12}O_6 + 6\ O_2 \rightarrow 6\ CO_2 + 6\ H_2O + 2880\ kJ$$

anaerobic respiration:
alcoholic fermentation, e.g. in yeast
$$C_6H_{12}O_6 \rightarrow 2\ C_2H_5OH + 2\ CO_2 + 210\ kJ$$
$$\text{alcohol}$$
lactic acid production, e.g. in muscle cells
$$C_6H_{12}O_6 \rightarrow 2\ C_3H_6O_3 + 150\ kJ$$
$$\text{lactic acid}$$

In anaerobic respiration, glycolysis takes place in the cytoplasm as we have seen before. During this process, the net gain is 2 ATP and 2 NADH + H$^+$. The reaction cannot continue in the manner described above since the ultimate oxygen acceptor is not available. This means that the electron carriers are reduced, and the Krebs cycle is unable to continue.

To allow glycolysis to proceed, two conditions must be met. Pyruvate (or pyruvic acid) cannot be allowed to accumulate since this would stop change the equilibrium of the reaction and a supply of NAD$^+$ is needed. Both conditions can be satisfied by reducing pyruvate into ethanol or lactic acid. NADH + H⁺ is oxidised into NAD⁺ in the process.

In the production of ethanol, CO_2 is released and the reaction is therefore not reversible. Lactic acid (or lactate) can be used to form pyruvate again (using energy) when oxygen becomes available.

EXERCISE

1. Lysosomes are vesicles containing digestive enzymes. Explain the role of the nucleus, rER and Golgi apparatus in producing lysosomes.

2. What is the role of the following in translation?
   a. GTP
   b. ribosomes
   c. polysomes
   d. codons

3. Which types of bonds are involved in each of the four levels of protein structure?

4. a. What is the function of the ATP and NADPH produced in non-cyclic photophosphorylation?
   b. What would be the purpose of cyclic photophosphorylation?
   c. What is the advantage of non-cyclic photophosphorylation over cyclic photophosphorylation for the plant?
   d. What is the purpose of the Calvin cycle?

5. a. Where in the cell does glycosis take place?
   b. Where in the cell does the Krebs cycle take place?
   c. Where in the cell is the electron transport chain found?
   d. Draw a diagram of the structure of a mitochondrion as seem with the electron microscope.
   e. How does the structure of the site for the Krebs cycle relate to its function?
   f. How does the structure of the site for the electron transport chain relate to its function?
   g. What would happen to the all parts of aerobic respiration of the outer membrane of the mitochondrion became permeable to protons (hydrogen ions)?

6. Compare and contrast the process of ATP production in chloroplasts and mitochondria.

Option C: Cells and energy

OPTION D: EVOLUTION

20

Chapter contents
- Origin of life on Earth
- The origin of species
- Evidence for evolution
- Human evolution
- Neo-Darwinism
- The Hardy-Weinberg Principle

Option D: Evolution

D.1 ORIGIN OF LIFE ON EARTH

D.1.1
Outline the origin and development of pre-biotic Earth, including cooling and formation of land, sea and a reducing atmosphere.

© IBO 1996

Many ideas of how we came about have been proposed over the past few centuries. Although no a single one of these has ever been scientifically proven, the '**theory of evolution**' (strictly speaking the 'hypothesis of evolution') is currently accepted by many people.

According to the evolutionary theory, the Earth has not always been as we see it today. Billions of years ago, the '**big bang**' occurred: explosions causing clouds of dust and gas to travel through the Universe. The process of expansion eventually stopped and clouds began to contract again. High temperatures at the centre of the contracting cloud caused nuclear fusion of hydrogen atoms to form helium and the Sun came into existence approximately 4.6 billion years ago.

Earth and other planets formed in similar ways. Approximately 3.8 billion years ago, Earth must have been a ball of fire and molten rock, at best with a thin layer of crust. As the Earth cooled down, this thin layer became thicker and liquid water began to condense from the clouds of water vapour. Land was formed.

The atmosphere at that time probably contained water vapour, methane, ammonia, hydrogen, carbon dioxide. It is speculated that the atmosphere was probably a 'reducing atmosphere', because metals in old rocks are found in their reduced form, e.g. iron as iron(II). Rocks that were formed later generally contain iron(III) which is the oxidised form. Experiments have shown that it is possible to form organic molecules in a reducing atmosphere but that this is very difficult to do when the atmosphere contains oxygen. (see also section D.1.2)

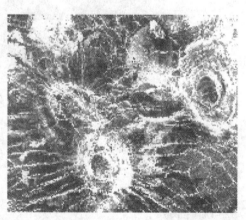

The surface of Venus (photo NASA) where present day conditions are similar to those on the primitive Earth

Frequent thunderstorms shot bolts of lightning through this mixture and because of the absence of ozone, UV light intensity was high. Life did not yet exist on Earth.

D.1.2
Outline the experiments of Miller and Urey into the origin of organic compounds.

© IBO 1996

Stanly Miller and Harold Urey worked on trying to confirm some of the above ideas. In 1953 Miller set up apparatus as shown below.

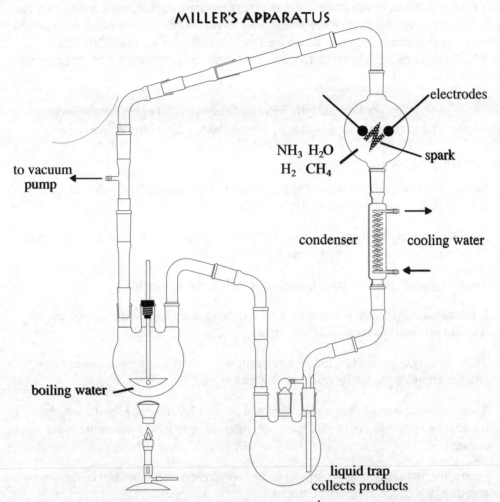

In the 'liquid trap' he collected amino acids, adenine and ribose after the apparatus had run for about a week. This experiment (Miller-Urey) shows that organic molecules can be formed under certain conditions. It has been repeated with some variations (e.g. in the presence of carbon dioxide) and other molecules have formed. If these indeed were the conditions on Earth, approximately 4 billion years ago, then the building blocks for molecules vital to life could have originated spontaneously. Note that this experiment contains a serious risk of explosion. You should not attempt to repeat it!

Meteorites have been found to contain organic material, including amino acids. In 1970, a meteorite was found to contain 7 different amino acids, 2 of which are not found in living things on Earth. So organic compounds can and do originate in Space.

Option D: Evolution

D.1.3 & D.1.4
Discuss the hypothesis that the first catalysts responsible for polymerisation reactions were clay minerals and RNA.
Discuss the possible role of RNA as the first molecule that could replicate.

© IBO 1996

One of the conditions of life is the process of self replication. In current cells, DNA can replicate but it needs the help of enzymes (proteins) to do this. The proteins are assembled according to the information carried on the DNA and transcribed into RNA. This resembles a chicken-egg situation: to make DNA you need proteins but to make proteins you need DNA.

One suggestion as to how the **catalytic action of clay** assisted in the formation of polypeptides from amino acids is made by **Katchalsky**, Cairns-Smith and Bernal.

The basis of this idea is as follows:

1. Some clays can grow by attracting molecules to themselves. They will then repeat a lattice-like organisation over and over again.

2. Amino acids may have stuck to the clay lattice and have been incorporated into it. They may have been attached to each other as well.

3. Some clay particles may have become a template for a protein.

4. If the protein product was a weak enzyme it may have speeded up the process of 'protein synthesis' with clay as a template.

5. Then the clay template for this particular protein would make more protein than another template whose product was not an enzyme.

6. Then nucleotides could have been attracted by the clay template, or the template with the attached proteins, and could have polymerised (into RNA) and come to act as a co-enzyme.

7. Again, the more successful template is the one where the enzyme and co-enzyme work together to produce more of themselves.

8. Eventually, the co-enzyme (the nucleotide polymer, RNA) could become the template for protein synthesis.

9. It is not clear how this process then continues so that DNA becomes involved in protein synthesis.

The series of events described above is not entirely random. The conditions on Earth made some processes more likely to occur than others.

D.1.5
Discuss a possible origin of membranes and prokaryotic cells.

© IBO 1996

Membranes are phospholipid bilayers with proteins (see Section 1.4). They are necessary to separate the cell from the environment and to control the passage of substances into and out of the cell.

Fox and his co-workers had some ideas about the origins of the cell membrane and decided to attempt to imitate the process in the laboratory. They heated amino acids without water and produced long protein chains. When water was added and the mixture was allowed to cool down, small stable **microspheres** or **coacervates** were formed. The microspheres seemed to be able to accumulate certain compounds inside them so that they became more concentrated that outside. They also attracted lipids and formed a lipid-protein layer around them.

If we assume that the microspheres/coacervates combined with self replicating molecules such as RNA, we are looking at a very primitive organism. This is thought to have happened about 3.8 billion years ago.

D.1.6
Discuss the endosymbiotic theory for the origin of the four eukaryotic kingdoms.

© IBO 1996

The above explains the sequence of events leading to the origin of the prokaryotic cell, about 3.8 billion years ago. The oldest fossils of eukaryotic cells have been found to be approximately 1.5 billion years old. One hypothesis on how eukaryotic cells arose comes from **Lynn Margulis** in what is called the endosymbiont theory (see Section 1.3).

Lynn Margulis suggests that mitochondra were originally independent prokaryotic aerobic organisms which developed a symbiotic relationship with another prokaryote, lacking a cell wall. The aerobic prokaryote was engulfed by the bacterium's cell surface membrane in the usual process of endocytosis, made easy by the absence of a cell wall in the bacterium. The aerobic prokaryote was not digested but continued to function inside the other cell. This association turned out to be profitable and was continued. The 'host' cell received energy that the aerobic prokaryote released, the 'mitochondrion to be' had all its other needs met by its host. A similar process occured later with the host cell (with mitochondria already present) and photosynthetic prokaryotes which became chloroplasts.

Support for the endosymbiont theory is found in the structure of DNA and ribosomes which is similar in prokaryotes and mitochondria and chloroplast but different in eukaryotes. See section 1.3 for details.

The four **eukaryotic kingdoms** are the Protoctista, Fungi, Plantae and Animalia (see section 13.1). Eukaryotic cells have some advantages over prokaryotic cells so the early eukaryotes survived and proliferated into the wide diverstity of species we know today.

Option D: Evolution

D.2 THE ORIGIN OF SPECIES

For many centuries, people accepted the species they saw around them and have assumed they had always existed. This is hardly surprising since in our daily lives we do not say evidence of any kind of evolution on a daily basis. However, in the 18th century, the finding of many strange species in other parts of the world together with the finding of fossils made people ready to depart from the idea of a 'steady state' and accept some concept of evolution.

D.2.1
State Lamarck's theory of evolution by the inheritance of acquired characteristics.
© IBO 1996

Although many people contributed to the concept of evolution and the theory of evolution as we now know it, some are recognised to have a major impact, even if we no longer agree with their hypotheses.

Jean Baptiste de Lamark (1744 - 1829) suggested that all species were created by God but that they undergo change over time. Lamarck is best remembered for his suggestions as to how this change came about. It is summed up in the phrase: 'inheritance of acquired characters'. What this means is that the behaviour of the individual determines the character that its offspring inherit.

D.2.2
Discuss the mechanism of, and lack of evidence for, the inheritance of acquired characteristics.
© IBO 1996

The example that illustrates this idea is the giraffe. According to Lamarck, the giraffe stretched its neck to reach the highest leaves because the leaves lower down had already been eaten. This constant reaching would stretch the giraffe's neck and result in its young being born with a longer neck.

At first sight, this idea might not seem unreasonable. However, if you then look at other situations, the concept seems not to work. For example, people have been docking (cutting off) ears and/or tails of many breeds of dogs for many generations. Yet no puppy of any of these breeds has yet been born with its ears/tail already the desired shape/size.

Another example can be found in humans. Circumcision is a well known and long established tradition among some people. Yet the baby boys are still born with their foreskins intact.

From your studies of genetics, you already know that Lamarck's idea cannot be correct. Most of these changes do not affect the genetic material. Even if they do, mutations to genetic material of somatic cells do not affect the offspring. Only when gametes are involved, does a change in the genetic material involve the next generation.

D.2.3
State the Darwin-Wallace theory of evolution as the natural selection and inheritance of favourable characteristics.

© IBO 1996

As Lamarck's hypothesis for the mechanism for change seemed unsatisfactory, **Alfred Russel Wallace** (1823 - 1913) and **Charles Darwin** (1809 - 1882) both came up with the alternative idea of 'natural selection' or 'a struggle for existence' as a mechanism driving the change process over time.

Both Wallace and Darwin had studied works of others (Lyell and Malthus) as well as having travelled to the far corners of the world (from England). Darwin's voyage on the HMS Beagle to South America and the Galapagos Islands is well documented. Wallace travelled to South America and also to Indonesia. Their previous reading and the information they gathered during their travels, made both Darwin and Wallace come up with the idea of natural selection as a force driving the process of evolution. In 1858, Wallace wrote a paper on this concept. Darwin published his controversial work *'On the Origin of Species'* in 1859. The text of this important work can be found on the internet at the address: http://earth.ics.uci.edu/faqs/origin.html

Using the example of the giraffe, which was used to illustrate the Lamarckian ideas, the giraffe is indeed always reaching for the leaves. The young giraffe that happens to be born with a gene for a longer neck, will therefore obtain the most food and will be successful in reproducing. It genes (including the one for a long neck) are passed on to its offspring. Competing with other giraffes with shorter necks, the long necked ones will have the advantage and obtain more food. They will not starve and produce many offspring until all giraffes have long necks. The mutation for a longer neck is random and just as likely to occur as the one leading to a shorter neck. The environment, with more competition for leaves lower down on the tree, will cause the longer necked mutant giraffe to have the advantage and the shorter necked mutant giraffe to probably die of starvation and not pass on its genes.

Evidence for the process of selection as a driving force for change can again be found in the breeding of dogs. Artificial selection by humans created many different breeds of dogs in a relatively short time. We can see the same in domesticated animals and plants used, for example, in agriculture.

Option D: Evolution

D.2.4
Describe experimental evidence for the process of natural selection, including bird predation on moths.

© IBO 1996

To see evidence of natural selection, we can look at fossil records (see section D.3.6). A good example is found in the evolution of the horse. Fossils have been found showing that 53 million years ago, the ancestor of the horse was a small herbivore probably living in the forest. It had 4 toes on its front feet and 3 toes on its hind feet. Over time, the animal started to live on grassy plains, grew bigger and the number of toes reduced until the horse just has one (the nail is the hoof) which allowed it to run faster. Many fossils have been found documenting this change. Being larger and faster is very important when you live on a grassy plain in sight of predators.

Another example can be found in the case of the Peppered Moth. *Biston betularia* (peppered moth) is found near Manchester. Before 1848 trees on which they rested were covered with off-white lichen. The moths were white and therefore camouflaged from predation by birds. Occasionally a black moth would appear. It, of course, was highly visible on the lichen covered trees and would have a very high chance of being eaten by a bird before reproducing.

Due to (coal based) industry, the trees became covered with soot and the white moths were easily spotted and eaten. The dark form (melanic form) now had the advantage and became predominant (95%) in certain areas in 1950.

Reduced use of coal has now made the trees green (covered in algae) and both forms are common. (This is called balanced polymorphism.)

The environment before 1848 was such that the white moth was the best adapted. However, when the environment changed, the black moth became the best adapted. This is a simple and short-time example of evolution.

D.2.5
Discuss other theories for the origin of species including special creation and panspermia.

© IBO 1996

Special creation involves the origin of humans. Many religions say that Man was created by God. As a result many people feel that the concept of evolution goes against their religion. Some people can accept the concepts of evolution when they see the hand of God in the process. Others do not find this acceptable. The IB syllabus requires knowledge about the ideas on evolution but does not require that you agree with them.

Panspermia suggests that life may have originated elsewhere and came to us from space. This may be possible but does not address the issue of the ultimate origin of life.

D.2.6
Discuss the nature of the evidence for all these theories and the applicability of the scientific method for further investigation.

© IBO 1996

The nature of the evidence of the theories is rather different. In section D.2.2, the evidence of Lamarckian ideas has been discussed and in section D.2.4 the same has been done for Darwinian ideas. Some evidence has been found to support panspermia as seen in section D.1.2 yet this is not overwhelming. There seems to be much more evidence to suggest that life originated on Earth. As the special creation theory seems to be founded in religion, it is difficult to find scientific evidence for it.

All these evolutionary theories are really hypotheses, since none of them has yet been proven . Some inconsistencies can be found for every theory, yet the hypothesis suggested by Darwin and Wallace seems to have the most convincing evidence and is supported by many people these days.

D.3 EVIDENCE FOR EVOLUTION

D.3.1
Describe the evidence for evolution as shown by the geographical distribution of living organisms, including the distribution of placental, marsupial and monotreme mammals.

© IBO 1996

One of earliest known pieces of evidence against the steady state hypothesis and supporting evolution came from information about the geographical distribution of living organisms (biogeography). It made sense to find different species in different locations when circumstances were different but sometimes the conditions of the various locations were similar yet completely different species were found.

An example of this can be found in the evolution of the mammal. Approximately 200 million years ago, all the continents were attached to each other. A mere 10 million years later, during the age of the dinosaurs, they started to break apart. Early mammals were already in existence but were small and most likely nocturnal.

Background information (not strictly on the syllabus):
Mammals developed approximately 180 million years ago and have the following characteristics:
- they maintain a constant body temperature.
- they have hair.
- they have mammary glands for providing their young with milk.

Mammals can be subdivided into three major groups:

1. **Monotreme mammals:** non-placental mammals which lay leathery eggs and nurse their young; e.g. duckbilled platypus.

Option D: Evolution

2. **Marsupial mammals:** mammals in which the female has a ventral pouch surrounding the nipples; the premature young leave the uterus and crawl into the pouch where they attach to the nipple until development is completed; e.g. opossum and kangaroo.

3. **Placental mammals:** mammals which carry the young in the uterus until an advanced stage of development; the placenta ensures efficient nutrition of the young; e.g. mice and humans.

Various hypotheses have been suggested to explain the evolution from reptiles to mammals. It seems simple to suggest one kind of ancestral mammal from which the three major groups evolved. Recent evidence suggests the step from reptile to mammal was made several times, independently. It also suggests that marsupials and placental mammals did share a common ancestor but the history of the monotremes is still unclear and this group may well have developed separately.

The great reptiles disappeared quite suddenly and the number of mammalian species greatly increased. They spread over the area that was available to them. As a result, we find marsupials in South America and Australia. These continents were attached via Antarctica and Australia did not separate from this continent until approximately 55 million years ago.

Evidence supporting this idea was found quite recently when a fossil marsupial was found in Antarctica but no fossils of other mammals have been found there.

In many places, the rise of the placental mammal was the downfall of the marsupial. Since Australia was not longer attached to other land masses, the placentals did not travel there and the marsupials reigned until invaded by *Homo sapiens*.

D.3.2
Outline the main stages in the process of fossilisation.

© IBO 1996

Fossils are any form of preserved remains from a living organism.
Some examples are:
- mammoths frozen in Siberia.
- mummies in acidic swamps in Scandinavia.
- insects in amber (resin which became hard).
- bones in rock.

Only in few cases are circumstances such that fossils are formed. This is why the fossil record is not complete. Most individuals do not leave a fossil after death.

It is also possible to find evidence of earlier life in 'rocks' which resemble the animal/plant/footprint/faeces.

If a (dead) organism gets buried in (sedimentary) silt, it will decay slowly and leave a space in the silt surrounding it. As the organism disappears, the space can be filled with minerals from the silt. They become solid and fill the exact gap the organism left behind (like making a cast). The silt around this then may solidify, becoming sedimentary rock. In the rock you find a different looking stone, looking like the dead organism. It is a fossil.

D.3.4
Define half life.

© IBO 1996

In determining the age of fossils (including preserved organic matter), people have made use of isotopes. Isotopes are atoms which have a mass different from the most other atoms of the element. This difference is caused by a different number of neutrons. Isotopes are often unstable and will spontaneously change into one or more atoms of other elements, often emitting some radiation. The time taken for this change is detemined by the kind of isotope. This process of radioactive decay follows rules. During a fixed interval (depending on which isotope is involved), half of the amount present will decay. As a result, at the end of the period, the radioactivity will be half of what it was before. This is called the **half life** of the isotope. Half lifes vary from fractions of seconds to thousands of years. The half life of ^{14}C is 5730 years.

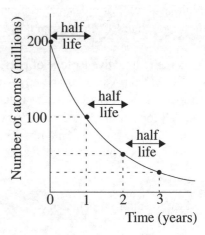

If you want to visualise half life you can do the following. Take a number of coins. Write the number down as you initial number at time 0. Toss the coins and remove all that land on 'heads'. Count the number of coins remaining and write down this number at time 1. Repeat until you have no coins left. Your graph will show how radioactive decay reduces the amount of isotope and the half life will be 1. (see section D.3.5)

D.3.3
Outline the method for the dating of rocks and fossils using radioisotopes, with reference to ^{14}C and ^{40}K.

© IBO 1996

'**Carbon dating**' has become an excellent method of determining the age of organic material. The principle of this method involves the radioactive isotope of carbon ^{14}C. A normal atom of carbon is referred to as ^{12}C. Its nucleus contains 6 protons and 6 neutrons. Around the nucleus, 6 electrons spin around. The nucleus of the isotope ^{14}C also contains 6 protons but it has 8 neutrons and again 6 electrons are found around the nucleus. This means that the isotope will chemically behave like a regular carbon atom but may have some different physical properties (e.g. a greater mass). It also happens to be somewhat unstable and spontaneously changes into ^{14}N (7 protons and 7 neutrons) and it emits some radiation.

Option D: Evolution

^{14}C is present in small amounts on Earth. The Sun's radiation causes a small but constant amount of ^{14}C to be present. As ^{14}C behaves chemically like the normal carbon, it is used in photosynthesis and enters the food chain. As a result, all living organisms contain ^{14}C in the same proportion as is found in the atmosphere. After the organism dies, the process of incorporating new carbon into the body (via feeding or photosynthesis) stops. After 5730 years, the amount of ^{14}C present in the remains is half of what is was at the time of death. After 11 460 years, it would be one quarter.

Carbon dating is an accurate and useful method to determine the age of organic remains. It is accurate for material up to approximately 20 000 - 50 000 years old (there are different opinions here). For older material, another isotope needs to be chosen: one that has a half life longer than ^{14}C. Potassium (^{40}K) has been found to be of use.

^{40}K is the radioactive isotope of potassium. It will decay to form ^{40}Ar which is stable. ^{40}K is released from volcanos and any Argon present at that time would be lost to the astmosphere as it is a gas. If the volcanic eruption becomes part of the sediment, then the relative proportions of ^{40}K and ^{40}Ar can be calculated. The half-life of ^{40}K is 1300 million years so this technique is mainly used for rocks over a million years old. The accuracy of this method is around 50 000 years.

D.3.5
Deduce the approximate age of materials based on a simple decay curve for a radioisotope.
© IBO 1996

As described above, the relative amount of ^{14}C present in organic material can tell us how long ago this material stopped incorporating new carbon into its system. That was the time it died. We can use a graph to help us determine the time. At time 0, the fraction of ^{14}C is 1, i.e. it is the original amount present, i.e the organism just died. When only half the original amount of ^{14}C is present, a time equal to the half life of the istopope has passed. In the case of ^{14}C, this is 5730 years. When only 0.125 of the orginal amount of ^{14}C is present, 3 half lives have passed since the time of death so the specimen died 17 190 years ago.

The Treasury building at Petra.

The age of such buildings can be estimated by carbon dating organic remains found in them.

D.3.6
Outline the palaeontological evidence for evolution using one example.
© IBO 1996

Biology

Several different kinds of evidence for evolution are recognised:
- **paleontological evidence** is based on fossils found.
- **biochemical evidence** is based on similarities and differences in the structure of macromolecules such as DNA or protein.
- **anatomical evidence** is based on the similarities and differences in the anatomy of species.

A classic example of paleontological evidence for macroevolution is found in the evolution of the horse. The first fossils of the ancestors of the modern horse are 53 million years old. From fossil records we know it had 3 toes on its hind feet, 4 on its front feet. Its eyes were halfway up its head (between nose and ears) and the structure of the teeth was such that *Hyracotherium* (also known as *Eohippus* which means dawn horse) must have been a browser (eating leaves) rather than a grazer (eating grass). Fossils have been found in successive younger strata (layers of sediment in which fossils may be found) suggesting the evolution to the modern horse. The horse grew progressively larger, one of its toes grew bigger and the others reduced, the eyes moved up in the head towards the ears (to be able to watch for predators while grazing) and the teeth became bigger and stronger to allow grazing. Many fossils have been found of lines that have since become extinct. These species probably were lived for some time but eventually were replaced by other species which were better suited to the environment. Only one line continued into our time which is that of *Equus*, the modern horse. The fossil record shows that a series of small changes occurred, some more successful than others and that natural selection eventually only allowed the modern horse to survive.

D.3.7
Explain the biochemical evidence provided by the universality of DNA and protein structures for the common ancestry of living organisms.

© IBO 1996

Evidence for evolution can be found in the biochemistry of organisms. As was already discussion in section 2.6.6, the genetic code is universal. This means that all organisms use the same four bases in DNA and DNA sections can be taken from one organism and introduced into another to create transgenic organisms (see section 2.7.3 and 10.6.7). In the same way, all species use the same 20 amino acids for proteins.

Although the same components are used to make DNA and protein in all organisms, the sequence of these components may be different. By comparing, for example, the amino acid sequence of haemoglobin in humans, cats and earthworms, it can be shown that humans and cats have greater similarities in the structure of their haemoglobin than humans and earthworms. This means that the common ancestor of humans and earthworms lived a longer time ago than the common ancestor of humans and cats.

Using DNA from termites that became trapped in resin (which over time turned into amber), it has been found that this 30 million year old DNA is very similar to that found in certain termites living today. However, when it is compared to that of other, less similar looking termites, the differences in the DNA are also greater.

Option D: Evolution

Generally the biochemical evidence is used together with other kinds of data such as paleontological data. Usually the two support each other.

D.3.8 & D.3.9
Explain how variations in specific molecules can indicate phylogeny.
Discuss how biochemical variations can be used as an evolutionary clock.

© IBO 1996

Phylogeny is the line of evolutionary descent. Biochemistry can be used to support other evidence about evolutionary relationships. The principle of this is simple. You study similar molecules in different species and determine how much difference there is between the molecules. The more difference there is, the longer the timespan since the two species had a common ancestor.

Commonly used in this technique are haemoglobin, cytochrome c and nucleic acids (especially rRNA). The bases for comparison between the molecules is either the primary structure (amino acid sequence) for proteins or the sequence of the nucleotides for nucleic acids.

Haemoglobin is suited to studying closer related organsims that have haemoglobin, for example, primates. Results suggest that humans are more closely related to chimpanzees than to gorillas or gibbons. This means that the ancestor common to humans and chimpanzees lived more recently than the one common to humans and gorillas.

Cytochrome c (a respiratory protein which is part of the electron transport chain) has been used to compare groups that are more different. It has been found that humans and chimpanzees have identical cytochrome c molecules and differ only in 1 amino acid from that of a rhesus monkey. Even humans and *Rhodospirillum* (bacteria) or yeast (fungus) have identical amino acid sequences in part of the cytochrome c molecule.

Again, as is the case with the other kinds of evidence for evolution, the results from comparative biochemistry alone do not prove anything. Yet they confirm data found using other methods and, together, make a convincing total.

D.3.10
Explain the evidence for evolution provided by homologous anatomical structures including vertebrate embryos and the pentadactyl limb.

© IBO 1996

Comparative anatomy is yet another way of providing supporting evidence for evolution. It concentrates on studying homologous structures. Two structures are homologous if they come from the same origin even though they may look different now and have different functions.

(Analogous structures have the same functions but come from different origins, e.g. the wing of a bird which used to be a limb and the wing of an insect which used to be a fold in the skin are analogous as they are both used for flying.)

An example of homologous structures are the pentadactyl limb. This limb has a basic pattern of bones including 5 digits (finger/toes). The pentadactyl limb is used differently by different mammals. As a result, the appearance varies but the same basic structure can be recognised.

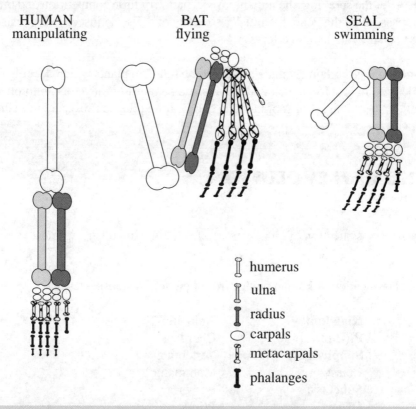

D.3.11
Describe modern examples of observed evolution including general biocide resistance, antibiotic resistance in bacteria, and heavy metal tolerance in plants.

© IBO 1996

Although evolution generally takes too long for us to observe, some examples exist of very rapid evolution. One of them is the colour change in the Peppered Moth in England, another is the rapid development of resistance in insects against insecticides.

During the day, the peppered moth sits on a tree trunk with its wings spread out. It becomes active during the night. Before 1850, most peppered moths found were light grey. It was rare to find a dark specimen (balanced polymorphism). The tree trunks were mostly covered with light coloured lichen which made the grey peppered moth hard to see. Dark specimens were easy to spot and usually eaten by birds. After 1850, industrialisation in England caused a lot of pollution. The lichens on the trees died and the tree trunks were covered with black soot. The light grey peppered moths became easy prey for the birds but the dark ones became more difficult to see. In a short time, the predominant colour of the peppered moth became black.

Another example was found at the beginning of the 20th century. Growers of citrus fruits had a problem with scale insects which feed on citrus trees. They used cyanide gas to kill the pest and were successful. However, in 1914, when they once again used the cyanide, some of the insects survived. They had a new gene which made them resistant to the cyanide. After the spraying, the mutant insects had very little competition and bred in large number, passing on the gene. Eventually so many of the scale insects were resistant to the cyanide that spraying had no effect.

Resistance against certain chemicals can also be found in plants which become resistant to herbicides and in bacteria which become resistant against (a range of) antibiotics. Especially in hospitals, strains of bacteria can be found that are resistant to many antibiotics.

D.4 HUMAN EVOLUTION

D.4.1
State the full classification of human beings from kingdom to sub-species.

© IBO 1996

The full classification of human beings from kingdom to subspecies is:

Kingdom:	Animalia
Phylum:	Chordata
Subphylum:	Vertebrata
Class:	Mammalia
Subclass:	Eutheria
Order:	Primates
Suborder:	Anthropoids
Family:	Hominidae
Genus:	*Homo*
Species:	*Homo sapiens*
Subspecies:	*Homo sapiens sapiens*

D.4.2
Describe the major physical features, namely the adaptations for tree life, which define humans as primates.

© IBO 1996

Major features which describe humans as primates are adaptations to tree life. They are the opposable thumb, acute vision and a large brain.

One of the most important features is the opposable thumb. Being able to move the thumb to a position opposite the fingers, greatly improves the ability to grasp which is important for tree dwellers. It also improves the ability to manipulate objects.

Living in trees means that it is possible to see further. Acute vision, combined with eyes that are placed so that they can see forward is a common trait of primates. Primate have a smaller field of vision which means they need to turn their heads to observe what is going on around them, but their vision is very good. As the field of vision of the two eyes overlaps a great deal, primates are very good at judging distances, which is vital when moving from branch to branch in the canopy of the trees. Colour vision enhanced the chances of finding food.

Primates have a relatively large brain. One can find a dog of the same mass as a rhesus monkey. Yet the monkey's brain will be much larger. This improves the ability to manipulate items and is related to the longer care of the young.

Primates care for their young for a long time. Primate babies are born very helpless. The upright position requires a modification to the pelvis and the birth canal which puts restrictions on the size of the baby's head. As a result babies are born helpless and require a long time of care. This gives them a good chance to learn many skills from their parents.

Primates also developed several different types of teeth. Having more than one type of tooth allows the individual to eat more different kinds of food, both of plant and animal origin.

D.4.3
Discuss anatomical and biochemical evidence which suggests that humans are a bipedal and neotenous species of African ape, that spread to colonise new areas.

© IBO 1996

The direct ancestor of *Homo sapiens* is believed to be *Homo erectus*, the upright human. Homo erectus is believed to have arisen in North Africa approximately 1.5 million years ago. *Homo erectus's* body was similar to our own, the skull was somewhat different (heavier). *Homo erectus* used tools that were purpose made (vs. the stone flakes used by its ancestor *Homo habilis*).

Homo erectus is believed to have originated in North Africa approximately 2 000 000 years ago and to have spread to Asia and Europe. However, it is believed that *Homo sapiens* arose from *Homo erectus* in one place in Africa and from there spread over the world and developed different regional features.

The alternative hypothesis is that *Homo sapiens* arose from *Homo erectus* at various points in the world which caused the differences we see between people today.

To test both hypotheses, biologists have traced the genetic material of the mitochondria. All the genetic material in a person's mitochondria is from his or her mother since the sperm cell only contributes half the genetic material of the nucleus and no other cell organelles. The mitochondral DNA seems to change faster than the nuclear DNA, making it more suitable for research over a short time period (in evolutionary terms).

Studying the mitochondria of different groups op people (Africa, Europe, Asia, Australia,

Option D: Evolution

etc.) the second hypothesis seemed extremely unlikely. It appears that one set of mitochondrial genes was at the origin of *Homo sapiens* and that the owner of these genes lived in Africa. This is based on the fact that there is more variation in mitochondrial DNA in Africa than anywhere else, suggesting that *Homo sapiens* existed in Africa approximately 200 000 years ago and only moved to other parts of the world 100 000 years ago. So *Homo sapiens* in Africa had more time to develop different kinds of mitochondrial DNA.

D.4.4
Outline the trends illustrated by the fossils of *Australopithecus* including *A. afarensis*, *A. africanus* and *A. robustus*, and *Homo* including *H. habilis*, *H. erectus* and *H. sapiens*.

© IBO 1996

Fossils have been found of various of the familiy *Hominidae* leading to several different speculations about their evolutionary relationships.

Those of the genus *Australopithecines* (southern ape) lived from 4 million years ago to 1 million years ago. Their brains were 500 cm^3 or less and they walked upright.

The earliest is believed to be *Australopithecus afarensis (A. afarensis)* (southern ape from the Afar desert) (4.0 - 2.8 million years ago). Remains have been found in Ethiopia and the south of Tanzania. One of the more complete fossil skeletons has been named Lucy. They were small individuals that walked upright. The hands were free and the brains were similar in size to that of a chimpanzee.

A later species is *A. africanus* (southern ape of Africa), from 3 - 2 million years ago. Remains have been found in South Africa. Some argue that *A. afarensis* and *A. africanus* are really the same species as the features are very similar. Both of these species showed bipedalism but did not have large brains. This development seemed to have arisen later.

A. robustus lived from 2.4 - 1.4 million years ago. Remains have been found in South Africa. *A. robustus* was large and heavily built.

Those of the genus *Homo* lived from 2 million years ago. Their brains are 600 cm^3 or more and they also walk upright.

The earliest is believed to be *Homo habilis (H. habilis)* (the handy man, remains often accompanied by simple tools). Although many different hypotheses are formed about the evolutionary relationships between the *Australopithecines* and *Homo* (and between the members of each family), many scientists agree that *H. habilis* probably arose from *A. afarensis*, around 2 million years ago in East Africa.

It is believed that *H. erectus* arose around 1 million years ago. This is likely to have happened in Africa. *H. erectus* migrated to other parts of the world. *H. erectus* had a larger brain than *H. habilis* and used more complex tools.

Approximately 500 000 years ago, *H. sapiens* arose from in Africa. Another wave of migration distributed *H. sapiens* all over the world. The earliest forms are called *H. sapiens neanderthalensis*. They have brains larger than 1000 cm^3. They lived in Eurasia.

Probably separate from the Neandertals, *H. sapiens sapiens* came to Europe from Africa. One of the earliest fossils found of this subspecies is called Cro-Magnon man. The physical resemblance with humans living today is very strong. They are suspected to be the first to use language.

Scientists do not all agree on the precise dates and durations. They also do not agree on which fossils make up a species. One school of thought has *H. heidelbergensis*, *H. neanderthalensis* and *H. sapiens* as 3 different species. Another approach is to consider them subspecies of *H. sapiens*. Neanderthals are then considered to be an example of 'archaic' *H. sapiens* or a subspecies by themselves *(H. sapiens neanderthalensis)* when modern man is called *H. sapiens sapiens*.

The diagram below is one suggestion of how the different species may have evolved. It is not the only explanation.

A PROPOSED MODEL OF HUMAN EVOLUTION

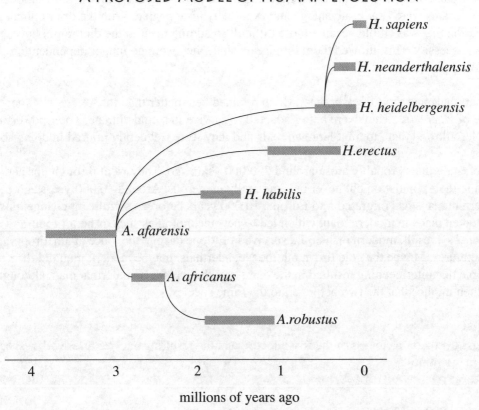

Option D: Evolution

D.4.5
Discuss the possible ecology of these species and ecological changes which may have prompted their origin.

© IBO 1996

Because of footprints found in volcanic ash in East Africa, dating from 3.7 million years ago, we know that the hominids alive at that time walked upright. This is supported by the structure of the skeleton of Lucy (*A. afarensis*) in Ethiopia. Lucy's skelton is more than 3 million years old. Living on the hot African plains, bipedalism meant that less surface area was exposed to the most intense rays of the sun, which helped in keeping cooler.

Lucy's brain was small, supporting the hypothesis that hominids first walked upright before developing a larger brain. As the climate grew cooler and drier, forests were replaced by savannas. Gradually, hominids climbed trees with decreasing frequency. As they were bipedal, this freed their hands to carry tools. The increase in the brain size allowed them to make full use of this. Eventually, *Homo* replaced *Australopithecus* as it was less dependent on trees.

One hypothesis is that the environment of early *Homo* was so diverse, they needed the larger brain to deal with all the different challenges. So individuals with the conventional smaller brains were at a disadvantage and eventually disappeared. With the drier climate, vegetation that was edible became more difficult so adding meat to the diet might have been a necessity. This allowed them to migrate since they were no longer dependent on (familiar) plants.

Neanderthals has short thick bodies, which retained heat better than the longer, slimmer shape of *H. erectus*. They were better adapted to survive in a cold climate. Their powerful muscles allowed them to hunt large animals and they were frequently injured in doing so.

The Neanderthals probably arose around 250 000 years ago. They went through ups and down in their population but never became really numerous. Around 40 000 years ago, modern humans had migrated into Europe. 10 000 years before, a significant change must have taken place in modern man, which led to new technologies such as better tools and shelter. As a result, modern man could survive in Europe despite the, once again, dropping temperatures. Maybe the cold drove out the Neanderthals, maybe modern man did. It is possible that interbreeding resulted in the demise of the typical Neanderthal man, although that then implies that the two belonged to the same species.

D.4.6
Discuss the incompleteness of the fossil record and the resulting uncertainties with respect to human evolution.

© IBO 1996

From 200 000 to 150 000 years ago, only a few Neanderthals survived the cold. As a result, only very few fossils are found. This has happened at other times and places too. Only a very few remains will ever be fossilised so if not many individuals were alive at any one time, the chance of leaving fossils becomes small. And then there is a large difference between leaving fossils and us finding them.

If you look at the appendix, you will find that the 'fossil remains' often consist of a skull and/or a few bones. A find of a limb bone led to some conclusions until other scientists argued that the bone was not human but belonged to a young horse.

D.4.7
Discuss the origin and consequences of bipedalism and increase in brain size.

© IBO 1996

From fossils such as 'Lucy' we know that bipedalism came before the increased brain size. It has been speculated that once our ancestors walked upright, their hands were freed from walking. They could therefore be used for other functions, such as carrying a tool. The increase in the size of the brain did not have a real advantage until the hands were freed from constant use in walking.

D.4.8
State that the evolution of speech and the development of the reflective mind (consciousness) occurred at some time in the Homo lineage.

© IBO 1996

The evolution of speech and the reflective mind (consciousness) occurred at the same time in the *Homo* lineage.

Many scientists have believed for a long time that only modern man had speech. Recent finds suggest that Neanderthal man may also have had the anatomy that allowed speech. They also hunted in groups, taking on large and dangerous animals which required excellent communication. They may have had speech.

One of the ideas associated with the increase of the brain size is to tie it in with the species diet. Brains are 'expensive', i.e. they require a large of amount of energy to operate. *Australopithecus* was mainly vegetarian, which requires large volumes of food to be consumed to fullfil the energy requirements. *Homo*, noted for its larger brain, was eating more meat. Any earlier mutation, having a larger brain, may have found it difficult to supply the brain with adequate amounts of energy.

D.4.9
Describe the origin and main trends in tool making, religion, art, agriculture and technology.

© IBO 1996

The oldest tools found are 2.6 million years old. The use of tools coincided with the increase in brain size. Having larger brains allowed them to work out plans involving tools, having tools allowed them to obtain more food, such as the nutritious marrow of bones left over from a large predator's kill; an easy risk-free source of food.

Whereas the use of tools is often subscribed to *Homo*, no *Homo* fossils have been found that are as old as the oldest tools found. Did *Australopithecus* use tools or is this another example of the incompleteness of the fossil record?

H. habilis is known to have used simple tools, flakes hammered out of pieces of rock having a sharp edge. *H. erectus* used much more sophisticated tools such as a hand axe. The flakes were much thinner and sharper than those used by *H. habilis*.

Neanderthals may have had a religion. They buried their dead with flowers and tools which could suggest that they believed in an afterlife. Neanderthals did not leave much evidence of art but some beads were found that must have taken a long time to make and had only a decorative function.

Cro Magnon (Modern Man) painted drawings of animals on the walls of their caves. They had elaborate art which possibly was related to a form of religion.

Cultural evolution (see below) eventually led to domestication of animals and agriculture around 10 000 years ago.

D.4.10 & D.4.11
Outline the difference between genetic and cultural evolution.
Discuss the relative importance of genetic and cultural evolution in the evolution of humans.

© IBO 1996

Genetic evolution involves the change of genetic material which is subsequently passed on. The change is random and whether the change is an improvement is dictated by the environment. The skills acquired are not passed on through the genes so the child of a great hunter would not know how to hunt unless it was taught.

Cultural evolution is the accumulation of useful skills and knowledge, and the discarding of harmful practices, passed down through thousands of human generations. It is based on the fact that we have elaborate language skills. Using language, accumulated experience can be passed from one generation to another. Using written language, we can even learn/ benefit from the experiences of a person we never met. The impact of this is incredible.

D.5 NEO-DARWINISM

D.5.1
State that mutations are changes to genes or chromosomes due to chance, but with predictable frequencies.

© IBO 1996

Mutations are changes to genes or chromosomes due to chance, but with predictable frequencies. Some of these are discussed below.

D.5.2
Outline phenylketonuria (PKU) and industrial melanism as examples of gene mutation and Klinefelter's syndrome as an example of chromosome mutation.

© IBO 1996

Gene mutations.
The well documented change in colour of the Peppered Moth has already been referred to earlier. The colour of *Biston betularia* is determined by the alleles present for one gene. Originally, there was a balanced polymorphism, with the dark allele being present rarely (as it was selected against). During and after the industrial revolution, the trees had fewer lichens growing on them and the selection started to favour the darker form of the species. So the percentage of the dark allele in the population increased (transient polymorphism). Since the airpollution is decreasing, natural selection now once again favours the light coloured morph and the allele for this is increasing.

PKU (**phenlyketonuria**) is a genetic disease caused by the presence of a homozygous recessive allele. A PKU individual cannot produce a certain enzyme. As a result, phenylalanine (an amino acid) cannot be changed into tyrosine (another amino acid). Therefore phenylalanine levels build up which is harmful to the brain (and some other organs). PKU individuals will become brain damaged unless they are given a special diet which contains little phenylalanine. Once adult, they can eat normally. This is why, in many countries, all babies are routinely tested for PKU so that they can be given the proper diet if necessary.

Both of the above examples are gene mutations. At some point in time, the 'normal' allele mutated and a new allele was created. The new allele (causing the dark moth or the PKU child) was not favourable but some individuals survived and passed on the allele.

It is also possible to have a deviation in the number of chromosomes rather than a change in the basepairs of one allele. These are called chromosome mutations and two examples are Down syndrome and Klinefelter syndrome.

Down syndrome is caused by a non disjunction during gamete production. It could take place both in oogenesis and in spermatogenesis although it is more common in females. The resulting gamete will have 2 copies of chromosome 21 (rather than one copy) and after fusion with a gamete of the other gender, the zygote will have a total of 47

Option D: Evolution

chromosome, 3 copies of chromosome 21. This causes the child to be retarded but the degree of the affliction is hard to predict.

If non disjunction occurs in the sex chromosomes during oogenesis, an XX gamete is produced. If this gamete fuses with a Y sperm cell, the resulting individual will also have 47 chromosomes, 3 sex chromosomes XXY. The individual will be a sterile male with some female traits (breasts) and sometimes reduced mental abilities. This is called **Klinefelter syndrome**.

D.5.3
Explain that variation in a population results from the recombination of alleles during meiosis and fertilisation.

© IBO 1996

Children of 2 parents are different from each other. Even sisters (or brothers) are not identical, unless they are identical twins.

During the process of gametogenesis, the chromosomes line up in pairs during Metaphase I (see section 10.1). Of each pair of chromosomes, one originates from the individual's father, the other from the mother. The process of pairing up is, of course, not random. Only homologous chromosomes pair up. Yet, the way they line up on the equator is random.

In the diagram below, it is shown how different gametes arise. The diagram only shows the cells after Meiosis I.

INDIVIDUAL'S CHROMOSOMES

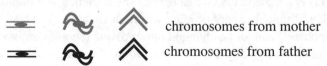

chromosomes from mother
chromosomes from father

Possible outcomes of meiosis I

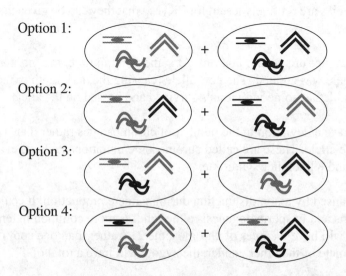

Biology

The diagram above shows just one aspect of variation caused by sexual reproduction. Another aspect is the fusion of the gametes. As both parents produce a number of genetically unidentical gametes, the amount of variation is increased as any of the gametes has an equal chance of fusion with any of the other parent's gametes. This greatly increases variation.

The third aspect of variation in sexual reproduction is crossing over (section 10.3). During crossing over, sections of homologous but non identical chromosomes break off and re-attach to the other chromosome. This way, any of the chromatids in the chromosomes above, could be partly striped (from the mother) while the rest would be dark (from the father). As a result, the chromatid would no longer be identical to its mate. Then the result would be 4 different gametes from 1 cell rather than 2 sets of identical gametes. Again, an increase in variation.

D.5.4 & D.5.5
Describe natural selection as leading to the increased reproduction of individuals with favourable variations.
State that adaptations (or micro-evolutionary steps) may occur as the result of an allele frequency increasing in a population's gene pool over a number of generations.

© IBO 1996

The above described ways of variation are all non-directional. That means that every change has an equal chance of occurring. The environment determines if the change is favourable or not. If the change is favourable, the individual is likely to grow bigger, obtain more food, have the best shelter etc. As a result, this individual is likely to pass on its genes because it is likely to produce offspring. Some of the offspring will inherit the favourable combination of traits (or mutation). They will be bigger, stronger, etc and have many offspring. This way, a new allele (from a mutation) can become more frequent in a population as time goes by.

D.5.6
Describe how the evolution of one species into another species involves the accumulation of many advantageous alleles in the gene pool of a population over a period of time.

© IBO 1996

A population in a habitat will interbreed, spreading the alleles over the offspring. If something happens and half the population move away to another habitat, they would no longer interbreed. Due to the different situation in another habitat, selection might favour a different set of characteristics than the original population. Each population may also undergo unique mutations, due to chance. After some time, the two populations could be so different that they could no longer interbreed to produce fertile offspring. They have become different species.

D.5.7
Discuss ideas on the pace of evolution including gradualism and punctuated equilibrium.

© IBO 1996

Option D: Evolution

Evolution is a slow process. In most cases it is too slow for us to witness in a life span or in a few generations using written or spoken records. So most of our ideas about evolution are supported by fossil records. These are incomplete because fossils are not formed under all conditions.

Darwin thought evolution to be a gradual process, a series of minor changes which, over time, led to a distinct difference between the individual and its ancestors. There is a lot to support this idea, although the fossil records do not always help (for reasons given above). Lately, a new idea has come up which, to a certain extent, explains the inconsistencies of the fossil record. In 1972, **N. Eldredge** and **S. Gould** suggested that evolution may occur in a short periods of rapid change, followed by long periods of no change.

The idea is that a large population which experiences different selection pressures will probably not change much. However, a small population, especially one that experiences a new environment, could undergo rapid changes due to selection pressure in a certain direction. This is called **punctuated equilibria**.

It is known that some species during certain times have evolved gradually, e.g. mammals in Africa. Others seem to follow the punctuated equilibria model. It is possible that the times of rapid change are caused by meteor impacts and/or volcanic eruptions which caused climatic changes.

D.5.8
State that a species is part of a potentially interbreeding population having a common gene pool.

© IBO 1996

A **species** is a part of a potentially interbreeding population having a common gene pool. (See section 13.1.).

For example, domestic dogs are of the same species and can interbreed. A domestic dog cannot, however, breed with a domestic cat. They are of different species.

D.5.9
Discuss the process of speciation in terms of migration, geographical or ecological isolation and adaptation leading to reproductive or genetic isolation of gene pools.

© IBO 1996

The process of **speciation** can be understood by thinking of the several important steps that occur. If a number of individuals of a population migrate to another habitat, geographical isolation is likely to occur. The small migrating group may have allele frequencies that are different from those found in the original population, which could contribute to a different genetic makeup of the new population (the founder effect). In addition, the new habitat will differ from the original habitat causing directional selection which will also increase the pace of evolution. All this may result in the two populations eventually becoming different species.

Ecological isolation may occur if populations do not use all of their ecosystem. The ecosystem may be divided into different habitats (which may be difficult for us to distinguish). If two populations use different habitats, they may not interbreed and undergo changes independently of one another. Eventually this could also lead to two different species. An example of this difficult to recognise process is found in the checkerspot butterfly in California. What appears to be one population is found on a large area of grassland. After years of careful studies, it has become evident that the area is divided into three sections, each supporting their own, isolated, population.

D.5.10
Describe an example to support macro-evolution (speciation) including ring species.
© IBO 1996

Many populations of a species may be found dispersed over a large area. Often the phenotypes of two populations which are far removed from each other are somewhat different. A **cline** is a gradual change in the frequency of the phenotypes over the geographical range of a species. A special type of cline is found in the **ring species**. Ring species form a circle around the North pole so that the 'ends' of the cline will meet again.

The herring gull is a ring species around the North Pole. Seven different populations are recognised, living around the North Pole. Each population will interbreed with its neighbours on either side, indicating that they cannot be considered as different species. However, the two 'ends' of the cline meet in Northern Europe. Here, the two populations (herring gull and lesser black-backed gull are found in the same area, sometimes nesting side by side. They normally do not interbreed, suggesting that they are different species.

D.6 THE HARDY-WEINBERG PRINCIPLE

D.6.1
Describe an adaptation in terms of the change in frequency of the alleles of a gene.
© IBO 1996

The adaptation of the Peppered Moth from the light coloured morph to the melanic morph has been discussed earlier. Before 1850, the melanic (dark) form was very rare but gradually became more common. The idea is that the industrial revolution caused so much pollution that it destroyed the light coloured lichens growing on trees. The light coloured moth, sitting on the tree, was no longer camouflaged and directional selection favoured the melanic form. Since efforts have been made to reduce pollution, the light coloured form is becoming more common.

In 1959, a study on the two morphs of the Peppered Moth started. 94% of the moths found were dark. This number gradually decreased. In 1969, it was around 90%, in 1979 around 79% of the moths were dark, and in 1989 only about 40%. In 1994 only 19% of the moths captured were dark.

Option D: Evolution

D.6.2

Explain how the Hardy-Weinberg equation: $(p^2 + 2pq + q^2 = 1)$ is derived.

© IBO 1996

In 1908 **Hardy** and **Weinberg** came up with a mathematical model for allele frequencies for a gene with 2 alleles.

Dominant allele is A and its frequecy is p (a number between 0-1)
Recessive allele is a and its frequency is q (a number between 0-1)

A gene must have an allele. The options are A or a. So if the allele is not A then it must be a. If a is not present, the frequency for A is 1. So, in that case, $p = 1$. If A is not present, then $q = 1$. Whatever the frequencies, $p + q = 1$, since the allele is either A or a.

Although this is an interesting concept, the real value of this observation came from the subsequent work of Hardy and Weinberg.

If the frequency of A is p, then the frequency of AA is p^2.
If the frequency of a is q, then the frequency of aa is q^2.
If the frequency of A is p and the frequency of a is q, then the frequency of Aa (which really is Aa + aA) is $2pq$.

Since any genotype is AA or Aa or aa, it means that

$$p^2 + 2pq + q^2 = 1 \quad \text{Hardy - Weinberg equation}$$

D.6.3

Calculate allele, genotype and phenotype frequencies for two alleles of a gene, using the Hardy-Weinberg equation.

© IBO 1996

In a certain population of *Drosophila* (fruit flies), 64 individuals are found to have red eyes (wild type) and 36 are found to have white eyes.

Find the allele frequency for each allele and the genotype and phenotype frequencies.

Total population is 100.

36 white eyed flies, i.e. $q^2 = \dfrac{36}{100}$, therefore $q = 0.6$

Since $p + q = 1$ and $q = 0.6$, $p = 0.4$

So the genotype frequencies are:
AA = p^2 = 0.4 × 0.4 = 0.16
Aa = $2pq$ = 2 × 0.4 × 0.6 = 0.48
aa = q^2 = 0.6 × 0.6 = 0.36

The phenotype frequencies can be found directly from the information.

- red eyed: $\frac{64}{100} = 0.64$

- white eyed: $\frac{36}{100} = 0.36$

D.6.4
State that the Hardy-Weinberg Principle can also be used to calculate allele, genotype and phenotype frequencies for genes with three or more alleles.

© IBO 1996

It is also possible to use the Hardy-Weinberg principle for genes which have 3 or more different alleles.

D.6.5
State the Hardy-Weinberg Principle and the conditions under which it applies.

© IBO 1996

The Hardy-Weinberg principle can be used to predict frequencies under certain conditions :
- large population.
- random mating occurs.
- no directional selection.
- no allele specific mortality.
- no mutations.
- no immigration or emigration.

D.6.6
Describe one exantple of transient polymorphism, and sickle cell anaemia as an example of balanced polymorphism.

© IBO 1996

If the above condition are met, the population has two different morphs. It is then said to show balanced polymorphism. Sickle cell anaemia is an example of balanced polymorphism. The incidence of the sickle cell allele in, for example, West Africa is relatively high. Despite the fact that the homozygous recessive individuals are in ill health (sickle cell anaemia) and are less likely to reproduce and pass on the allele, the heterozygous individual shows little signs of the disease. This individual has an advantage in being less susceptible to malaria which is common in West Africa. This combination of factors has caused the allele for sickle cell anaemia to be more common in West Africa than in other, malaria free, parts of the world.

Industrial melanism in the Peppered Moth is an example of transient polymorphism. Due to directional selection which first favoured the dark morph and later the light morph, the

Option D: Evolution

percentage of dark vs. light coloured moth has dramatically changed in only a few decennia (see D.6.1).

D.6.7
Describe how heterozygosity is advantageous for a gene pool.

© IBO 1996

Heterozygosity (being heterozygous) is an advantage for the gene pool. Having many individuals with many heterozygous traits, maintains recessive alleles in the gene pool. In the case of sickle cell anaemia in West Africa, the advantage of the heterozygous over the homozygous (dominant or recessive) is clear. Related individuals often share the same heterozygous genes. Inbreeding will therefore reduce the amount of heterozygosity which often results in 'inbreeding depression'. It is well known that conservationists are often worried about the effect of inbreeding in a protected species of animals when the number of individuals is low. Its effects have been seen in breeding programmes such as pedigree dogs. The reverse is also true where plant breeders often find that the hybrid F_1 is easier to grow because of 'hybrid vigour'.

EXERCISE

1. a. Who first suggested the endosymbiont theory?
    b. Outline the main concept of this theory.
    c. Give 4 arguments to support the endosymbiont theory.

2. a. Who suggested the inheritance of acquired characters?
    b. Outline what is meant by this concept.
    c. Outline 2 examples which show that this concept does not apply.
    d. Discuss the genetic background to this concept.

3. At an archeological site, some organic material is found. It is suspected it may be very old. When carbon dating is used, the organic material is found to contain less than one percent of the original amount of ^{14}C. The radioactive half life of ^{14}C is 5730 years.
    a. Plot a decay curve for ^{14}C.
    b. Use the curve to estimate how old the material must be.

4. The evolution of humans is still hotly debated. Many models of human evolution have been made. Some of the ones that are, or have been, more common are given below.

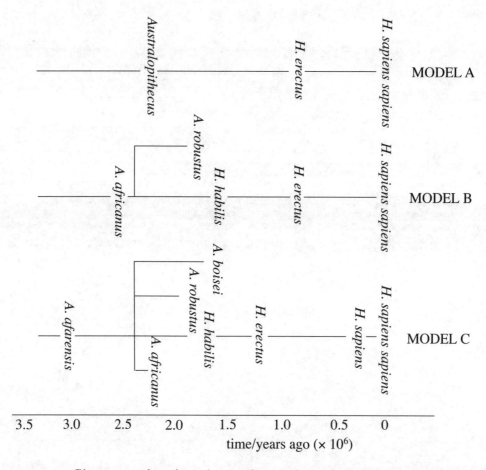

a. Give one explanation why *A. afarensis* is not included in model A or B.
b. What is one of the main differences between the *Australopithecus* genus and the *Homo* genus?
c. Why is it more likely that there were two waves of migration from Africa : one by *H. erectus* and later one by *H. sapiens*?

5. Refer to the diagram in section D.5.3.

a. Given the genetic make up of the individual with 6 chromosomes, how many different gametes could this individual make without crossing over?
b. If you had an individual with 4 chromosomes, how many different gametes could it produce without crossing over?
c. How many different gametes could a human produce without crossing over?

6. In a population of *Drosophila* (fruit flies), the frequency of the recessive allele for vestigial wings is 30%. Predict how many flies have normal wings in a population of 125.

Option D: Evolution

OPTION E: NEUROBIOLOGY AND BEHAVIOUR

21

Chapter contents
- Introduction and examples of behaviour
- Perception of stimuli
- Innate behaviour
- Learned behaviour
- Social behaviour
- The autonomic nervous system
- Neurotransmitters and synapses

Option E: Neurobiology and Behaviour

E.1 INTRODUCTION AND EXAMPLES OF BEHAVIOUR

E.1.1
State that behaviour is the response of animals to the stimuli they receive.

© IBO 1996

Behaviour is the response of animals to the stimuli they receive. For example, animals pull away when they touch a hot object.

E.1.2
State that innate behaviour arises as part of normal development whereas learned behaviour is influenced by conditions experienced during development.

© IBO 1996

Innate behaviour arises as part of normal development whereas learned behaviour is influenced by conditions experienced during development. For example, feeding is a part of the innate behaviour of all animals. However, captive animals often learn that they will be fed immediately after their owner returns from work, has a shower etc. When they receive these signals, they can become excited, salivate etc. Much of human behaviour, such as driving, a car is learned.

E.1.3
Explain the role of natural selection in the development of behaviour patterns.

© IBO 1996

Innate behaviour patterns (instinct) are genetically inherited and stereotyped responses to environmental stimuli. The behaviour patterns are adaptive and suit the organism to its environment. Possessing a cerain gene makes it more likely that a specific behaviour pattern will develop. This is the basis of the role of natural selection.

Hedgehogs have the innate behaviour pattern of rolling up into a ball when they feel threatened. Being nocturnal, many hedgehogs are killed at the roads at night. They cross a road, a car approaches, the hedgehog rolls up and is run over.

There is some evidence that some hedgehogs have a different pattern of behaviour. When a car approaches, they quickly run to the side of the road and hence are not killed. If this behaviour is genetic, the hedgehogs with the allele(s) which dictate this behaviour are likely to survive longer and hence produce more offspring. The selection process would favour the hedgehogs with the alleles for running away.

E.1.4
Explain, using species of birds or mammals (other than human), one example of each of the following types of behaviour: migration, grooming, communication, courtship and mate selection, and parental care.

© IBO 1996

Five categories of behaviour are considered to be very important for most animals:

a. migration.
b. grooming.
c. communication.
d. courtship and mate selection.
e. parental care.

- **Migration.** Many animals migrate over shorter or longer distances. Ducks and geese are well known examples of migratory birds. Often the areas that are preferred for breeding are not desirable during other times of the year (winter). Hence migration. Invertebrates also migrate. Monarch butterflies from the eastern part of the USA and Canada migrate to Mexico to overwinter and return in spring. However, the lifespan of the butterfly is such that only a few individuals make the return journey and it is mostly the next generation(s) that complete the trek. Since migration might involve long distances, it is important to have some system of orientation. The sun and the stars are used but so also are the Earth's magnetic field and, for marine animals, possibly wave action.

- **Grooming.** Grooming has one obvious function. It keeps the coat clean and removes parasites. Baboons and chimpanzees are two species where grooming also has a social role: it reinforces the relationships within the group. Not only primates groom: silver backed jackals are an example of other groups in which grooming plays a role.

- **Communication.** Social animals have ways of communicating. Communication can take place with sounds, movements, colour, and scents. A warning call may alert the members of the group to danger. Showing teeth in dogs and wolves indicates a threat, while rolling over indicates submission. A direct stare followed by a frown challenges a male baboon.

- **Courtship and mate selection.** During courtship, animals produce specific signals for communications. The signals may serve to ensure that only individuals of the same species mate (e.g. specific blinking pattern in each species of fire flies) or to attract the attention of the opposite sex (e.g. display of colourful feathers in the peacock). The signals may be visual as colour and/or movement or involve smells (pheromones). Mate selection is based on some characteristic of the mate which suggests that offspring from this mate will have a greater chance of survival than that of others. If the mate is large, the offspring may inherit this trait and be able to be at an advantage in intraspecific competition. If the mate has a large territory, the young may have the most food and therefore the best start to life. A high status may also be passed on to offspring in a society. Often, but not always, the males will display the courtship signals and fight for the females. It is to the advantage of the female to wait for the outcome of a fight and then go with the winner.

- **Parental care.** One reproductive strategy is to produce large numbers of offspring hoping that some of them will survive. Another strategy is to produce limited numbers of offspring and to take care of them so that most will survive. Generally, the greater the parental care, the fewer offspring. Birds and primates both have high levels of parental care, involving both parents.

Option E: Neurobiology and Behaviour

E.1.5
Explain the need for quantitative data in the studies of behaviour.

© IBO 1996

It is very easy, in the study of animal behaviour, to go into purely descriptive studies and draw conclusions from them. However, the study of animal behaviour needs to follow the same rules as other scientific research. If, for example, you want to show that one difference in the environment will cause a significant difference in behaviour, you should do a statistical analysis.

In the study of animal behaviour it is common to use codes for certain kinds of behaviour. For example, 1.1 is standing and eating, 3.0 is sleeping. Behaviour can be recorded either every interval (1 minute or 5 minutes) or by using a computer to type in the new code when the behaviour changes and having the computer calculate the duration of each behaviour after a certain interval (1 hour or 5 hours).

Comparing two groups, it is possible to use a statistical analysis to show a significant difference in certain behaviours.

E.1.6
Analyse data relating to any of the above.

© IBO 1996

Any results from experiments can be used. For example a choice chamber (damp/dry or dark/light) and some insects (woodlice).

To set up an experiment of this type, place some woodlice in an enclosure that provides two types of environment. For example, light and dark. Observations are then made at regular intervals of the numbers of individuals in the light and the number in the dark. These are then analysed numerically using, for example, significance testing.

Some examples of this sort of study were discussed in Chapter 10.

E.2. PERCEPTION OF STIMULI

E.2.1
State that sensory receptors act as energy transducers.

© IBO 1996

Sensory receptors act as energy transducers. State that sensory receptors act as energy transducers. (A transducer is a device for converting a non-electrical signal into an electrical one.) The result is an action potential. See section 14.1.

An example is the conversion of light energy into electrical signals by the cells in the retina in the eye.

E.2.2
State that human sensory receptors are classified as mechanoreceptors, chemoreceptors, thermoreceptors, electroreceptors or photoreceptors.

© IBO 1996

Sensory receptors are classified as
a. chemoreceptors.
b. electroreceptors.
c. mechanoreceptors.
d. photoreceptors.
e. thermoreceptors.

These terms are considered in the next section.

E.2.3
Describe what is meant by each of the terms in E.2.2 with reference to one named example of each classification of receptor.

© IBO 1996

Chemoreceptors have special proteins in their membranes. These proteins can bind to a particular substance and this will result in a depolarisation of the membrane leading to an action potential being sent to the brain. Chemoreceptors are responsible for our sense of smell and taste but also detect the blood pH.

Electroreceptors are found, for example, in sharks. Muscle contractions generate electrical fields which are conducted by the water. The shark can sense these fields with its electroreceptors and detect its prey. As air is a poor conductor, terrestrial organisms would not have much use for this sense organ.

Mechanoreceptors are those that are sensitive to some kind of movement. In fish, it is found in the 'lateral line system', which detects vibrations in the environment.
We have a similar system in our inner ear to inform us of our body's position and

Option E: Neurobiology and Behaviour

movement. Three fluid filled semi circular canals are connected to the area. At the end of the canals, as well as in the common area, we find a system of hair cells. Any change in speed or direction will move the fluid in at least one of the semicircular canals, bending the hairs. This causes an action potential to be sent to the brain.

Photoreceptors are the rods and cones in our eyes (section E.2.4). Rods and cones contain photopigments that are broken down when exposed to light. This causes the cells to send an action potential to the brain.

Thermoreceptors are found, for example, in the skin. Cold receptors, just under the surface of the skin, will send an action potential when the temperature drops. Warm receptors, located a little deeper, will send an action potential when the temperature increases. The temperature centre in the hypothalamus of the brain also contains thermoreceptors to monitor the temperature of the blood.

E.2.4
Draw the structure of the human eye.

© IBO 1996

In the eye, the retina contains the light sensitive cells. All the other structures shown in the schematic diagram below, help to project a sharp image onto this retina.

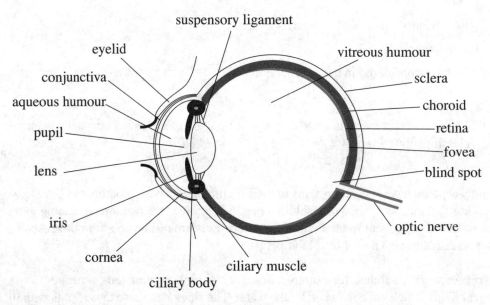

Structure and function of the eye.
The eye is protected inside a bony socket of the skull and is moved by a set of 6 muscles. Around the eye, we find eyelids and tear glands to keep the eye moist and clean, eyelashes to keep dust out and eyebrows as a protection against sweat running down from the brow. The eye is supported by hydrostatic pressure of the aqueous and vitreous humours.

conjunctiva:	thin transparent layer continuous with the epithelium of the eyelids.
cornea:	transparent front of the sclera; the curved surface is very important in refracting the light towards the retina.
aqueous humour:	clear solution of salts.
pupil:	variable opening in the iris to allow light to enter the eye.
lens:	transparent, elastic bi-convex structure which focusses the light onto the retina.
iris:	coloured part of the eye; circular and radial muscles which control the size of the pupil.
ciliary body:	contains blood vessels and ciliary muscle.
ciliary muscle:	bundle of circular muscles which control the shape of the lens.
suspensory ligaments:	attach the lens to the ciliary muscle.
vitreous humour:	clear gelatinous substance which fills the eyeball.
sclera:	white protective covering of the eye.
choroid:	black layer which prevents internal reflection of light and contains blood vessels to supply the retina.
retina:	contains rods and cones and nerve cells for vision.
fovea:	'yellow spot' contains cones only, spot of most accurate vision.
blind spot:	point where optic nerve leaves the eye; not light sensitive.
optic nerve:	carries impulse to the brain.

Pupil size.
In bright light, the circular muscles of the iris contract and the pupil becomes smaller. This reduces the amount of light entering the eye and prevents damage to the retina. In poor light, the radial muscles relax, increasing the size of the pupil and the amount of light entering the eye.

Thickness of the lens.
Light reflecting from an object far away, enters the eye as parallel rays. Light reflecting from a near object, comes in as diverging rays which need to be refracted more to be focussed on to the retina. Hence in the latter case, the ciliary muscles contract, cease pulling on the suspensory ligaments and allow the lens to attain its own more convex shape.

Option E: Neurobiology and Behaviour

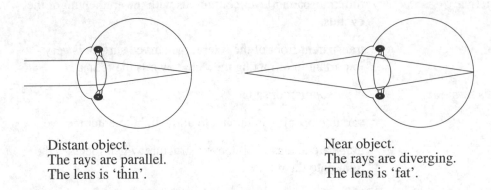

Distant object.
The rays are parallel.
The lens is 'thin'.

Near object.
The rays are diverging.
The lens is 'fat'.

With age the lens becomes less elastic and it grows progressively more difficult to focus on near objects. Around the age of 45-50 most people need glasses to read.

Long sight is caused by the eyeball being to short or the lens not being elastic enough. Close objects cannot be seen clearly: the light is focussed behing the retina. Converging lenses in glasses will help the lens.

Short sight is cause by the eyeball being to short or the lens being too elastic. Distant objects cannot be seen clearly: the light is focussed in front of the retina. Diverging lenses in glasses will compensate for this.

E.2.5
Draw the structure of the human retina.

© IBO 1996

Structure and functioning of the retina.

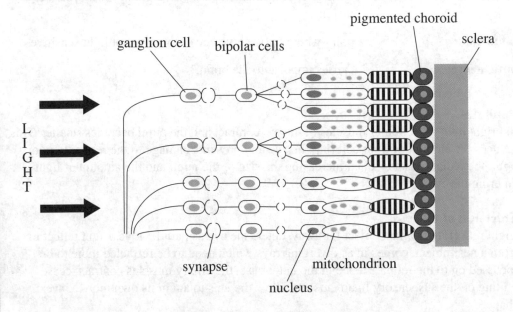

E.2.6
Outline the principle of trichromacy in relation to colour vision.

© IBO 1996

The basis of human colour vision is the fact that we have 3 different kinds of cones. The rods contain the pigment **rhodopsin** which is broken down by light. This will generate an action potential to the brain. The cones have one of 3 similar pigments (called **iodopsins**). They absorb light of a certain colour, red, green or blue. In the same way as in the rods, this will break down the pigment and cause an action potential to the brain.

A certain amount of overlap in the absorption of colours exists, especially in red and green light which may trigger responses of both kinds of cones.

If white light hits the fovea, all three kinds of cones will send equally strong messages to the brain. The brain will interpret this as white light. If blue light hits the fovea, the 'blue' cones will send a strong signal to the brain. The other will send little or no signal. The brain will interpret this as 'blue'. So the initial discrimination of colour occurs in the retina but the final colour perceived involves interpretation of action potentials by the brain.

E.2.7
Explain briefly how visual stimuli are perceived and processed in the retina including the roles of rods and cones, pigments, diffuse and monosynaptic bipolar cells, neurons of the optic nerve and visual cortex.

© IBO 1996

The light enters the eye and is refracted by the cornea and the lens. It passes through the clear vitreous humour and then reaches the retina. The light will have to pass between the nerve cells (ganglion cells and bipolar neurons) to reach the rods and cones which form the layer of light sensitive cells.

The rods and cones contain photosensitive pigments which is broken down under the influence of light. This causes one or more impulses to be sent to the brain. In the absence of light, the pigments are rapidly reformed.

The cones are mostly found in the fovea. Cones contain 3 different pigments which are sensitive to different colours of light. Cones are individually linked to the bipolar neurons and are therefore less sensitive to lower light intensities but give a more accurate picture in bright light.

Rods are found all through the retina except in the fovea. The pigment in rods is called rhodopsin or visual purple. Several rods are linked to one bipolar neuron. This means that their impulses are 'added up' and this explains their higher light sensitivity. However it also reduces accuracy.

When the light has caused an action potential in the rods or cones, it is passed on to the bipolar neurons. Here, the presence of action potentials of other cells may inhibit or

Option E: Neurobiology and Behaviour

further excite the bipolar cell. An action potential may then be passed on to the ganglion cell and from here to the optic nerve.

Action potentials travel through different fibers of the optic nerve, depending on the place in the retina from which they originate. Some of the fibres of the optic nerve of the right and left eye will cross in the optic chiasm, others, the ones from the 'outside' of the retina do not. As a result the complete picture of the left object is transmitted to the right side of the brain and vice versa.

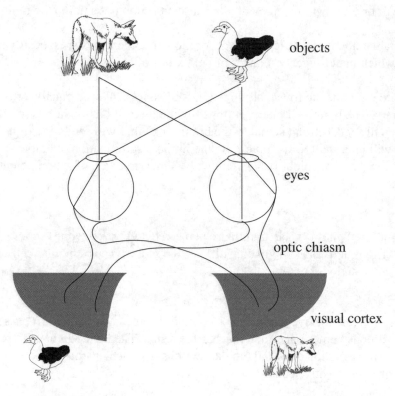

E.3 INNATE BEHAVIOUR

E.3.1
Define innate behaviour.

© IBO 1996

Innate behaviour: behaviour which normally occurs in all members of a species despite natural variation in environmental influences. Most migration in birds is an example (see section E3.11).

E.3.2
Outline three examples of human spinal reflexes including the pain withdrawal reflex.

© IBO 1996

Spinal reflexes are reflexes that involve the spinal cord (and not the brain). They are part of the innate behaviour. They are very simple forms of response, involving only two or three nerve cells.

One example of a spinal reflex is the knee jerk reflex. When someone taps your knee so that the tendon is stretched, the stretch receptor in the muscle (the receptor) will send an action potential to the spinal cord. Here it is passed to an effector neuron which will make the muscle contract. As a result, your lower leg moves forward.

See the diagram in section E.3.3. The knee jerk spinal reflex does not involve an association neuron, the receptor (sensory) and effector (motor) neuron are directly connected.

Another example is the withdrawal reflex caused by pain. If you prick your finger, the receptor neuron will send an action potential to the spinal cord. It again is relayed (via an association neuron) to an effector neuron. The axon of this particular effector neuron will end in the biceps of the upper arm. Contracting the biceps causes the finger to be drawn away from the pain causing object. See diagram in section E.3.3.

The main advantage of having reflexes is that they are fast. If you are burning your finger, you do not want to take a lot of time over processing the information and deciding on a course of action. The other advantage of reflexes is the simple adjustment of the body to changing situations which do not require any sophisticated thinking.

E.3.3
Draw the structure of the spinal cord and its spinal nerves, to show the components of a reflex arc including receptor, effector and association neurons.

© IBO 1996

The structure of the spinal cord and its spinal nerves can be seen in the diagram below. The spinal reflex arc as described above also involves the second diagram below.

Option E: Neurobiology and Behaviour

SCHEMATIC DIAGRAM OF A CROSS SECTION OF THE SPINAL CHORD

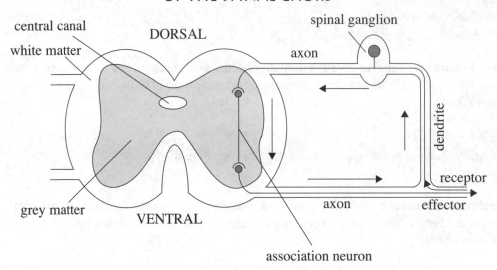

Receptor neurons are often called sensory neurons, effector neurons are often called motor neurons.

THE 'KNEE JERK' REFLEX

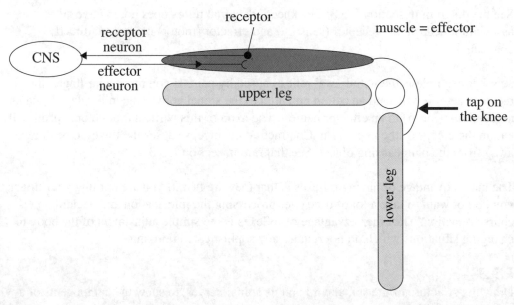

Spinal nerves are mixed; dorsal roots contain sensory neurons and ventral roots contain motor neurons.

E.3.4
Outline three cranial reflexes including the pupil and the Hering-Breuer reflexes.

© IBO 1996

Biology

The pupil reflex is a cranial reflex. When bright light is perceived, the iris will immediately contract, reducing the amount of light upon the retina so that it will not be damaged. This is an example of a cranial reflex.

Another example is the blinking reflex. When an object comes close to your eye, you will blink or close your eye to avoid or reduce damage if the object were to hit you. This reflex can be 'undone'. Student A will put on a pair of safety goggles. Student B will (gently) toss of ball of loosely scrunched up paper in her face. Student A will blink even knowing that the paper ball will not hurt her eyes because s/he is wearing goggles. If the experiment is repeated several times, student A can learn to go against his/her reflex and not blink. In this case, learned behaviour interacts with a reflex.

The **Hering-Breuer reflex** is a way to protect against overinflation of the lungs. The stretch receptors in the walls of the bronchi and bronchioles will send an action potential via the vagus nerve to the respiratory centre in the medulla oblongata. This will inhibit the inspiratory areas and stop further inspiration of air. Expiration (a passive process) will follow.

E.3.5
Draw the gross structure of the brain, including the medulla oblongata, cerebellum, hypothalamus, pituitary gland, and cerebral hemispheres.

© IBO 1996

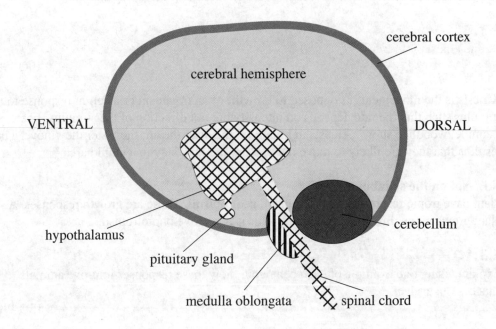

SCHEMATIC DIAGRAM OF THE GROSS STRUCTURE OF THE BRAIN

Option E: Neurobiology and Behaviour

E.3.6
State that 12 pairs of cranial nerves are connected to the human brain.

© IBO 1996

Twelve pairs of cranial nerves are connected to the human brain. They are involved in, for example, moving the eyeball and the sense of smell.

E.3.7
Outline the neural pathways involved in the pupil reflex and describe how this reflex is used to test for brain stem death.

© IBO 1996

The pupil reflex is the constriction of the pupil (caused by relaxation of the muscles in the iris) due to bright light coming in. As the retina could be damaged by too much light, this reflex is important.

The brainstem is responsible for this reflex. The absence of a pupil reflex can be caused by the brainstem no longer functioning properly.

E.3.8
Define taxis.

© IBO 1996

Taxis is the locomotion of an organism in a particular direction in response to an external stimulus (e.g. positive phototaxis to light in *Euglena* is a positive chemotaxis to gradients of dissolved substances such as predators to the chemicals their prey excrete).

E.3.9
Define kinesis.

© IBO 1996

Kinesis is the movement (as opposed to growth) of an organism or a cell in response to a stimulus such that the rate depends on intensity but not direction of the stimulus. For example, woodlice show a kinesis to humidity: the drier the air, the faster they move. The result is that they are likely to move out of a dry area but stay in a humid area.

N.B. Not on the syllabus.
Plant have tropic responses (phototropism, geotropism). These are growth responses. A plant grows in the direction of the light. This is a positive phototropism.

E.3.10
Explain, using one example of each behaviour, how these responses improve animals' chances of survival.

© IBO 1996

Euglena needs light to photosynthesise. A positive phototaxis helps the organism to find light and allows it to photosynthesise.

Woodlice have gills to assist in gaseous exchange. They would be damaged in dry conditions. By moving faster when the air is dry, they are likely to get out of these conditions.

E.3.11
Discuss the importance of innate behaviour to the survival of animals.

© IBO 1996

Innate behaviour is sometimes called instinctive behaviour. The organism does not have to learn it (it is genetically progammed) and usually shared by all member of the species.

Innate behaviour can be vital for survival. Kittiwakes (a sea bird) build their nests on narrow ridges. When the young hatch, they do not move about much. If they did, they would most likely fall off the ridge. The young do not have a chance to learn this behaviour; having it as innate behaviour has survival value.

Innate behaviour is suitable in situations where:
- the stimulus is always the same.
- it is important to respond quickly.
- it is disastrous when the response is wrong.

Learning is more suitable for:
- changing situations.
- social organisms (learn from others).
- animals that live a long time (will encounter many different situations).

Option E: Neurobiology and Behaviour

E.4 LEARNED BEHAVIOUR

E.4.1
Define conditioning.

© IBO 1996

Conditioning is the modification of behaviour in an animal as a result of detection of correlations between external events.

E.4.2
Outline the experiments of Ivan Pavlov into conditioning of dogs.

© IBO 1996

In the 1920, the Russian psychologist **Ivan Pavlov** did the first studies into conditioning. Dogs, like most mammals, will start to produce saliva at the smell or taste of food. When Pavlov for some time rang a bell before feeding the dog, eventually the dog began to produce saliva at the sound of the bell. Pavlov called the food the unconditional stimulus. The bell became the conditional stimulus: the dog had been conditioned to associate it with food.

E.4.3
Define operant conditioning.

© IBO 1996

Operant conditioning is a learning procedure in which a reinforcement follows a particular response on a proportion of occasions.

E.4.4
Outline the experiments of Skinner into operant conditioning.

© IBO 1996

The term operant conditioning was used by Professor Skinner who developed the '**Skinner box**', a special cage in which an animal could press a lever and obtain a reward (food). Operant conditioning is also called 'trial and error learning' because the animal (e.g. a rat) will at first accidentally press the lever and after some time learn that pressing the lever will bring the reward. Operant conditioning is easier and faster if the action (pressing the lever) is logically related to the reward (obtaining food).

E.4.5
Define imprinting.

© IBO 1996

Imprinting is an attachment to an object encountered during a short period after birth, usually a parent.

E.4.6
Outline Konrad Lorenz' experiments on imprinting in geese.

© IBO 1996

It is important that an organism knows members of its own species from others. The type of learning called **imprinting** is common in birds. They will imprint on the first moving object they see after hatching. This is most likely to be their mother. They will follow her around and therefore be under her protection. Konrad Lorenz ensured that he was the first moving object that some young geese saw and, as a result, they followed him around.

E.4.7
Define insight learning.

© IBO 1996

Insight learning is a form of intelligent activity and a function of cognitive effort, which contrasts with more passive trial and error mode of learning. Insight learning is also known as 'reasoning'. It involves the combination of data on the current situation with experiences held in memory. This way a problem not previously experienced can be solved without going through a trial and error procedure.

E.4.8
Discuss theories relating to the neural basis of memory and learning.

© IBO 1996

Memory involves learning from experience, being able to store the information and being able to recall the required information when needed. Although much is still unclear about this process. It seems to involve the build-up of a neural net in which synapses between groups of neurons are either strengthened or weakened. It may involve Ca^{2+} ions moving through membranes and the development of receptor sites between neurons.

E.4.9
Discuss how learned behaviour improves chances of survival.

© IBO 1996

Whereas innate behaviour does not change when the environment changes, the opposite is true for learning. In the process of learning, an animal will change its behaviour according to changing circumstances. This has survival value. An animal may learn to recognise that the warning call or behaviour from individuals of another species is likely to be followed by the appearance of a predator. This may very well save its life. Any nature programme on, for example, the Serengeti will show this behaviour.

Option E: Neurobiology and Behaviour

E.5 SOCIAL BEHAVIOUR

E.5.1
List three examples of animals that show social behaviour.

© IBO 1996

Social behaviour is easily observed in animals that live in groups:
- bees.
- ants.
- wolves.
- lions.

All four of these examples are of animals who gather or hunt food cooperatively. To the extent that each seeks shelter or territory, they act as a group in this respect also.

E.5.2
Describe the social organisation of honey bee colonies.

© IBO 1996

The honey bee lives in groups that often exceed 20 000 individuals. Most individuals in the colony are workers and there is one queen. The queen will lay fertilised diploid eggs that hatch into larvae. If the larvae are fed a special diet (more protein), they develop into queens. Otherwise they develop into workers. After about a week, the larva will become a pupa and two weeks later will emerge as an adult bee.

The various tasks in the colony are divided according to the age of the worker. Young workers start off as nurses, taking care of the larvae for about two weeks. They then are assigned house keeping duties, cleaning and guarding the hive and removing sick or dead bees. They gradually start to go outside, making short trips and will finally, for about 6 weeks, go out to collect honey and pollen.

When the colony gets too big, the queen will prepare to leave. As a result, some of her larvae will be fed the special diet and they will become queens. Some unfertilised eggs will hatch into drones (males). The old queen will leave with half the colony and soon after the new queen will fly out. She will be followed by the drones. She will be fertilised and store the sperm to use over the next few years.

The new queen returns and starts laying eggs, fertilising them with the stored sperm. She will not fly again and she is looked after by the workers. The drones also return. They have no further role and when food becomes scarce, they are driven out or killed.

E.5.3
Discuss the role of altruistic behaviour in social organisations using two examples.

© IBO 1996

Altruistic behaviour is behaviour that benefits others and involves risk or cost the performer. The best know and most wide spread example of altruism is in the behaviour of parents to their young. Another example is the worker bee, never reproducing, spending her life looking after a colony.

A lioness feeding her young will not directly benefit from this. She gives up nutrients from her body or puts herself at risk when going out to catch a prey for them. She may live in a small group and even feed another lionesses cubs. If, however, the male of the group is replaced by another male, he is likely to kill all the young.

There is a certain logic to this, seemingly very different, behaviour. **Richard Dawkins** came up with the idea that the individual is not important but his genes are. All that a gene wants to do is multiply and distribute as many copies of itself as possible. It has been described as the '**selfish gene**'. The lioness will therefore care for her offspring because they carry some of her genes. She will even care for cubs of the other lioness because they are likely to be related (the mothers are probably sisters or half sisters) and therefore will share genes. The new male is not related to the cubs and will kill them. This makes the females start a new cycle and soon the male will have his own cubs in the pride, carrying his genes.

This also explains the behaviour of the bees. Since males are haploid, all sperm cells are identical. Egg cells are made via meiosis so some variation exists. Workers are therefore genetically very similar. They share more genes with the queen and the other workers that they would with their own offspring if they were able to produce them. Using the concept of the 'selfish gene', the interest of the worker's genes is better served by promoting the growth of the colony by caring for the queen's eggs than by producing their own offspring.

The idea of the selfish gene is very interesting and it explains behaviour that seems to go against the interest of the individual. However, it does not explain everything and should not be over rated.

E.5.4
Discuss the effects of alcohol abuse.

© IBO 1996

Alcoholic drinks such as beer, wine, cognac, etc. are drinks that contain a certain amount of ethanol (C_2H_5OH). Alcoholic drinks have standard amounts in which they are served: beer is served in large glasses, gin in very small. This relates to the the alcohol content and as a result each 'portion' will roughly contain the same amount of alcohol: the stronger the drink, the smaller the glass.

Many people enjoy drinking alcohol. Many do not because it is against their religion, or they do not like it or other reasons. It is a personal choice how much alcohol you are willing to drink. Unfortunately, some people drink too much, either regularly or on occasion.

Option E: Neurobiology and Behaviour

Alcohol is a depressant and reduces orientation, coordination and judgement. It may reduce inhibitions and change your mood. However, even if you perceive yourself as funnier than when you are sober, this view may not be shared by others. Even the smallest amount of alcohol will have the above effects and a large amount will be very noticeable. When judgement is impaired by alcohol, people are often extremely unpleasant because they do not heed to the mores of civilisation. They can be agressive or overly and inappropriately affectionate. The often used excuse: "you will have to forgive my behaviour because I was drunk" is of course equally unacceptable. Excessive acute alcohol abuse (caused, for example, by some bet in a pub) can lead to acute alcohol poisoning which may be fatal.

Frequent and long term alcohol abuse leads to severe consequences. It increases the chance of cancer (esp. oesophagus and pancreas), damages the immune system, reduces sexual and reproductive functions, causes high blood pressure and causes irreparable liver damage (cirrhosis) which may be fatal.

The social effects of alcoholism are financial (drinks are expensive), the ability to hold a job (many jobs cannot be done when 'under the influence', e.g. taxi driver) and the general effects of impaired judgement. Soccer hooliganism has been reduced since alcohol was banned from the stadium.

E.6 THE AUTONOMIC NERVOUS SYSTEM (HL ONLY)

E.6.1
State that the ANS consists of sympathetic and parasympathetic motor neurons.

© IBO 1996

The **autonomic nervous system** (ANS) consists of sympathetic and parasympathetic neurons.

E.6.2
State that the roles of the sympathetic and parasympathetic system are largely antagonistic.

© IBO 1996

The roles of the sympathetic and parasympathetic system are largely antagonistic. This means that they act in opposite senses. Examples of this are discussed in section E.6.4.

E.6.3
State that the autonomic nervous system serves the heart, blood vessels, digestive system and smooth muscles.

© IBO 1996

The autonomic nervous sytem in vertebrate animals controls the internal organs without

Biology

any conscious effort on the part of the animal. The heart beats, blood vessels contract and dilate, the digestive system operates and smooth muscles contract automatically under the control of the autonomic nervous system.

E.6.4
Explain the effects of sympathetic and parasympathetic system by reference to the heart, salivary glands and iris of the eye.

© IBO 1996

The **sympathetic system** prepares the body for action while the **parasympathetic system** is active at times of rest.

	sympathetic system	parasympathetic system
heart frequency	speeds up	slows down
salivary glands secretion	a little	a lot
iris	radial muscles contract	radial muscles relax
pupil	dilates	constricts

E.6.5
Discuss relationships between the influence of the conscious part of the brain and automatic reflexes as shown by bladder/anus control, meditation and yoga.

© IBO 1996

Although the automanic nervous system mainly works under involuntary control, it is possible to achieve voluntary contol of some of it when this is desired. The autonomic nervous system innervates the muscles of the bladder and rectum. Yet we learn to control these in our early years (although this process takes some time). In the same way, but with even more effort, is it possible to control your cardiac frequency. It has been shown that by long practice in meditation or yoga that people can have voluntary control over systems that normally are controlled by the autonomic system.

YOGA

Option E: Neurobiology and Behaviour

E.7 NEUROTRANSMITTERS AND SYNAPSES (HL ONLY)

E.7.1
State that synapses of the peripheral nervous system are classified according to the neurotransmitter used including acetylcholine and noradrenaline.

© IBO 1996

Synapses of the peripheral nervous system are classified according to the neurotransmitter used including acetylcholine and noradrenalin.

E.7.2
Explain how presynaptic neurons can either encourage or inhibit postsynaptic transmission by depolarisation or hyperpolarisaton of the postsynaptic membrane.

© IBO 1996

Below you will find a brief explanation of the working of a synapse. This is also explained in section 14.1.

In the synapse, the arrival of an action potential causes a change in membrane permeability for Ca^{2+}. As a result, Ca^{2+} flows into the synaptic knob. The consequence of this is exocytosis of transmitter substance in vesicles. The neurotransmitter then diffuses across the synaptic cleft (20 nm) and attaches to receptors in the post synaptic membrane. The receptor sites change their configuration and open the Na^+ channels (in an excitatory synapse) which causes an action potential in the neuron. In an inhibitory synapse, the configuration change in the receptors opens the K^+ and Cl^- channels; K^+ moves out and Cl^- moves in, increasing polarisation of the neuron and increasing the distance from the threshold value.

SCHEMATIC DIAGRAM OF A CHEMICAL SYNAPSE

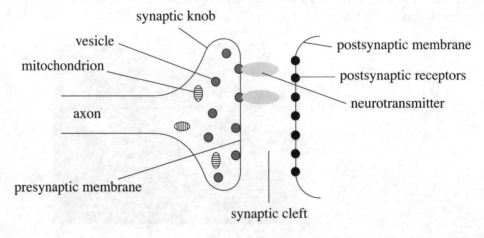

After the post synaptic membrane has been affected, enzymes break down the neurotransmitter. (Acetyl)cholinesterase (found on the post synaptic membrane) changes acetylcholine into choline and ethanoic acid. These diffuse back to the synaptic knob, are absorbed and recycled into acetylcholine.

So an action potential arriving at a synapse can either depolarise the postsynaptic membrane (encourages post synaptic transmission) or polarise the postsynaptic membrane (inhibits post synaptic transmission).

It is possible that an action potential of one presynaptic neuron arriving at a synaptic cleft will not release sufficient neurotransmitter to lead to depolarisation that will pass the threshold potential in the postsynaptic membrane. Action potentials of several presynaptic neurons arriving at the same time might be sufficient to cause an action potential in the postsynaptic membrane. However, if some of the presynaptic neurons are inhibitory neurons, an action potential from them will hyperpolarise the postsynaptic membrane, making it more difficult to pass the threshold potential so less likely that an action potential will be passed on.

It is the combination of which neurotransmitter and the properties of the postsynaptic membrane which will decide whether depolarisation or hyperpolarisation takes place.

E.7.3
Outline the way in which pain is sensed and how endorphins and enkephalins can act as pain-killers.

© IBO 1996

Pain receptors are nerve endings surrounded by a thin layer of support cells. Pressure on the support cells will trigger an action potential.

Enkephalins and **endorphins** are small polypeptide chains. They act by inhibiting association neurons that transmit pain to the brain.

A sensory neuron forms a synapse with an associate neuron. Near this synapse, another synapse is found, coming from the pain control centre in the brain. When the pain control centre sends out an impulse, the synapse will release endorphins/enkephalins which will hyperpolarise the postsynaptic membrane of the association neuron, inhibiting an impulse from being passed on. Enkephalins and endorphins are destroyed very rapidly so their effect does not last long.

E.7.4
Outline the symptoms of Parkinson's disease and the involvement of dopamine.

© IBO 1996

Parkinson's disease is caused by an absence of dopamine in certain parts of the brain. It affects muscle movements. Symptoms are hand tremors, shuffling gait and muscle weakness.

Option E: Neurobiology and Behaviour

Without dopamine, the impulses cannot be tranmitted properly which causes the above described problems. A certain drug (levodopa) is available which the brain cells can use to make dopamine. This reduces the symptoms for several years. Surgically inserting dopamine producing cells in the brain has been moderately succesful but controversial since the source of the transplant is human embryonic tissue.

E.7.5
Explain that psychoactive drags affect the brain and personality by either increasing or decreasing synaptic transmission.

© IBO 1996

Psychoactive drugs affect the brain and personality. They do this by increasing or decreasing synaptic transmission. The drugs can bind to the receptor site on postsynaptic membranes, mimicking the transmitter or blocking the binding of the tranmitter substance. It can also reduce the effect of the enzyme which normally breaks down the transmitter substance, causing an increase in the effect of the neurotransmitter.

Some examples are
- **nicotine** - it mimicks the working of acetyl choline.
- **curare** - blocks the action of acetyl choline.

E.7.6
Discuss the effects of excitatory psychoactive drugs including nicotine, caffeine, cocaine and amphetamines.

© IBO 1996

Effects of excitatory psychoactive drugs.
Nicotine causes the release of adrenalin from the adrenal glands and therefore increases blood pressure and cardiac frequency. It affects the mood, acting like a stimulant and causing a feeling of euphoria.

Caffeine increases heart rate and urine production. It causes some mood elevation and increases alertness.

Cocaine raises cardiac frequency and body temperature and dilates pupils.
It causes euphoria and excitation - 'crack' : a smokable form of cocaine; the effects are more intense but shorter lived.

Amphetamines increase cardiac frequency, respiration, blood pressure. Increases alertness, reduces appetite. - 'ecstasy' is a derivative of amphetamines.

E.7.7
Discuss the effects of inhibitory psychoactive drugs including benzodiazepines (Valium™ and Temazepam™, etc.) and cannabis.

© IBO 1996

Effects of inhibitory psychoactive drugs.

Benzodiazepines (e.g. brands like Valium™, Librium™ or Temazepam™) relax muscles, decrease circulation, respiration and blood pressure. They reduce anxiety and elevate the mood.

Cannabis increases cardiac frequency. It also causes euphoria and sensory distortions.

EXERCISE

1.
   a. What is innate behaviour?
   b. What causes innate behaviour?
   c. Give an example of innate behaviour.
   d. What is the role in natural selection in the development of behaviour patterns?

2.
   a. What is the name of the light sensitive layer in the eye?
   b. What are the names of the light sensitive cells in this layer?
   c. Which of these kinds of cells will work better in low light conditions? Explain.
   d. Outline how light is perceived and changed into an action potential.
   e. What is the role of the visual cortex in the perception of visual stimuli?

3. Animals depend on innate and learned behaviour for survival. Give 2 examples of different situation where innate behaviour is vital for survival.

4.
   a. Give an example of conditioning
   b. Give an example of operant conditioning
   c. Give an example of imprinting
   d. Give an example of insight learning
   e. How does insight learning differ from operant conditioning? How does this affect survival?

5.
   a. What is meant by altruistic behaviour?
   b. Give 2 examples of altruistic behaviour in different situations.
   c. What is one explanation for altruism?

6.
   a. What is the general effect of the sympathetic system?
   b. What is the result of sympathetic stimulation of the heart, salivary glands and iris?
   c. Explain the survival value of the above described responses.

Option E: Neurobiology and Behaviour

7. a. Explain the different effects of an excitatory and inhibitory presynaptic neurone on the postsynaptic membrane.

b. How does this affect the transmission of other, simultanous, messages at this synapse?

OPTION F: APPLIED PLANT AND ANIMAL SCIENCE

22

Chapter contents

- Science and the world food problem
- Applied plant science
- Applied animal science
- Science applied to horticulture
- Modern methods and techniques
- Wider biological and ethical issues

Contributed by Barbara Free

Option F: Applied plant and animal science

F1 PLANT AND ANIMAL SCIENCE AND THE WORLD FOOD PROBLEM

F.1.1
Discuss the imbalance between food production and need on a global scale in terms of the social, economic and climatic conditions.

© IBO 1996

The production of food is not uniform across the world, nor is it fixed from year to year. Factors such as climate and general technological development affect a country's ability to produce food. Some areas of the world are very densely populated, for example, Japan and Hong Kong. These countries have comparatively small areas devoted to farming. They struggle to feed their populations and often rely on imports and fishing. Other areas, such as Canada and Australia, are sparsely populated. Here the problem is that there are few people to work on farms as not only is the population sparse but it tends to be highly urbanised. Farming in these countries tends to be large scale and highly mechanised. These countries often produce an excess of food which is exported.

Food shortages and **famines** result from several causes. The most common is climatic changes which result in the failure of rain with inevitable crop failures. Some developing countries also experience economic difficulties brought on by large debts which consume substantial amounts of the gross national product in just repaying the interest. Agriculture seldom flourishes in such situations. War is also a frequent cause of a country's failure to produce enough food to feed its citizens. Apart from direct damage, wars are costly and take farmers from their fields. Ethiopia, a frequent recent victim of famine, has suffered from all of these scourges.

Whether or not food should simply be transferred free of charge (aid) to countries experiencing famine is one of the world's most pressing moral issues. Farmers in developed countries fear that free distribution of food will reduce prices, undermining their incomes, threatening their futures and the future of world food production.

In 1999 the world produced more than enough food to feed all its population. Starvation and malnutrition are a result of the uneven distribution of supplies.

F.1.2
Describe the imbalance referred to in F.1.1 in relation to one named developed and one named developing country.

© IBO 1996

Canada is a developed country. In 1997, Canada had population of 30 287 000. This was 3.3 persons per square km. 37% live in cities and 63% in rural areas. The population is not increasing rapidly.

Some major agricultural products are:

Wheat (23 million tonnes), barley (14 million tonnes), maize (7 million tonnes). There are many other staple products. There are 13 million live cattle and a fish catch of 1 million tonnes annually.

Canada is a major food exporter, in the 1990s exporting $Can24 000 000 000 worth of food annually.

Indonesia is a developing country. In 1997, Indonesia had population of 199 544 000. This was 104 persons per square km. 78% live in cities and 22% in rural areas. The population is increasing quite rapidly.

Some major agricultural products are: rice (51 million tonnes), sugarcane (32 million tonnes), cassava (16 million tonnes) There are many other staple products. There are 14 million cattle and an annual fish catch of 4 million tonnes.

Indonesia is self sufficient in food but with few food exports.

In comparing these two countries, there are a number of clear differences. Canada has a much smaller population density than Indonesia and thus has larger area per person available for farming even taking into account the mountainous areas that cannot be farmed at all.

Just looking at the principal plant product in each country, Canada produces 760 kg of wheat per person annually and Indonesia produces 255 kg of rice per person annually.

Comparing animal products, there are 43 live cattle for every 100 Canadians and 7 live cattle for every 100 Indonesians. The situation is somewhat more even if you make a similar comparison of fish catches.

Whilst neither country is currently experiencing famine, it is clear that Canada produces much more food than is needed by its people and can export. Indonesia is in a more precarious position in respect of its supplies and would be more at risk if it faced, for example, an unusually dry year.

As a final comparison, Canada increased its food production by 27% between 1980 and 1990 whereas Indonesia improved its production by 65%. This is a characteristic of countries that actually do 'develop'!

F.1.3
Describe how increases in human population require improvements in food production, transport and storage.

© IBO 1996

A country should plan for future. The English economist and demographer **Thomas Malthus** (1766-1834) said that populations always grow exponentially whereas food supplies grow linearly.

Option F: Applied plant and animal science

Graphically, this appears as:

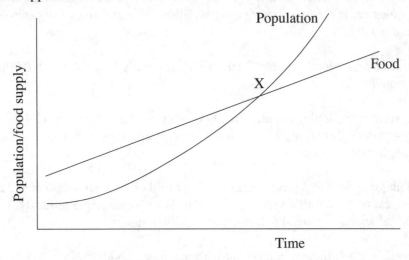

Malthusian theory suggests that, however efficient farming becomes, the nature of the straight line and exponential graphs, ensures that there will always come a moment (X) when population will outstrip food supply. There were a number of results that followed from this theory. It had been believed that high fertility and birth rates contributed to national wealth. Family planning could be said to have its origins in the theory. It also had the effect of limiting traditional charities because they were thought to encourage large families amongst the poor.

Whatever the truth of this complex issue, it seems prudent that countries should seek to ensure that their population growth is less than exponential and the growth in food production is better than linear. Also, storage of food is important to cover periods in which production may be reduced by poor rains, frosts etc.

F.1.4
Explain the need for international collaboration to solve the world food problem.

© IBO 1996

As we have seen, neither the world's population, nor its food production is uniformly distributed across the globe, This means that food needs to be transferred from one place to another to cover short term shortages. At the time of writing, war in the Balkans has significantly reduced production in that area. There are large numbers of displaced people and the world community has had to fly in food supplies in the hope that, when peace is restored, this area will be returned to its previous status as an over-producer of food. These efforts are coordinated by private charities as well as organisations such as the United Nations.

F.1.5
Analyse data relating to any of the above.

© IBO 1996

We have already looked at some data relating to populations and food production in different countries. On the global scale, there are fears that climate changes resulting from the greenhouse effect may limit our ability to increase food production.

Population statistics are:

	1950	1991	2025 projection
Total world population (millions)	2516	5384	8646
Developed countries	832 (33%)	1219 (23%)	1412 (16%)
Developing countries	1684 (67%)	4165 (77%)	7234 (84%)

(Eurostat demographic statistics 1993)

As far as world food production is concerned, if 1979-81 production is used as a basis of 100, production in subsequent years was recorded as:

Year	1979-81	1991	1992	1993	1994
Production index	100	126	129	131	135

(United Nations food & agriculture organisation)

These data indicate a variety of trends. Populations are increasing more rapidly in developing than developed countries. Food production, currently sufficient, is increasing, but is it increasing at a sufficient rate? Only time will tell!

F.2 APPLIED PLANT SCIENCE

F.2.1
Define leaf area index.

© IBO 1996

Leaf area index: the ratio of the total area of leaves of a plant to the area of soil available to it is known as the leaf area index.

Densely planted crops such as beans which have large leaf area and which are planted close to each other have a large leaf area index. Thinly planted crops such as asparagus which require a lot of root space but small leaves have a small leaf area index.

Since plants trap the sun's energy in their leaves, crops with a high leaf area index tend to produce more usable foods than those with low indices.

Option F: Applied plant and animal science

F.2.2
Define harvestable dry biomass.

© IBO 1996

Dry biomass: the total commercially usable mass of organic matter in a plant or a crop, organism or ecosystem (excluding water) is known as the dry biomass. Water is excluded from this measure as it is not usable organic matter.

The measure is useful when applied to food crops as it represents the amount of usable food. 1 kg of dried wheat contains more 'food' than 1 kg of jelly fish (which are mainly water).

F.2.3
Define net assimilation rate.

© IBO 1996

Net assimilation rate (NAR): the net increase of biomass per unit leaf area per unit time.

$$NAR = \frac{\text{increase in dry biomass per unit time}}{\text{leaf area}}.$$

If a crop has a high net assimilation rate, it means that it is an efficient converter of solar energy into organic matter.

F.2.4
Describe how plant productivity can be measured in terms of relative growth rate, leaf area index, harvestable dry biomass and net assimilation rate.

© IBO 1996

All the factors described above affect the total primary production of an environment. Productive plants have high leaf area indices and high net assimilation rates. As a general indicator of the general productivity of different types of plant environments, Tropical forests produce about 1800 grams of dry biomass per square metre annually compared with 1250 grams of dry biomass per square metre in temperate forests.

Other environments produce the same sort of range of biomass. All figures are given in grams of dry biomass per square metre annually.

Swamp	2500
Cultivated land	650
Desert	70

Aquatic environments can also be surprisingly productive:

Reefs	2000
Estuaries	1800
Lakes	500
Deep sea	25

Of course, when it comes to producing useful food, not all this biomass is 'harvestable'. When it comes to a crop such as wheat, we have an edible part, the grain, and an inedible part, the straw. The straw is not useless and is used as animal feed. Consequently about 10% of the straw turns up later as meat and dairy products. Annual production of wheat is running at about 600 million tonnes grown on an area of 230 million cultivated hectares.

F.2.5
Outline how the following factors affect plant productivity: light, water, concentration of carbon dioxide, temperature, availability of nutrients, disease, predators and genotype.

© IBO 1996

Limiting factors are also discussed in option C. As discussed above, a plant's productivity depends on a number of factors that are particular to its genotype. Basically, some plants are inherently more productive than others. However, that is not all. The environment of a particular plant may place a limit on its productivity.

An example is water supply. If a crop does not receive adequate water from rain or irrigation its productivity will be limited by this alone. Fertilizer will not be effective in increasing the yield.

There are other potential limiting factors such as lack of light, absence of essential nutrients, disease, predators etc.

F.2.6
Explain how plant productivity can be optimised using greenhouses.

© IBO 1996

The effects of greenhouses are:
- to increase the ambient temperature (even in the absence of artificial heat). This happens because the glass transmits the heat energy from the sun more readily than is does the lower energy heat re-radiated by the soil and the plants. Some of the heat is thus 'trapped'. This also allows 'out of season' production which allows growers to sell produce at higher prices than they get 'in season'.

- Moisture is retained, reducing the need for irrigation. Natural rain can still be fed to the plants.
- Protection from wind and frost.
- More sophisticated greenhouse operations use hydroponic culture in which plants are grown without soil, their roots receiving nutrients from a special solution designed to optimise their cropping.

Appropriate ventilation is an essential part of an effective greenhouse operation.

Option F: Applied plant and animal science

F.2.7
Describe how intensive monoculture can lead to nutrient depletion and pest invasion and therefore requires fertilisers and pesticides.

© IBO 1996

Monoculture: the growing of the same crop on the same land year after year.

There are some advantages to monoculture including the selection of the optimal crop for a particular soil type or geography. Thus forage crops (such as clover) are often grown on steep slopes. They give an acceptable yield and at the same time protect the soil against erosion. Also, the need to re-seed the land is avoided.

Crops such as grapes, whose vines live for many years, have to be cultivated in a monoculture environment.

The major disadvantages of monoculture are:
- depletion of nutrients. Many plants deplete some nutrients more than others. Tomato plants are particularly heavy users of nitrogen. Gardeners who grow their tomatoes on the same part of the garden year after year will find their crops decreasing. This is because the soil is depleted of nitrogen in the first year and subsequent plantings leave the plants short of nitrogen.
- pests. Many pests such as aphids are active during summer producing eggs that survive winter in the soil. Monoculture makes it easier for such pests and diseases to survive and reinfest each planting.

These problems can be at least partly controlled by using fertilizer to replace soil nutrients and pesticides to control pests.

A classic example of the dangers of monoculture is the Irish potato famine of 1845-49. The Irish climate (cool and damp) is particularly suited to growing potatoes. As a result, the population had become significantly dependent on the vegetable as a source of food. Extensive crop failures caused by the fungus *Phytophthora infestans* (blight) led to a serious famine. The fungus arrived from North America by accident. The consequences were serious. 1.5 million irish people migrated to America and the Irish population fell from 8.4 million in 1844 to 6.6 million in 1851.

F.2.8
Explain how intensive monoculture can lead to increased crop production in terms of efficient land use, timing of intervention and harvesting.

© IBO 1996

'Intensive agriculture' is a term used to describe methods that aim to use science to optimise cropping. It is much more common in developed countries. Typical methods involve use of irrigation, fertilization using chemicals specific to the particular crop and the use of chemicals to control pests and diseases and the use of specialised machinery. One of the surprising consequences of intensive farming is that prices have been reduced by the increased supply, reducing profitability.

F.2.9

Explain how a knowledge of genetics has led to an improvement in yield of a cereal crop, using wheat as an example.

© IBO 1996

Wheat is a cereal grass of the *Gramineae* family. Originally a wild grass, wheat has been bred to produce hundreds of different varieties. These have been bred to increase yield, improve the range of climates in which the crop can be grown and increase the ease with which the grain can be separated from the less useful parts of the plant. Modern wheats can be grown in climates from those in the southern United States (hot and dry) to sub-arctic conditions.

There are three main varieties of wheat in common use:
- *Triticum aestivum*. This is used to produce flour for bread making, cakes and biscuits.
- *Triticum durum*. This is mainly ground into semolina and is the basis of pasta.
- *Triticum compactum*. This is principally used in the production of confectionery and biscuits.

The history of domestication of wheat is complex. One of the earliest stages was to cross the diploid eincorn wheat (*Triticum monococcum*) with a species of grass (*Aegilops speltoides*). This resulted in a doubling of chromosome complement and tetraploid wheats such as emmer (*Triticum dicoccon*). As well as improved yield, threshing and winnowing (separation of the grain) is much easier in these crosses.

F.2.10

Explain how plant science has been applied to the improvement and cultivation of another crop plant.

© IBO 1996

Another major world crop is rice. Large parts of Asia are heavily dependent on this crop. It is estimated that 95% of the 400 million tonnes produced annually are for human consumption.

Wild rice (Indian rice), *Zizania aquatica*, is not related to rice (*Oryza sativa*). Today it is considered a delicacy and is cultivated for the 'health food' market.

Recent breeding has been aimed at improving yield and disease resistance. 'Miracle rice', a product of the 1960s has short stalks (which minimise loss by seed dropping) and disease resistance.

Option F: Applied plant and animal science

F.3 APPLIED ANIMAL SCIENCE

F.3.1
State that hominid ancestors have used animal products for over half a million years.
© IBO 1996

Hominid ancestors have used animal flesh as food for a very long time. Evidence derived from the fossil beds in Olduvai gorge in Tanzania shows both vegetarian (*Australopithecus boisei*) and omnivorous (*Homo habilis*) ancestors of man. Conclusions about diet are based on the size of the molars which are much larger in the vegetarian species. The earliest *Homo* fossils have been dated to about 1.5 million years (by using the ratios of potassium and argon isotopes).

Evidence from cave paintings at Altamira in Spain and Lascaux in France, which are thought to date from 15 000 BC, suggests a heavy practical and cultural dependence on hunting animals. The paintings show hunt scenes but also animals in a more abstract vein that seems to indicate a more mystical relationship between man and the natural world.

Cultural evidence of man's long term dependence on animal products abound. It is common for myths to include man/beast crosses (minotaur) and also to show a clear understanding of ecological principles. The shark, feared by most 'advanced' cultures, is sacred to many fishing cultures. The presence of sharks indicates good stocks of other fish and hence 'good fishing'.

F.3.2
Outline how people have used selection in the domestication of animals to produce breeds more suitable for ploughing, transport, food, and fur and skins for clothing, and shelter.
© IBO 1996

The horse (*Equus caballus*) was one of the first and most useful of domesticated animals. All horses are of the same species, with differences between them being as a result of careful interbreeding. This has lead to the development of types of horse that have different characteristics.

'Draught horses' are large and strong, comparatively slow and have large feet. They are particularly suited to ploughing which requires pulling a heavy weight fairly slowly over slippery ground.

'Race horses' have been bred for speed. The science/art of racehorse breeding is particularly advanced largely because of the large amounts of money generated by betting on races. Good racehorses are amongst the most valuable animals on Earth and enjoy 'supermodel' status.

Cattle are all thought to belong to the species *Bos taurus* (short horn and Jersey) or *Bos indicus* (Brahman). There are crosses of these two such as Santa Gertrudis. One of the major reasons for crossing cattle is to increase resistance to disease and pests. Farmers in tropical Queensland (Australia) farm a cross between Malayan and Brahman cattle as these are resistant to the paralysis tick which is fatal to other breeds.

F.3.3
Outline how a knowledge of genetics applied to domesticated animals has led to an improvement, using as examples either milk yield in cattle or egg yield in poultry.

© IBO 1996

Improvements in the milk yield per cow have been marked during the latter part of the 20th century.

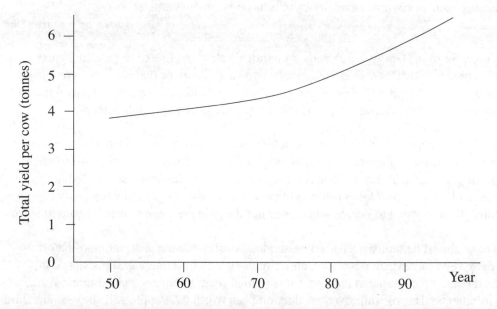

These have been achieved by improved feed, health and breeding for increased milk production.

Similar improvements have been made in the production of beef.

Chickens originated in Asia and were first of interest because their aggressive nature made them suitable for exhibition fights (cockfights). There are many breeds of chicken with a variety of characteristics. The White Plymouth (a cross between the Dominique with the Black Cochin) is mainly raised for its meat. By contrast, the Leghorn (a mediterranean breed) is farmed for its eggs (which are white shelled). The Cornish, an English breed, is a poor 'layer' but produces good meat. Good use of cross breeding to produce hybrid vigour has been largely responsible for improved production of both eggs and poultry meat.

The modern trend towards 'traditional' foods has ensured that original breeds are still farmed, one of the major reasons for retaining these is their importance in retaining a diverse gene pool for future breeding.

Option F: Applied plant and animal science

F.3.4
State that the intensive culture of farmed animals leads to an increase in the yield of protein.

© IBO 1996

Farmers who use 'intensive' methods generally produce greater yields of saleable meat. They use many methods to achieve this; selective breeding, improved feed, use of growth stimulators and antibiotics.

F.3.5
Evaluate intensive farming techniques in terms of yield and ethical issues.

© IBO 1996

Jonathan Swift in *'Gulliver's Travels'* (a political satire and one of the world's great books) observed that "whoever could make two ears of corn or two blades of grass to grow upon a spot of ground where only one grew before would deserve better of mankind and do more for his (*sic*) country than the whole race of politicians put together".

Swift is observing that, in his time, the activities of scientists (who are also satirised in the chapter '*A voyage to Laputa*', which highlights their obsessions) are more useful than those of politicians. Today, scientists have 'the stage' and they have succeeded in 'making two ears of corn or two blades of grass to grow upon a spot of ground where only one grew before'. We now need to decide whether or not the gain has been worth the consequences.

We have almost certainly over-used pesticides (resulting in the destruction of natural predators and increased resistance), anti-biotics (which have increased resistance in bacteria that infect humans) and fertilisers (which seem to have damaged marine environments). Has the improved production been worth it? People still starve in the third world and 'advanced societies' are afflicted with the 'diseases of affluence'.

That said, across the world, the lifestyle and lifespan of *Home sapiens* has improved. What we need to decide is whether or not the costs are worthwhile. Is an extra day on the average life of the average human worth the extinction of the giant panda?

F.3.6
Explain how veterinary techniques have been applied to improve the health and fecundity of animals including artificial insemination, vaccination and nutrient supplementation.

© IBO 1996

Intervention in the raising of animals has included selective breeding and artificial insemination. These ensure that animals have desirable characteristics. Other major techniques include the use of vaccination and good nutrition to enhance both the health and longevity of livestock. Also, artificial insemination reduces the transmission of some diseases. Vaccination is compulsory for participation in, for example, horse shows. Nutrient supplements: you only need to look in any tack shop or stand at a horse show to find 'the latest' in food supplements.

F.3.7
Discuss the use of antibiotics and growth hormones in livestock production.

© IBO 1996

The administration of antibiotics and growth hormones to livestock both increases their health and growth rates. This means that animals thus treated grow faster and produce a greater return to farmers. The concerns about these practices centre around fears that the use of antibiotics in cattle have speeded the development of antibiotic resistance and have thus reduced their effectiveness in the treatment of human disease. Also, the use of hormones in cattle is thought to be having a damaging effect on human fertility. The use of oestrogens and their ingestion by males may be contributing to a reduction in human male fertility. In Holland and some other countries, all milk is tested for the presence of antibiotics and the farmer is fined if antibiotics are discovered in his milk.

F.4 SCIENCE APPLIED TO HORTICULTURE

F.4.1
Draw the generalised structure of dicotyledonous insect-pollinated flowers, as seen with the naked eye and hand lens.

© IBO 1996

The structure of flowers was discussed in section 16.1.1

F.4.2
Describe the differences between wind-pollinated and insect-pollinated flowers.

© IBO 1996

Methods of pollination vary. Wind pollination (anemophily) depends on wind to spread pollen. It is thought to have evolved in plants that originated in rain forest and which subsequently faced a more arid and hostile environment in which pollinating insects were rare. Such plants produce huge quantities of pollen. For example, a single male sorrel plant produces 400 million pollen grains. These plants have small petals to allow the wind to disperse the pollen as much as possible.

By contrast, insect pollinated plants tend to have large, coloured and scented petals often designed to fool the pollinating insect into thinking that it has found a mate. Thus the petals of such flowers have the same flamboyant colours as the wings of insects.

F.4.3
Define pollination.

© IBO 1996

Pollination: the transfer of pollen grains from the anther to the stigma. Sometimes plants pollinate themselves (**self pollination**) and sometimes they pollinate others (**cross pollination**). Cross pollination spreads a particular plant's genes.

Option F: Applied plant and animal science

F.4.4
Outline the differences between pollination and fertilisation in flowers.

© IBO 1996

The process of fertilization is the fusion of male and female gametes. Pollination is the tranfer of pollen while fertilisation is the fusion of gametes. Pollenation does not always lead to fertilisation.

For example, you can pollinate a flower with pollen of a different species but usually fertilisation then does not happen.

F.4.5
Explain how commercial production of cut flowers and flowering plants requires a scientific understanding of flower structure, pollination and fertilisation as well as the techniques of vegetative propagation.

© IBO 1996

Commercial production of cut flowers is a large international business. Annually the business is running at about $US 5 billion. It has been estimated that half of the exported cut flowers originate in the Netherlands.

The **tulip** is one of the best loved of cut flowers. It produces seeds but is most commonly propagated by digging up (lifting) the bulbs, storing through winter, dividing and re-planting.

The **geranium** is generally propagated by '**cutting**'. A part of the plant is cut off, placed in the soil when it will generally produce roots and a plant identical to the parent. See also F.4.7. Whilst horticulturists use cutting frequently, it is a natural process by which plants damaged by wind, animals etc. can produce new individuals from broken parts.

The **croton**, a tropical plant is generally propagated by cutting the stem and wrapping the cut in moist sphagnum moss covered in plastic (to retain moisture). Roots are produced and the part of the plant above the cut can be removed and conventionally planted. This process is known as '**layering**'.

Normal seed propagation is also common. This produces individual plants with variable characteristics. If the horticulturist wishes to produce a uniform product it is more likely that asexual reproduction of the types described would be used.

F.4.6
Describe the role of auxins in terms of apical dominance and how pruning can result in a bushy decorative plant.

© IBO 1996

Auxins are plant hormones that control growth. They are present in greatest concentration in the tip of most plants. Growth is thus most rapid here (**apical dominance**). This leads to plants with a tall structure (such as a forest tree). If the growing tip is cut off (pruned), this removes the growing tip and induces the production of side shoots (new growing tips). The result is a much bushier plant and a more ornamental appearance. Pruning is routinely used in the care of roses which are pruned almost down to soil level each autumn. The effect is vigorous growth and healthy flowering in the next spring and summer.

F.4.7
Describe how plant growth substances can be used commercially to promote rooting, to induce flowering at the required time and to produce fruits without seeds.

© IBO 1996

When using vegetative reproduction techniques such as cuttings commercially, it is important to ensure a high percentage of success. Some plants root easily (coleus and willow) whereas others benefit from receiving some help. This 'help' can include the careful control of the environment, humidity etc. More sophisticated methods involve the use of substances normally found in the plant; plant hormones (such as auxins). Some success has been achieved indolebutyric acid, a synthetic auxin included in commercially available 'hormone rooting powders'. To maximise the chance of successful rooting, care should be taken to cut a healthy part of the plant to cut. The cut is dipped in water and then in hormone rooting powder. This is then planted in a rich moist potting medium. Care should then be taken not to overwater or otherwise mistreat the cutting.

The main technique to induce out of season flowering is to use artificial light to alter day length. Many plants native to places where the day length varies from season to season flower when the correct day length for spring occurs. If lights (and shades) are used to mimic the day length of the season in which a plant normally flowers, it can be induced to flower artificially.

In some cases, pollination, not followed by fertilisation is a sufficient stimulus to produce enlargement of the pistil to form a seedless, or pathenocarpic fruit. This process has been used in certain varieties of citrus fruits, grapes and bananas to produce a seedless product.

F.4.8
Outline how tissue culture techniques can be used commercially in the production of decorative house plants.

© IBO 1996

Tissue culture involves taking a small (often very small) part of a growing plant and attempting to grow a new plant from this fragment. These are transferred to a medium containing auxins, cytokinins and nutrients. They are then incubated and can produce very large numbers of identical offspring in a short time. This has obvious commercial advantages.

Option F: Applied plant and animal science

F.5 MODERN METHODS AND TECHNIQUES OF PLANT AND ANIMAL SCIENCE – HL ONLY

F.5.1
Discuss gene manipulation involving sense/antisense technology, with reference to 'Flavr-Savr™ tomatoes.

© IBO 1996

The **Flavr-Savr™ tomato** is a genetically engineered plant and was the first such plant to be approved for sale (in 1995). The aim was to produce a plant that gives fruit that stays ripe (i.e. red and firm) for as long as possible as well as being able to be ripened on the plant, producing a better flavour.

Traditional commercial tomato production has the fruit picked green when it is less liable to bruising (allowing mechanical picking). The fruit is then packaged and exposed to ethene gas which initiates ripening in time for the fruit to be marketed.

The Flavr-Savr™ tomato contains introduced genes to prevent over ripening. The over-ripening gene is blocked. This is because the antisense RNA strand is introduced so that the double-stranded part prevents the gene from being expressed. The tomato also contains genetic material that makes it resistant to disease.

F.5.2
Discuss the science behind the use of plant growth regulation and hormone weed killers.

© IBO 1996

Synthetic auxins can be used to, for example, removed dicotyledonous weeds from monocotyledonous cereal crops or lawns. They do this by interfering with the natural auxins resulting in disturbed growth. Sometimes this results in very rapid growth and premature death. Chemically most of these substances are phenoxyacetic acids. 2,4,5-T was one of the most effective but has now been banned because it contained traces of the toxic substance dioxin.

F.5.3
Explain the techniques used in modern plant production including micropropagation and cloning, growth media, aseptic techniques, auxins, kinetin and gibberellins.

© IBO 1996

Micropropagation and **cloning** was discussed in section F.4
Growth media are used in tissue cultures. The success of the technique depends on placing the plant fragment in a bath of appropriate composition (growth hormones, nutrients, water, etc.). The medium must, of course, be aseptic to avoid rotting or bacterial infestation.

Auxins: plant growth hormones discussed in previous sections.
Kinetin (6-furfurylaminopurine): is synthesised in the roots and in conjunction with

gibberellins: control growth and cell differentiation.

Gibberellins have the general structure shown:

gibbane skeleton

gibberellic acid

The gibberellins control cell elongation and, hence stem length. They also have a role in the germination of some seeds, particularly cereals. You are not required to memorise their structures.

F.6 WIDER BIOLOGICAL AND ETHICAL ISSUES

F.6.1
Discuss the biological and ethical issues surrounding transplanting animal organs into humans.

© IBO 1996

Transplanting of organs from human to human is quite common in modern surgery. It is undertaken when a person's organ(s) fail. Transplants can be from the same individual (autograft - skin grafts), from a closely related person (isograft - kidney transplant), from an accident victim (allograft - all types of transplants), and from animals or mechanical devices (xenograft - a developing field).

The major difficulty in transplantation is rejection. The body treats the transplant as a pathogen and rejects it. This is not a problem in the case of skin grafting because the transplant is not foreign tissue. In the case of transplants from other individuals, rejection can be minimised by tissue typing (ensuring that they donor organ is similar to the tissue of the receiver) and by the use of immuno-suppressive drugs.

Most people find acceptable the earlier types of transplant where parts of one's own body are used to benefit ourselves. Also, the donation of a kidney from one twin to another has tended to be a consensual process. The use of organs from accident victims is more controversial as it requires rapid removal of the organ after death making the 'definition of death' a particularly important issue.

When it comes to using animal organs in people, we have a new ethical issue. Firstly, is it appropriate to keep animals purely for the purpose of using them to replace human parts? We now know that, biologically at least, we are animals, but many people still make a

Option F: Applied plant and animal science

distinction.

Another problem also arises in that it is quite difficult to predict the wider impact of using animal material in humans. Might there be unexpected results such as increased resistance to antibiotics?

F.6.2
Discuss the biological and ethical issues surrounding 'organic' versus 'non- organic' farming methods.

© IBO 1996

Organic farming uses traditional fertilizer, compost animal manure etc. to ensure continued cropping. Large scale farms generally use inorganic fertilizers produced by chemical industries. The outputs of modern farms are generally larger than the more traditional types of cultivation. However, there are problems. Small amounts of deleterious chemicals are leaking into the food chain with possible long term effects on human and animal health. Also, mineral fertilizers tend to leach into waterways more than do organic types. They are then transported to the sea. Many marine environments are fragile and have been seriously damaged by the presence of nitrogenous compounds. Coral reefs, for example, require a nutrient free environment if they are not to be swamped by algae. The rate of destruction of the World's reefs which are very biologically productive, as well as being places of great beauty, is a major cause for concern.

F.6.3
Discuss the biological and ethical issues surrounding the eating of meat, fish, eggs and dairy products.

© IBO 1996

The need for a balanced diet has already been discussed. Difficulties arise mainly as a result of human over population. This leads to excessive exploitation of natural resources. This can be obvious such as over-fishing (the over fishing of herring in the North Sea has resulted in the almost complete elimination of the fish and the collapse of the industry).

Many people have ethical objections to killing animals for food that are not related to fears of exploitation. Such people become vegetarians or vegans.

F.6.4
Discuss the biological and ethical issues surrounding the use of animals for education and research.

© IBO 1996

Animals are used in both education and research. Veterinary students, for obvious reasons, dissect animal carcasses. Other students of animal biology do the same. To what extent should this happen?

The same can be said for the use of animals in research. This includes using animals to test products such as drugs, cosmetics, skin care products etc. Where should we 'draw the line

Biology

as to what is and is not acceptable?

F.6.5
Discuss the biological and ethical issues surrounding biological and chemical pest control.

© IBO 1996

In the period after the second World War it appeared that chemical pest control (using substances such as DDT) was the answer to crop pests such as locusts and to insect born diseases such as malaria. The chemical worked for a time, infestations declined and productivity increased. American biologist and author Rachel Carson (1907-1964) in the celebrated book '*Silent Spring*' exposed the dangers of chemical over use and the damaging effects DDT and other chemicals were having on the wider biosphere. Carson faced well organised opposition from established science and from the chemical companies. However, her book is now seen as the 'first shot' in the environmental revolution and she is now seen as one of the twentieth century's most influential people. This can be seen as a biological issue. Many people also have ethical concerns about tampering with the natural balances of the biosphere.

Biological pest control generally uses natural predators, artificially introduced, to control pests. Rabbits breed rapidly and, having been introduced for sport shooting to Australia, has done enormous damage across the continent. Control by introduced diseases such as myxomatosis have been the only way of controlling the animals. However, the disease results in a painful death and is not a desirable means. Research is underway to produce a more effective and humane alternative.

Today, we make extensive use of technology in farming, land management and forestry.

Option F: Applied plant and animal science

EXERCISE

1. Consider these statistics on India (1997), its population and food production.

   Population: 967 613 000 (27% urban, 73% rural) projected to double in 43 years.
   Density: 306 persons per square km.
   Major plant products: sugar cane (255 million tonnes annually), cereals (214 million tonnes annually).
   Live cattle: 196 million.

   a. Calculate the amount of cereals produced per person annually.
   b. Find the number of live cattle per 100 Indians.
   c. India currently produces enough food to feed its population. Will it continue to do so?.

2. Use the data in section F.1.5 to draw graphs showing the expected world population and predictions for food production. Are we heading for a 'Mathusian catastrophe'?

3. Greenhouse cultivation is common in Holland but comparatively rare in Australia. Explain why this might be the case.

4. Discuss the effect that the introduction of a 'super-cow' might have on society. The cow is a combination of mutation and interbreeding. It has so much muscle bulk that it cannot be born naturally and must be delivered using Caesarian section.

5. A new anti-biotic can double the egg production of chickens. Write a newspaper article opposing the use of this anti-biotic.

6. Investigate the cut flower industry in your country. Which flowers are most popular, where are they grown and how they are propagated?

7. Investigate the production of the commercial decorative plant, the African Violet.

8. Investigate the effect of a major dam project (Aswan, Kariba etc.) on the long term ecology of the region. Did it achieve the aims of its builders?

OPTION G: ECOLOGY AND CONSERVATION

23

Chapter contents
- The ecology of species
- The ecology of communities
- The ecology of ecosystems
- Biodiversity and conservation
- Microbial ecology (HL only)
- Reducing harmful impacts of humans

Option G: Ecology and conservation

G 1.1 THE ECOLOGY OF SPECIES

G.1.1
Explain the factors that affect the distribution of plant species including temperature, water, light, soil pH, salinity and mineral nutrients.

© IBO 1996

The presence or absense of a certain plant species in a certain environment depends on several factors. The characteristics of the abiotic environment is one of them. Factors affecting plant distribution are:

temperature, water, light, soil pH, salinity and **mineral nutrients.**

Pine trees from the Austrian Alps photosynthesize best at 15°C but the Hammada bush (found in the Israeli desert) has an optimum temperature of 44°C for photosyntesis.

Xerophytes (plants that live in dry areas, e.g. cacti, pine trees) have several adaptations to conserve water. Hydrophytes (waterplants, e.g. duckweed, water lily) do not have these adaptations.

The pH of the soil affects its capacity to retain minerals. In acid soil, H^+ ions replace the positive ions clinging to the clay particles and the positive ions leach out of the soil. Calcium becomes more soluble (and hence more available to plant roots) as pH increases whereas iron becomes less soluble as pH increases. Alfalfa, clover and other leguminosae require high calcium, i.e. alkaline soil. Rhododendron and azalea require high iron, i.e. acid soil.

The salinity of the soil is the salt content. The salt in the soil will dissolve in water. This water then has a relatively high osmotic value. Since roots take up water by osmosis, salt water is more difficult to take up and many plants cannot survive this. The plants which are adapted to high salinity often have some of the characteristics of xerophytes.

The mineral content of soil depends on several factors. One factor is the rock from which the soil was formed. Another factor is the size of the particles which make up the soil. Minerals (and water) drain rapidly through sand (large particles) and are held by clay (small particles). Since clay particles carry a negative charge, they bind positively charged particles such as Ca^{2+}, K^+ and Mg^{2+}.

The most important factors affecting mineral content of the soil are biological factors. An undisturbed environment will recycle nutrients via soil, plants, animals and microorganisms. Harvesting can quickly deplete the soil when the nutrients removed are not replaced (fertiliser). In some countries too much manure is put on the land resulting in a very high mineral content. The minerals will leach into the streams and lakes and cause eutrophication (=rapid growth of algae which depletes the oxygen in the water when they die and are broken down by bacteria).

G.1.2
Explain the factors that affect the distribution of animal species including temperature, water, breeding sites, food supply and territory.

© IBO 1996

The presence or absence of a certain animal species in a certain environment depends on several factors. They can be biotic and abiotic. Factors affecting animal distribution are: temperature, water, breeding sites, food supply and territory.

When cells freeze, they are damaged by the formation of ice crystals. When the temperature becomes too high, enzymes are denatured. ('Too high' is different for different enzymes. Some algae live in hot springs at 80°C.) Some animals have thermoregulation, others do not.

1. Thermoregulation: the ability to maintain a body temperature either at a constant level or within an acceptable range.
2. Advantage: optimum temperature makes enzymes more effective and allows species in more different environments.
3. Disadvantage: costs a lot of energy so more food is required.
4. Homeotherms: regulate body temperature.
5. Poikilotherms: do not regulate body temperature.
6. Heterotherms: regulate body temperature part of the time.

Ways to warm up the body:
- **Endothermic:** generation of heat through internal metabolic processes.
- **Ectothermic:** use of external heat sources by behavioural means.

Ectothermic regulation: reptiles lie in the sun in the morning to warm up and look for a shady place when it becomes too hot.

Endothermic regulation: the hypothalamus is the control centre.

EXAMPLE 1.
The Owlet moth is a heterotherm: its body temperature is equal to the surroundings when at rest. Preparing for flight, they become endothermic so that the flight muscles can function properly. A counter-current blood flow retains heat in the body.

EXAMPLE 2.
The blue fin tuna is endothermic (unusual for fish). The swimming muscles produce heat. Heat is then retained in the core of the body by an efficient counter-current system in the blood.

Birds and mammals.
A decrease in body size means an increase in metabolic rate. The small shrew consumes 100 times as much oxygen per gram of body mass as the elephant.

Option G: Ecology and conservation

G.1.3
Analyse the significance of the difference between two sets of data using the Students' t-test given the appropriate formula and tables.

© IBO 1996

When you take samples of 2 populations which are different in one aspect (amount of light OR temperature OR predation) you want to know if this factor has affected these populations so that they are different (in weight, size, number of fruits, etc).

The size of your sample determines the reliability of the results.
If your sample size is small, the difference between the samples will have to be bigger to conclude a significant difference between the populations than when you have large samples.

The **Students' *t*-test** will show the significance of the difference between two sets of data.

The following formula is used:

$$t = \frac{|\bar{x}_1 - \bar{x}_2|}{\sqrt{\frac{s_1^2}{n_1} + \frac{s_2^2}{n_2}}}$$

where
- $\bar{x}$ = mean
- s = standard deviation
- n = number of entries in a set of data
- s^2 = variance
- $|\bar{x}_1 - \bar{x}_2|$ = the positive difference between the two means

Large values of t indicate little overlap and almost certainly a difference between two sets of data. Small values of t suggest a lot of overlap and probably no difference between the populations.

To find out if there is a significant difference between the populations from which we took the samples, you need to go through several steps.

Step 1:
Formulate the H_0 and H_1 hypothesis.

H_0 states: there is no difference between the populations.

H_1 states: there is a difference between the populations (two-tailed test) OR population 1 is bigger/stronger/more seeds etc. than population 2 or vice versa (one-tailed test)

You need to know whether to use a one-tailed test or a two-tailed test. If the question is about a difference between the populations, you use a two-tailed test, if the question is about one population being significantly larger (better, more offspring, etc.) than the other, you use a one-tailed test. The difference is in which column to use from the table. At any degree of freedom, the required t value for a two-tailed test is larger than for a one-tailed test.

Step 2:
Find the t value from a table and compare it with the calculated t value.
To find the t value from a table you need to know the 'degrees of freedom'. In these calculations, the degrees of freedom will be $n_1 + n_2 - 2$.
If $n>30$ you can use infinite degrees of freedom (d.f. $= \infty$) for IB Biology.

You also need to know with what degree of confidence you reject H_0. For IB Biology, we use $p < 0.05$. This means that if we reject H_0 based on the differences in the 2 samples, there is less than 5% chance that these differences were caused by a coincidence in the sample taking and that the populations actually were not different. In Biology this is usually considered acceptable.

Step 3:
If calculated t value exceeds the t value from the table, then you are entitled to reject H_0 and have to accept the alternative hypothesis which suggests that the populations are not the same.

Critical values of Students' t test

level of significance for:

degrees of freedom	one-tailed test $p = 0.05$	two-tailed test $p = 0.05$
1	6.314	12.706
2	2.920	4.303
3	2.353	3.183
4	2.132	2.776
5	2.015	2.571
6	1.943	2.447
7	1.895	2.365
8	1.860	2.306
9	1.833	2.262
10	1.812	2.228
11	1.796	2.201
12	1.782	2.179
13	1.771	2.160
14	1.761	2.145
15	1.753	2.131
16	1.746	2.120
17	1.740	2.110

Option G: Ecology and conservation

18	1.734	2.101
19	1.729	2.093
20	1.725	2.086
21	1.721	2.080
22	1.717	2.074
23	1.714	2.069
24	1.711	2.064
25	1.708	2.060
26	1.706	2.056
27	1.703	2.052
28	1.701	2.048
29	1.699	2.045
30	1.697	2.042
∞	1.645	1.960

EXAMPLE

In a test of the effectiveness of a new fertilizer, two groups of tomato plants of the same variety were planted. One was treated with the new fertilizer and the other was not.

The yields from each of the plants in the two groups are given below:

With Fertilizer	Without Fertilizer
3.5	4.1
5.1	9.0
6.7	4.9
10.6	7.7
10.4	1.8
4.1	2.6
14.3	5.8
4.9	7.0
12.4	9.7
14.0	9.8
9.1	6.2
9.0	5.9
10.9	
5.2	

Use the t-test to decide if the evidence of this experiment produces a significant increase in yield.

SOLUTION
Step 1:
Formulate the H_0 and H_1 hypothesis.

H_0 states: there is no difference between the yields. That is $\mu_1 = \mu_2$ the means are equal.

H₁ states: there is a difference between the yields and that the fertilized plants produce more tomatoes than the unfertilized plants. That is $\mu_1 > \mu_2$.

Step 2:
The statistics on the two data sets are:

	With Fertilizer	Without Fertilizer
Mean	8.586	6.208
Standard Deviation	3.684	2.604
Number of plants	14	12

The 'degrees of freedom' $n_1 + n_2 - 2 = 14 + 12 - 2 = 24$

$$t = \frac{|\overline{x_1} - \overline{x_2}|}{\sqrt{\frac{s_1^2}{n_1} + \frac{s_2^2}{n_2}}}$$

$$= \frac{|8.586 - 6.208|}{\sqrt{\frac{3.684^2}{14} + \frac{2.604^2}{12}}}$$

$$= 1.9196$$

Step 3:
The calculated value of t is 1.9196. The table value for 24 degrees of freedom is 1.711 (for the one tailed test). So, as the calculated value (1.9196) is greater than the value obtained from the table, we can reject H₀ and conclude that the new fertilizer is effective in increasing yield.

We now show you how this can be done using the TI–83 graphics calculator:
Step 1: Enter the data as two lists
press STAT, then select **1:Edit** from the STAT EDIT menu.
Enter each value in the first sample (press enter after each entry). Once you have finished with the first List, use the arrow key to move to the next column and repeat the previous step for the second set of data.

Step 2: press STAT, then use the arrow key to hghlight **TEST**.
Then, select **4:2–SampleTTest** (press ENTER).
In this case, as we have entered the data as List 1 and List 2, we simply use the arrow key to move down and select the alternative hypothesis (in this case, that $\mu_1 > \mu_2$).

Option G: Ecology and conservation

Step 3: Keep using the down arrow key to highlight the **Calculate** option and then press ENTER:

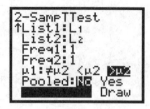

Step 4: A summary of results are displayed on the next screens

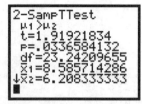

From the first part of the screen we see that the calculated value of t is 1.919 (>1.711). Also, the p–value is 0.0336 (< 0.05). Both results clearly indicating that we would reject the null hypothesis in favour of the alternative.

G.1.4

Explain what is meant by the niche concept, in terms of an organism's spatial habitat, its feeding activities and interactions with other organisms.

© IBO 1996

An organism's **niche** describes its way of life. It includes its habitat, feeding activities and interactions with other organisms.

The vulture is a large african bird that feeds off the corpses of dead animals. Its lifestyle consists of circling the kills of predators, waiting until they have eaten and then landing to 'clean up'. The vulture is thus one of nature's garbage collectors and, whilst we might not find its lifestyle very attractive, it performs a useful function in seeing that the environment is not littered with rotting meat. The vulture's niche in the world consists of these important parts of its lifestyle and contribution to the environment and, hence, to the welfare of other animals.

G.1.5

Explain the significance of the principle of competitive exclusion.

© IBO 1996

The **competitive exclusion principle** states that no two species can share a niche. They will compete until only one species is left.

This was illustrated in an experiment by **G.F. Grause** (1934, USSR) in which two *Paramecia* species were both grown separately and in competition. These graphs illustrate the results:

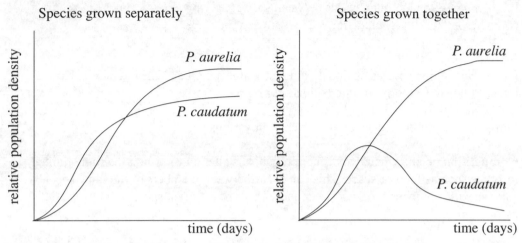

This means that even if organisms of both populations can exist in a certain environment, one will be able to survive better than the other. In this case *P. aurelia* can tolerate higher levels of waste products than *P. caudatum* and this will be the 'fitter' to survive. *P. caudatum* will not survive at all. Two species cannot share the same niche at the same time.

Competition is intense in forests

Option G: Ecology and conservation

G.2 THE ECOLOGY OF COMMUNITIES

G.2.1
Explain the following interactions between species, giving two examples of each: competition, herbivory, predation, parasitism and mutualism.

© IBO 1996

Individuals of one species obviously interact with each other. This interaction can be intense (ants in a colony) or rare (spiders for reproduction).

Individuals of different species also interact in various ways.

Competition occurs when two or more organisms attempt to exploit the same limited resource such as food or space. This can occur between members of the same species (intraspecific competition) or between individuals of different species (interspecific competition).

An example of the latter kind is competition for food between 5 different species of warbler (a small bird) in a spruce tree in North America. This has been solved by niche differentiation (or resource partitioning). Each species has a preferred area of the tree for feeding so that these species are not really sharing the same niche.

Another example of interspecific competition can be found between different species of duckweed. *Lemna gibba* and *Lemna polyrrhiza* can be found in similar ponds and lakes and will compete mainly for light. *L. gibba* grows more slowly than *L. polyrrhiza* but since is has tiny air filled sacs, it will float on the surface and shade other species and win the competition for light.

Herbivory is the eating of plants. This has been compare to parasitism since neither herbivores nor parasites intend to kill their source of food. Examples of herbivory are any herbivorous animal (cattle, goats, zooplankton)

Predation is the eating of live organisms. In some definition this includes plants as well as animals. Venus flytrap is a carnivorous plant which will capture insects and digest them to supplement its minerals. Of course, the lion is a well known example of a predator and its prey consists of zebra, gazelles, wildebeest, etc. (Often, the lion will steal the prey from, for example, hyenas.)

Parasitism is a long lasting relationship between individuals of different species where one individual benefits (the parasite) and the other is harmed (the host). The other difference from predation is that the parasite is considerably smaller than its host.

Examples are fleas on a dog and the fungus which causes athlete's foot.

Mutualism is the long lasting relationship between individuals of different species where both benefit. Examples are lichens which are a symbiosis of an algae (photosynthesis) and a fungus (water and minerals). Other examples are the bacteria living in the rumen (stomach) of cows which helps them digest plant material and the bacteria in our gut which also aid in digestion and produce vitamin K for us.

G.2.2
Define gross production.

© IBO 1996

Gross production is the amount of organic matter produced by photosynthesis in plants. (Approx 2% of the light striking a forest will be used for photosynthesis.)

G.2.3
Define net production.

© IBO 1996

Net production is the part of gross production not used in plant respiration.

G.2.4
Calculate the above values from given data.

© IBO 1996

This results in the conclusion that
Gross production – respiration = net production

If a plant's gross production is 2 kg over a month and 0.95 kg is lost through respiration, then the net production is 2 – 0.95 = 1.05 kg.

In plants on average net production is about half of gross production (i.e. 50% of the captured energy is used in cellular respiration).

G.2.5
Explain the differences in photosynthetic efficiency between tropical forests, temperate forests, deserts and polar ecosystems.

© IBO 1996

The limiting factors of photosynthesis are
- light.
- water.
- carbon dioxide.
- temperature.

In tropical forests, the limiting factor is probably carbon dioxide (or light at the forest floor) since the other factors are near optimum. In temperate forests, temperature will be the limiting factor for most of the year and cause the overall photosynthetic efficiency to be lower than in tropical forests. In deserts, photosynthesis is low due to a high temperature and a lack of water. (CAM plants are more efficient since they obtain carbon

Option G: Ecology and conservation

dioxide at night and utilise it during the day while the stomata are closed to conserve water). In polar ecosystems, photosythesis is also low. Temperate climates can limit photosynthesis but also water which can be unavailable since it is in the form of ice. Light intensities are low and during winter the day is short.

G.2.6
Discuss the difficulties of classifying organisms into trophic levels.

© IBO 1996

In section 4.1.8 and 4.1.9 we discussed the flow of energy through a food chain and a food web. The first organisms in either one must always be the producer or autotroph. It is relatively simple to deceide if an organism is a producer or not. It becomes more difficult when we try to place predators as secondary or tertiary consumers. Since few predators prey on only one species, any one predator can be a secondary consumer (eating a herbivore) today and a tertiary consumer (eating another carnivore) tomorrow. Omnivores are always primary as well as secondary consumers. This shows again the discrepancy between the nicely ordered food chain and the reality of an ever changing food web.

G.2.7
Explain the small biomass and low numbers of organisms in higher trophic levels.

© IBO 1996

Food chains on land are rarely longer than 4 links. Those in the oceans may have up to 7 steps. (In both cases parasites and detritivores are not included.) This is caused by the fact that most energy contained in one trophic level is not used to create biomass in the next level. The metabolism of the consumer requires most of the energy and very little is converted in biomass.

On average the biomass found at a trophic level is 10% of that found at the previous trophic level. Consider the following example :
10 000 kg of waterplants can produce 1000 kg insects which produce 100 kg of small carnivorous fish which produces 10 kg of bass which produces 1 kg of human.
To sustain a reasonable biomass at the end of a long food chain, you need very large amounts of producer biomass. Since the organisms at the end of the food chain are usually large, this means that you can only afford to sustain a small number of individuals at the highest trophic level.

Biology

G.3 THE ECOLOGY OF ECOSYSTEMS

G.3.1
Draw the water cycle in a named terrestrial ecosystem including precipitation, evaporation, condensation, transpiration, and drainage or capillary action in soils.

© IBO 1996

Since the Earth does not exchange materials with space, matter needs to be recycled. Energy is constantly received from the Sun and also radiated into space but matter is recycled in a number of ways.

The **water cycle** is unlike other cycles (carbon and oxygen cycle, nitrogen cycle) since it involves no chemical changes to the water. Only changes of state are involved.

The water cycle includes the following elements:
    precipitation, evaporation, condensation, transpiration, drainage.

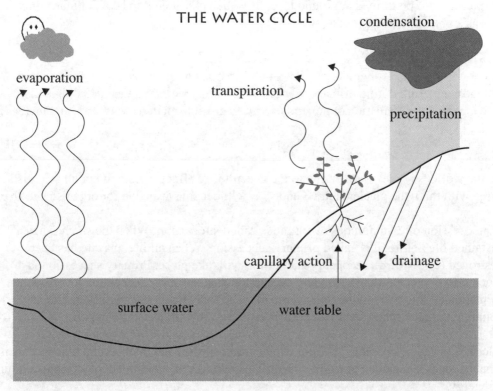

Water vapour condenses in the atmosphere and returns to the earth as precipitation e.g. rain. The water will drain into the soil and to lakes or the sea. Capillary action of soil particles will allow some of the water to move up again towards the roots of plants. Some of the water will be absorbed by these roots, the remainder will add to the water table and will eventually become part of the surface water. Wet soil and surface water will evaporate water (using the heat of the sun) and produce water vapour. The process of tranpiration in plants also evaporates water absorbed by the roots.

Option G: Ecology and conservation

G.3.2
Explain how oxygen is interconverted between oxygen, carbon dioxide, water and ozone.
© IBO 1996

The **Carbon-Oxygen cycle** involves the two basic processes of photosynthesis and cellular respiration.

Plants take up carbon dioxide and produce organic (carbon based) compounds. In this process water molecules are broken down and oxygen is released.

When the oxygen is in the atmosphere, UV radiation can break apart the molecules and create oxygen atoms. These will recombine to form ozone (O_3).

During cellular respiration, carbon (from organic compounds) is combined with oxygen from the atmosphere, producing carbon dioxide.

Ozone can also be formed at ground level. It is then considered to be a pollutant since it acts as a greenhouse gas (see Section 4.4).

G.3.3
Explain the effects of living organisms on the abiotic environment with reference to the changes occurring during primary succession to climax communities including soil development, accumulation of minerals, reduced erosion, altered river flows, and increased rainfall.
© IBO 1996

As the world becomes warmer due to the **greenhouse effect** (although not all scientists agree with this), the glaciers retract and sites will become available for organisms to live.

A progression of communities over time is called **succession**. When the site has never sustained life before, it is called primary succession. When all life in a site has been destroyed (e.g. by fire), secondary succession will take place. Primary succession is a much slower process than secondary succession since primary succession involves the formation of soil. Succession will eventually lead to a climax community. A climax community has reached stability and is in equilibrium with its environment.

Succession can be prevented by removing certain species and maintaining others. This is what happens in a garden or a golf course. This requires the investment of a certain amount of energy.

The first step in primary succession is the formation of soil. Weathering (heat, water, freezing) will break rocks into smaller ones. The first organisms in primary succession are often lichens. They cling to irregularities in the rocks with root-like rhizoids and secrete acid to dissolve the rocks. When lichens die, their remains are added to the soil. When a little soil has accumulated, mosses often follow lichens and will shade them, causing them to die. This way, more organic material is added to the soil.

Eventually, the mosses will be replaced by ferns, grasses and shrubs and eventually trees (confireous, possibly followed by deciduous). The roots of all these plants will break rocks apart adding to the formation of the soil.

The rocks broken apart will contribute to the mineral content of the soil but especially the organisms living on the soil will change its mineral content (see section G.1.1). Nitrogen fixing bacteria (e.g. those in symbiosis with leguminosae) will change nitrogen from the air into nitrates to be used by the plants (see section G.5.6).

Exposed soil will erode quickly. It is easily washed away by rain or blown away by wind. Plants cover the soil and keep it in place with their roots, almost completely preventing erosion.

The plants will also move water from the soil into their leaves where transpiration takes place. This will assist the water cycle since it increases the amount of water in the air and hence the amount of rain.

G.4 BIODIVERSITY AND CONSERVATION

G.4.1
Discuss current estimates of numbers of species of organisms living in the world.

© IBO 1996

In 1991, 1.7 million species of plants and animals had been identified but many scientists believe that as many as 40 million different species may inhabit our planet. Many of the known animal species belong to the phylum of the Arthropods (approx 750 000). Scientists now suspect that species are becoming extinct before they have even be classified.

The approximate number of species known are:
- Viruses   1 000
- Prokaryota   30 000
- Protoctista   30 000
- Fungi   70 000
- Plantae   250 000
- Animalia   1 030 000

G.4.2
Outline the factors that caused the extinction of two named animals and one named plant species.

© IBO 1996

According to the theory of evolution, many species that once lived on Earth are no longer around. They have become extinct. There are many ways in which a species can become extinct.

Option G: Ecology and conservation

The dodo is one of the species to recently become extinct. Dodos used to live on the small island of Mauritius. Mauritius lies in the Indian Ocean, 2000 km east of Africa.

Mauritius was discovered in the early sixteenth century and quickly became a place for ships to stop and collect fresh food and water. Since dodos were not afraid of people, they could be approached easily. Many dodos were killed by sailors, although they were rarely used for food.

Eventually people settled on Mauritius and brought with them cats, dogs, pigs and rats (unintentionally). These animals ate the dodo or its eggs or chicks and the dodo became extinct by 1680.

The mammoth and the woolly rhinoceros became extinct during the Stone Age. At the end of the last ice age, these animals found that their fur coat was too thick and they could not cope with the warmer climate. However, some people are starting to wonder about the human contribution to this process. People were hunter-gatherers at that time and started to hunt in groups and even use fire to drive animals into traps or ravines. This caused slaughter at a larger scale than had occurred before as well as, sometimes, dramatic habitat destruction. This way, people may have been part of the reason why the mammoth and the woolly rhinoceros became extinct.

The rhinoceros is not yet extinct but may soon be. In 1970, there were an estimated 65 000 white rhinos, in 1990 3000 only existed.

Rhinos are hunted for their horn which is believed to have magical or medicinal properties. In some nature reserves, rhinos are shot by tranquiliser darts and the game warden will remove the horn (with a chain saw). Although this is not pleasant, it will protect the animal from poachers. The rhino is too big to be hunted by predators and the loss of the horn does not interfere with its life.

Along the same lines, people (used to) spray seal cubs with a dye that would permanently stain their white fur. The cubs became more vulnerable to predators but no longer were suitable for fur coats.

The above examples are caused by direct and deliberate actions of humans. Possible even bigger is the effect of habitat destruction which takes place at a large scale today and which threatens the existence of many species.

The giant panda lives in China and feeds almost exclusively on bamboo. The bamboo forest grows smaller all the time as humans use more land for agriculture. The panda seemed doomed, but the government of China (and of course WWF) has invested money and energy into a panda protection plan and the future of the panda seems secure, by establishing panda farms which essentially are nature reserves.

Currently about 4500 animal and 20 000 plant species are endangered and may become extinct.

G.4.3
Compare the relative biodiversity of biomes including tropical rainforest, temperate forest, desert and tundra.

© IBO 1996

A **biome** is a major ecological community in a certain climate, formed by a certain plant cover and its associated organisms. The biome is often given the name of its predominant vegetation.

Within climatic zones, different conditions exist which can support the argument that the area includes different biomes. Any biome will fall into one of two categories: terrestrial biomes or aquatic biomes.

The most commenly accepted terrestrial biomes are: tundras, coniferous forests (or taigas), deciduous forests, tropical rain forests, grasslands and deserts.

Conditions vary a great deal in the different biomes and the diversity of species reflects this.

In the tundra, the ground is permanently frozen (permafrost). The growing season is short and there is little precipitation. Mosses and lichens are the predominant plants, insects are abundant which attracts migrating birds. Mammals are rare although migrating groups of herbivores are found. Species diversity is quite low.

Many of the same conditions are found in the desert. Little precipitation and harsh conditions lead to a low diversity of species. In the desert it is not the cold but the heat which make it hard to survive.

Tropical rainforest is at the other extreme. Species diversity is greater than in any other biome, caused by a long growing season and a lot of precipitation. Plant life concentrates in the higher regions of the rainforest since the canopy is so dense that little light penetrates to the ground.

The 'jungle' is found around rivers or where vegetation has been disturbed. Only then will light reach the ground and allow vegetation to grow. When disturbed areas are left alone, 'jungle' will eventually reach the climax community of 'tropical rainforest'.

The temperate forests are found south of the taiga or coniferous forest. Its species diversity is greater than that found in the tundra or desert but not as much as is found in the tropical rainforest. Temperatures are more moderate and precipitation is sufficient to sustain the growth of (deciduous) trees (e.g. oak, maple, birch). The temperate forest is more open and not as tall as the tropical rainforest.

Aquatic communities are divided into marine and freshwater biomes.

Option G: Ecology and conservation

G.4.4
Calculate the index of diversity using the Simpson formula and outline its significance.
© IBO 1996

The index of diversity gives a measure of how likely two organisms found in an area are to be of the same species. We calculate this by using the Simpson formula:

$$D = \frac{N(N-1)}{\sum n(n-1)}$$

where N is the total number of individuals in the area, n is the number of individuals per species and D is the diversity index.

EXAMPLE
Two islands each have populations of four species:

Island 1.

Species	n	$n(n-1)$
A	345	118680
B	260	67340
C	342	116622
D	598	357006
Totals	1545	659648

Island 2.

Species	n	$n(n-1)$
A	50	2450
B	20	380
C	40	1560
D	1250	1561250
Totals	1360	1565640

The diversity index for island 1 is: $D = \frac{N(N-1)}{\sum n(n-1)} = \frac{4(4-1)}{659648} = 0.1819152 \times 10^{-4}$

and for island 2: $D = \frac{N(N-1)}{\sum n(n-1)} = \frac{4(4-1)}{1565640} = 0.76645972 \times 10^{-5}$ so, using this measure, the second island has smaller species diversity than the first.

A high value for D suggests a stable and ancient site. A low D value could suggest pollution, recent colonisation or agricultural management. The index of diversity is usually used in studies of vegetation but can also be applied to animals or diversity of all species.

One of the consequences of the pollution caused by the Gulf war, was that the diversity of marine species around Bahrain dropped dramatically.

Biology

G.4.5
Explain the use of biotic indices and indicator species in monitoring environmental change.

© IBO 1996

It is important to know when and how much the environment changes. This is especially true if human actions might have caused the change. Careful environmental monitoring can warn us in time of undesired changes and we may be able to take some action which prevents the damage from becoming worse.

Here we will consider monitoring eutrophication. **Eutrophication** means nutrient enrichment (of an acquatic environment).

Eutrophication can be measured chemically, for example, by measuring the rate of oxygen depletion by organisms in the biochemical oxygen demand (BOD). It reflects the activity of micro-organisms in decomposing organic material.

However, it is often easier to measure eutrophication biologically by using **indicator species**. In Britain, professionals in the water industry generally use the Trent Biotic Index (TBI) and the Chandler Biotic Score (CBS).

TBI monitors presence/absence of key species together with species richness. Eutrophic waters show high abundance but low species diversity. In addition, certain species can tolerate higher levels of pollution than others. All this has given rise to the TBI tables which link the presence of certain species etc. to certain levels of pollution.

G.4.6
Discuss reasons for the conservation of biodiversity including ethical, ecological, economic and aesthetic arguments, using rainforests as an example.

© IBO 1996

Why should we care about biodiversity? Although this issue is not restricted to any one place or situation, the disappearance of the tropical rainforest is a good example.

- **Ethical arguments:** "we did not inherit the Earth from our ancestors but we are keeping it for our children".
- **Ecological arguments:** cutting down the rainforest will remove nutrients from the area. The remaining soil is not capable of sustaining many good crops and will soon be left by farmers for another area. Erosion will cause further destruction.
- Researcher predict that if we continue without change, that by the year 2000 nearly 1 billion species will have become extinct and that nearly half the world's species will disappear in the next 500 years. Although there have been times when many species became extinct in a relatively short time, it has never happened on this scale.
- **Economic arguments:** these are probably the most powerful arguments. Many of our medicines orgininate from plants. Many species of plants may have cures for diseases that we cannot cure at this moment. Losing these species might affect human health.

Option G: Ecology and conservation

Also if we destroy the tropical rainforests, we can no longer use mahogany for furniture or the environment for recreation and tourism.

The rosy periwinkle was on the verge of extinction until its medicinal properties were discovered. It contains compounds now used in chemotherapy against a few types of cancer.

- **Aesthetic arguments:** 'extinction is forever'.

G.4.7 & G.4.9
Explain the advantages of conservation of endangered species in situ (terrestrial and aquatic nature reserves).
Explain the use of ex situ conservation measures including captive breeding of animals, botanic gardens and seed banks.

© IBO 1996

As the Chinese found out, pandas are very hard to breed in captivity. They therefore set up 'panda farms' which essentially are nature reserves in areas where pandas normally live. Breeding has been more successful here.

The conservation of species is not an easy task at any time. Plants are easier to conserve since they can be kept as seeds. Conserving an animal species means keeping alive a certain population of animals and having them breed successfully. Many animals do not thrive in captivity and even if they do reproduce, there are often signs of reduced welfare. All this reflects on the animal's health. The more conditions differ from the animal's normal habitat, the more difficult it becomes to keep them.

G.4.8
Discuss the management of nature reserves including control of alien species, restoration of degraded areas, promotion of recovery of threatened species and control of exploitation by humans.

© IBO 1996

management of nature reserves
- control of alien species.
- restoration of degraded areas.
- promotion of recovery of threatened species.
- control of exploitation by humans.

Almost every nature programme on television about nature reserves (e.g. Serengeti) will have information on this.

G.4.9
Explain the use of ex situ conservation measures including captive breeding of animals, botanic gardens and seed banks.

© IBO 1996

To see a species in its natural environment, you might have to travel a long way and even then it may be hard to find. Plants might live in less accessible places and animals might prefer to hide from you. So people have captured the animals and put them in a small cage so that they could be seen easily. This was then called a zoo. Was this an improvement on shooting the animal and using its skin for a rug or did it only just prolong the animal's misery?

Capturing animals is hard work. They tend not to cooperate and often many are killed in the process. It might be easier to shoot the mother and then take the young. They then need to be transported which also many do not survive, especially if they need to be smuggled. Finally, many species die younger in captivity (sea mammals) or fail to reproduce.

Some of this is also true for plants which can be kept in botanic gardens. To maintain a climate in which the plant will grow, often a lot of energy might have to be spent. Any plant that would not normally grow in the area of the botanic garden, is not part of the climax community for that area and hence will be replaced by others unless action is taken to prevent this.

Of course plants are easily kept as seeds for a long time and seed banks are set up to do exactly this and preserve as many species as possible with a minimum investment of energy. These seedbanks are a valuable back up and play an important role in preservation of certain genes.

However, zoos and botanical gardens do raise awareness of regions other than the one in which people live and might make them more concerned about global issues. Although we have seen elephants roam in the Serengeti on television, seeing a real living elephant still inspires a sense of awe. Zoos have also changed very much over the last few decades and are now concerned about the living conditions of the animals and play an important role in breeding programmes. The lion in the small cage is a thing of the past.

G.4.10
Discuss the types of action that can take place at national level including data collection, monitoring and legislation.

© IBO 1996

Action at national level: data collection, monitoring, legislation.

In the Netherlands, a government institution monitors air pollution. Spread out over the entire country, you find 'sniffing poles' which continuously collect data on several variables. This information is then collected and studied.

Based on this (and other) information, the government has passed a law which states

Option G: Ecology and conservation

national speed limits. Because, especially truckdrivers, were frequently caught speeding, the government passed another law which dictates that all trucks have a system of speed control which makes it impossible for them to go faster than a certain speed. This reduces air pollution. Cars fitted with a catalyst which makes the exhaust fumes less polluting are taxed less than cars without.

G.4.11
Discuss the importance of surveys of biodiversity, environmental monitoring and environmental impact assessment.

© IBO 1996

The importance of surveys of biodiversity, environmental monitoring and environmental impact assessment is to note changes on time, as they happen. No one noticed that the numbers of salmon in the Rhine were decreasing until they had entirely disappeared. By continuous monitoring, you can detect changes early and, hopefully, still do something about them.

G.4.12
Discuss the role of international agencies and measures in conservation including IUCN, the Rio Convention on Biodiversity, CITES, WWF and Red data books.

© IBO 1996

The **IUCN** (International Union for the Conservation of Nature and Natural Resources) is a union involving 200 governments and 300 private organisations. It focuses on conservation of species and habitats.

Red Data Books are published by IUCN. They contain lists of endangered species.

The Convention on Trade in Endangered Species (**CITES**) is an offshoot of IUCN. Member states of CITES have to accept and implement the ruling of the entire convention regarding trade in organisms or their products, for example, on banning trade in ivory which intended, and succeeded, in reducing the poaching of elephants.

The **Rio Convention** on Biodiversity was organised in 1992 in Rio de Janeiro by the United Nations Conference on Environment and Development (also called the Earth Summit). One of the results of the Convention was the Biodiversity Treaty. Countries signing the Treaty committed themselves to the concept of the richer countries giving money to the poorer countries for conservation of biodiversity in these countries.

The **World Wildlife Fund** aims at the conservation of biodiversity by encouraging sustainable use of resources. Contrary to those mentioned above, WWF is not linked to any government of the country.

Greenpeace is also a non-governmental organisation. Greenpeace is also concerned with biodiversity; its methods to achieve its goals are sometimes more controversial than WWF. The Internet is a good source of information on these groups.

G.5 MICROBIAL ECOLOGY (HL ONLY)

G.5.1
State that all chemical elements that occur in organisms are part of biogeochemical cycles and that these cycles involve water, land and the atmosphere.

© IBO 1996

All chemical elements that occur in organisms are part of biogeochemical cycles. The cycles involve water, land and the atmosphere. (See section G.3.1 for the water cycle and G.5.6 for the nitrogen cycle.)

G.5.2
Explain that all such cycles summarise the movement of elements through the biological components of ecosystems (food chains) to form complex organic molecules and subsequently into simpler inorganic forms which can be used again.

© IBO 1996

Any diagram of a cycle summarises the movement of elements through the biological components of ecosystems (food chains) to form complex organic molecules and subsequently into simpler inorganic forms which can be used again (See section G.5.6 for the nitrogen cycle).

The above statement does not apply to the water cycle since water is not changed chemically during the cycle.

G.5.3
Explain that chemoautotrophs can oxidise inorganic substances as a direct energy source to synthesise ATP.

© IBO 1996

Chemoautotrophs can oxidise inorganic substances as a direct energy source to synthesise ATP. This means that chemosynthetic bacteria use carbon dioxide as a source of carbon but obtain energy from a chemical reaction (not from light).

Examples of this are the iron bacteria, colourless sulphur bacteria and nitrifying bacteria. Iron bacteria (e.g. Leptothrix) essentially change Fe^{2+} into Fe^{3+} and energy.

$$4\ FeCO_3 + O_2 + 6\ H_2O \rightarrow 4\ Fe(OH)_3 + 4CO_2\ (+ \text{energy})$$

Colourless sulfur bacteria (e.g. Thiobacillus) changes sulfur into sulfate and energy.

$$2\ S + 3\ O_2 + 2\ H_2O \rightarrow 2\ H_2SO_4\ (+ \text{energy})$$

Nitrifying bacteria fall into two groups.

Option G: Ecology and conservation

1. oxidising ammonium (e.g. Nitrosomas) into nitrite + energy.

$$2NH_3 + 3\,O_2 \rightarrow 2\,HNO_2 + 2H_2O\ (+ \text{energy})$$

2. oxidising nitrite to nitrate (e.g. Nitrobacter)

$$2\,HNO_2 + O_2 \rightarrow 2\,HNO_3\ (+ \text{energy})$$

In all the above examples, oxygen is the electron and proton acceptor, as it is in photosynthesis. The difference is that the source of energy to drive these reactions is NOT light energy but energy from the oxidation of inorganic substances.

G.5.4
State that chemoautotrophy is found only among bacteria.

© IBO 1996

Chemoautotrophy is found only among bacteria.

G.5.5
State that electron donors (substrates) include hydrogen sulfide, ammonia, nitrite ions and sulfur and that electron acceptors (oxidising agents) include oxygen, sulfate ions and nitrate ions (cross reference 9.1.1 and C.5.1).

© IBO 1996

In the process of chemosynthesis, electron donors include:

hydrogen sulfide (H_2S), ammonia (NH_3), nitrite ions (NO_2^-) and sulfur (S).

Electron acceptors (oxidising agents) include:

oxygen (O_2), sulfate ions (SO_4^{2-}), nitrate ions (NO_3^-).

G.5.6
Draw a nitrogen cycle including the process of nitrogen fixation (free-living, symbiotic and industrial), denitrification, nitrification, feeding, excretion, root absorption, and putrefaction (ammonification).

© IBO 1996

Plants need to absorb nitrogen since it is found in amino acids. Weathering rock does not contain nitrogen so a different source must be found. Air is mostly nitrogen but plants cannot take up free nitrogen. Plants take up nitrogen in the form of nitrates (NO_3^-) or ammonium (NH_4^+). In the cells nitrates are reduced to ammonium ions which are then combined with carbon containing compounds to make amino acids.

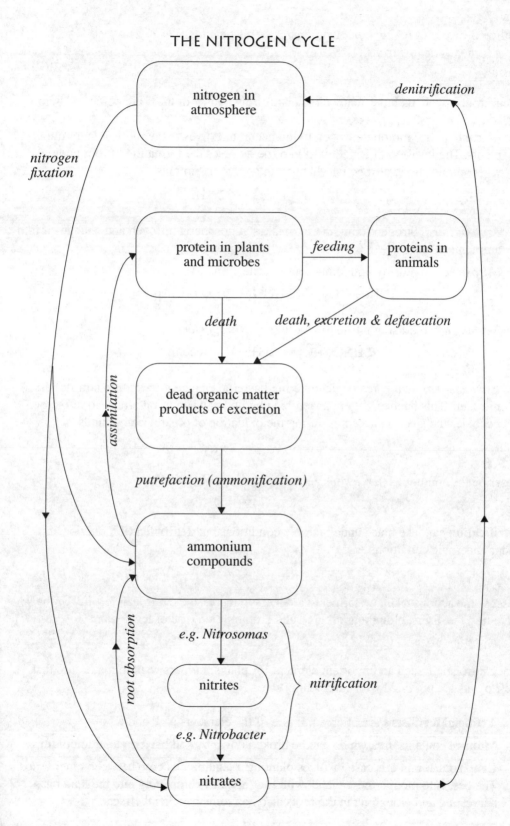

Option G: Ecology and conservation

G.5.7
Outline the roles of *Rhizobium, Azotobacter, Nitrosomonas, Nitrobacter,* and *Pseuclornonas denitrificans.*

© IBO 1996

The nitrogen cycle includes many chemoautrotrops. Some of them are described below.

- *Rhizobium:* a symbiotic nitrogen fixing bacteria. It lives in the roots of leguminous plants. The invasion of Rhizobium into roots causes the formation of root nodules where atmospheric nitrogen is changed into ammonium ions.

$$N_2 + 6H^+ + 6e^- \rightarrow 2 NH_3$$

- *Azotobacter:* Nitrogen fixing in that it takes atmospheric nitrogen and changes it into ammonium ions.
- *Nitrosomas:* oxidising ammonium into nitrite + energy.

$$2NH_3 + 3 O_2 \rightarrow 2 HNO_2 + 2H_2O \text{ (+ energy)}$$

- *Nitrobacter:* oxidising nitrite to nitrate (e.g. *Nitrobacter*)

$$2 HNO_2 + O_2 \rightarrow 2 HNO_3 \text{ (+ energy)}$$

- *Pseudomonas denitrificans:* denitrifying bacteria, change nitrates into atmospheric nitrogen. This happen only in anaerobic conditions when nitrate (instead of oxygen) becomes the electron acceptor during the oxidation of organic compounds.

G.5.8
Describe the conditions that favour denitrification and nitrification.

© IBO 1996

Nitrification can take place under aerobic conditions but **denitrification** only occurs under anaerobic conditions.

G.5.9
Discuss the actions taken by farmers/gardeners to increase the nitrogen fertility of the soil including fertilisers, ploughing/digging and crop rotation (use of legumes).

© IBO 1996

Since nitrogen is such an important element for plants, farmers want to make sure that their plants are provided with sufficient nitrates.

- **Artificial fertilisers** are often a mixture of the elements N, P and K.
- **Manure** contains urea which can be broken down by soil bacteria to ammonium.
- **Crop rotation** in this case involves planting a leguminous crop between other crops. The best is to plough these plants with root nodules completely into the land but harvesting and ploughing in the roots also has some beneficial effects.

- **Deep ploughing** will make available some minerals not previously used but often upsets the structure and water draining properties of the soil.

G.5.10
Outline the consequences of releasing raw sewage and nitrate fertiliser into rivers including eutrophication, algal blooms, deoxygenation, increase in biochemical oxygen demand (BOD), and subsequent recovery.

© IBO 1996

Using too much fertiliser or manure or having sewage drain into rivers and lakes can cause eutrophication of the water.

Eutrophication: process by which pollutants cause a body of water to become overly rich in organic and mineral nutrients, so that algae grow rapidly and deplete the oxygen supply (when decaying).

Algal bloom: extensive growth of algae in a body of water, usually as a result of the phosphate content of fertilizers and detergents.

The resulting increased biochemical oxygen demand and deoxygenation of the water cause many heterotrophic organisms to die.

Recovery of the system can be aided by removing phosphate rich sediment and nutrient rich algae after the orginal source of pollution has been reduced.

G.5.11
Analyse quantitative data on biogeochemical cycles.

© IBO 1996

Biogeochemical cycles circulate the essential elements of living things. We have already considered the water and nitrogen cycles. Data on a cyclical system can be either researched or measured experimentally.

If you wish to find some data on cycles, some websites (at the time of publication) affording this were:

http://irina.colorado.edu/lectures/Lec11.html
http://www.newi.ac.uk/bartlett/ECOLOGY/nutrient.htm
http://www.univ-brest.fr/IUEM/BIOFLUX/SINOPS/observer.htm
http://www.monterey.edu/academic/institutes/essp/esse/climate/climatebiogeo.html
http://www.arts.ouc.bc.ca/geog/G210/210~2~2~7.html

Option G: Ecology and conservation

G.6 REDUCING HARMFUL IMPACTS OF HUMANS ON ECOSYSTEMS

G.6.1
Describe the role of atmospheric ozone in absorbing ultra violet (UV) radiation.

© IBO 1996

The **ozone layer** in the stratosphere (10 - 45 km) stops 99% of the UV radiation from the Sun. Ozone is formed spontaneously when UV light strikes the atmosphere. When UV light strikes an ozone molecule, it separates into 3 oxygen atoms which are very reactive. They will recombine with forming ozone, releasing the absorbed energy as heat (see also Section 4.4).

Ozone is necessary at high altitudes but a pollutant near the Earth's surface (the troposphere, 0 - 10 km). Here it is a component of smog and a greenhouse gas. Ozone in the lower atmosphere is converted back to oxygen in a few days and does not replenish the ozone layer in the stratosphere. In plants ozone inhibits photosynthesis, probably by altering the permeability of cell membranes.

G.6.2
Outline the effects of UV radiation on living tissues and biological productivity.

© IBO 1996

The main effect of **UV radiation** on living tissues is found in the harm it does to DNA. High levels of UV radiation alter the structure of DNA which eventually may cause (skin) cancer.

UV light can be absorbed by organic molecules causing them to dissociate forming atoms or groups with unpaired electrons. These substances are very reactive and can cause unusual reactions to take place.

Phytoplankton are very sensitive to UV light. These organisms at the base of aquatic food chains function less well in even only moderate levels of UV light. Higher levels cause them to die. Since phytoplankton also contribute significantly to the total oxygen production on Earth, damage to these organisms will have far reaching consequences.

Terrestrial plants have been shown to have a lower yield when UV levels are increased. Nitrogen fixing bacteria have been shown to be killed by high levels of UV light.

G.6.3
Outline the chemical effect that chlorine has on the ozone layer.

© IBO 1996

The major cause of the depletion of the ozone layer are CFCs. This abbreviation stands for Chlorofluorocarbons. These molecules are very useful in refrigerators, airconditioners, fire extinguishers and as propellants in aerosol sprays. They were used in large amounts and were thought to diffuse harmlessly into the stratosphere where they were broken down by sunlight.

This is partly true except that in the process of breaking them down, a chlorine atom is produced. The effect of chlorine on the ozone layer is very serious. The chlorine will react with an ozone molecule and break it apart so that it will not reform. One chlorine atom can destroy 100 000 ozone molecules.

G.6.4
Discuss methods of reducing the manufacture and release of ozone-depleting substances including recycling refrigerants, reduction of gas-blown plastics and CFC propellants.

© IBO 1996

In 1990, 93 nations signed an international agreement to phase out the use of CFCs by the year 2000. It will take a century before the ozone layer will be complete again since CFCs are broken down only slowly.

When a refrigerator, freezer or airconditioner is disposed of, it is usually not because the CFCs have disappeared. So it is possible to collect these unwanted items and recycle the CFCs. In many countries, systems have been set up to do this.

Alternatively, producers are looking for alternatives to CFCs and 'green' refrigerators are now available. Although the alternatives tend to be more costly, it does not seem to have a very large effect on the total price.

In many countries, CFCs are no longer allowed to be used in aerosol sprays and alternatives have been found. This is usually indicated on the aerosol can so that the consumer can chose to buy a product for that reason.

G.6.5
Outline the origin, formation, and the biological consequences of acid precipitation on plants and animals.

© IBO 1996

Acid rain
Rain is normally slightly acidic (pH 5.6) due to the presence of carbonic acid (from dissolved CO_2). Over the last few decades, the pH of the rain has become lower than this and a pH between 4.0 and 4.5 has unfortunately become usual. In very severe cases, levels as low as pH 2.1 have been measured.

The presence of sulfuric acid (H_2SO_4) and nitric acid (HNO_3) have caused this increase in acidity. These gaseous oxides will dissolve in the water droplets and come down as precipitation.

Sulfur oxides are produced by volcanic eruptions, combustion of high sulfur coal and oil and the smelting of sulfur containing ores.

Option G: Ecology and conservation

Nitrogen oxides are also produced by volcanic eruptions, petrol combustion in cars and by generating electricity in certain ways.

Many plants cannot tolerate acid rain. Coniferous forests will be damaged, especially the highest parts of each plant. Plants living on limestone ($CaCO_3$) have some advantage in that the limestone will somewhat buffer the pH of the water and nutrients will leach less and remain available to the plants. Those less fortunate may show reduced germination, decrease in seedling survival, reduced growth and less resistance to disease.

Lakes and streams can be severely affected by this. Especially lakes at high elevations will suffer the consequences. A study in 1977 showed that many high altitude lakes had a pH of 5.0 or less and that most of these no longer sustained life. Pictures of healthy trout and those coming from acidic streams show a clear difference in size and general appearance.

G.6.6
Discuss ways to reduce emissions of gases that promote acid precipitation including flue-gas (emissions resulting from the burning materials) and fuel desulfurisation (removal of sulfur), and the use of alternative energy sources.

© IBO 1996

Reducing emission of gases that promote acid precipitation can be done by, for example, reducing flue-gas (emissions resulting from burning materials) as well as fuel desulfurisation (removal of sulfur) and the use of alternative energy sources.

G.6.7
Discuss the advantages and disadvantages of the use of renewable energy sources in various parts of the world including solar, hydroelectric, tidal, geothermal, wind and oceans (tidal, currents and temperature differentials).

© IBO 1996

Energy can be divided into two groups: renewable and non-renewable.

Renewable energy comes from a source which is not going to run out (for a long time). **Non-renewable energy** comes from finites sources.

Examples of renewable energy are: solar energy, hydroelectric, tidal, geothermal, wind and oceans (tidal, currents and temperature differences).

Solar energy uses the Sun's energy to heat water or produce electricity. Solar panels are still fairly expensive but are fast becoming cheaper. Solar energy is best used in areas with a lot of sunshire such as Bahrain.

Hydroelectric power uses the energy of falling water to drive a generator and produce electricity. This method requires large water reservoirs and can change the landscape. Bahrain cannot use this source of energy but Switzerland can (and does).

Tidal energy utilises the up and down motion of the water over 12 hours. The movement of the water can be converted to electricity.

Geothermal sources use the heat of the Earth. Water is pumped deep into the Earth and will come up hot. Areas with volcanic activity (e.g. Iceland) can easily tap into this source of energy.

Wind energy is used in areas where it often is windy. Windmills have been used for centuries to move water or mill grain. They are now also used to generate electricity. They are often large and noisy and many are needed. They also are exposed to great forces and break down easily.

Oceans have waves, tides and currents (see tidal energy). They also have deeper warmer layers (see geothermal energy).

G.6.8
State that biomass can be used as a source of fuels such as methane and ethanol.
© IBO 1996

Biomass can be used as a source of fuel. This has been known for a very long time (burning of wood). However, new sources have been found, for example, in methane produced by cattle (manure) and ethanol from plant sources.

G.6.9
Explain the principles involved in the generation of methane from biomass, including the conditions needed, organisms involved and the basic chemical reactions.
© IBO 1996

The generation of methane from biomass involves the methanogenic bacteria. These strict anaerobic prokaryotes live in swamps, marine and fresh-water sediments and in the animal gut especially those of ruminants. These bacteria are chemosynthetic and produce ATP by converting CO_2 and H_2 to methane. Doing so, they complete the process of fermentation of organic materials under anaerobic conditions where CO_2 and H_2 were formed by other anaerobes.

When trying to produce methane from cow manure, the bacteria are already present (from the cows' guts). A closed off space (without oxygen) is needed to allow the methanogens to survive.

G.6.10
Outline the damage caused to marine ecosystem by the overexploitation of fish.
© IBO 1996

The North Sea is currently one of the most over-fished areas in the world. Since many countries border the North Sea, they all claim the right to fish there. The problems this causes have been recognised for some years. Although the equipment has improved

Option G: Ecology and conservation

greatly (better ships, better nets, sonar to locate the schools of fish), they have not caught more fish. The average size of the fish has gone down, suggesting that they do not have time to grow before they are caught. This then affects the larger carnivores which normally live on these fish.

G.6.11
Discuss international measures that would promote the conservation of fish.

© IBO 1996

International measures which promote conservation of fish all relate to restrictions on the number of fish caught. Currently all EU countries bordering the North Sea have a quota for each species of fish. They can catch a certain amount per year. Strict control is required to ensure that these quotas are not exceeded. Some countries sell some of their quota to fishermen in other countries which causes frictions. Even without this additional trouble, views from fishermen and scientist about the amount of fish that can be caught often differ

EXERCISE

1. Students measured the length of the leaf of the wooddaisy in two different sites. They found the following results :

site 1 (in cm)	site 2 (in cm)
9	17.5
6.5	8.5
7	17.5
5.5	12.5
7.5	16
5	10
10	20
8	17
6.5	19
5	21
6.5	12.5
6	13
8	14
6.5	10
7	14
6	12
13	13
12	11
9	14
9	12

   a. Are the leaves of the wooddaisy significantly different in site 1 compared to site 2 ($p<0.05$)? Show your working.

Site 1 was an open area, site 2 had many large bushes growing in it.

    b.   Explain the difference found in the size of the leaves using this information.

2.   Compare and contrast herbivory and parasitism.

3.   Explain the process of primary succession as it affects the soil.

4.   Students sampled two fresh water ponds at different sites.

They found the following results :

	site 1	site 2
water beetle	19	
water snail	66	
mosquito fish	19	
leech	11	
diptera larvae	11	
bivalve	786	
Ostracada	5	16
tadpoles	8	32
true worms	74	39
water hoglouse		157
swimming mayfly nymph		102
dusky mayfly nymph		16
non biting midge larva		614
'Hawker' dragon fly		16

    a.   Which site has a higher diversity? Explain.

    b.   Which site would you expect to be more polluted? Explain.

5.   a.   Why would you expect a food chain starting with chemoautotrophs to be shorter than one starting with photoautotrophs?

    b.   What are the various ways of increasing nitrogen fertility of the soil? What are the advantages/disadvantages of each?

6.   Discuss the following aspects of the ozone layer:

    a.   How is is formed?

    b.   How is it beneficial to life on Earth?

    c.   What causes ozone depletion?

    d.   How can ozone depletion be reduced?

Discuss the following aspects of acid rain:

    e.   What causes it?

    f.   How is it harmful?

    g.   How can it be reduced?

Option G: Ecology and conservation

OPTION H: FURTHER HUMAN PHYSIOLOGY

24

Chapter contents
- Homeostasis
- Digestion
- Absorption of digested food
- The functions of the liver
- Transport
- Gas exchange

Option H: Further human physiology

H.1 HOMEOSTASIS

H.1.1
State that homeostasis involves maintaining internal environment at a constant level or between narrow limits, including blood pH, water potential, oxygen and carbon dioxide concentrations, blood glucose and body temperature (cross reference 5.5.2).

© IBO 1996

Homeostasis involves maintaining an internal environment at a constant level or between narrow limits, including blood pH, water potential, oxygen and carbon dioxide concentrations, blood glucose and body temperature.

H.1.2
Explain that homeostasis involves monitoring levels of variables and correcting changes in levels by negative feedback mechanisms.

© IBO 1996

Homeostasis involves monitoring levels of variables and correcting changes in levels by negative feedback mechanisms.

An example of negative feedback is the regulation of our body temperature, which is discussed in the next section.

H.1.3
Describe the control of body temperature, including the roles of sweat glands, hairs, skin arterioles and shunt vessels, shivering, hormones (thyroxine, TRI-I, TSH), anterior pituitary gland, hypothalamus and thyroid.

© IBO 1996

Regulation of body temperature is also covered in section 5.5.2

Sweat glands (exocrine glands) secrete sweat into ducts leading to the outside surface of the body. The sweat will evaporate, withdrawing heat from the skin.

Hairs capture a layer of air. Air is a good insulator. Raising the hair will increase the thickness of the insulating layer, reducing heat loss.

Skin arterioles, together with shunt vessels, regulate the amount of blood going to the skin. This determines the skin temperature and hence the amount of heat lost.

Shivering is the contraction of voluntary muscles. Various groups of muscle fibres within a muscle contract and relax out of phase. This can lead to an increase of heat production up to five times the basal level.

Thyroxine (produced and secreted by the thyroid) increases the metabolic rate and as such the heat production. Thyroxine is released only when the anterior pituitary produces and releases **Thyroid Stimulating Hormone** (TSH). The anterior pituitary only does this when **Thyroid Releasing Hormone** is produced in and released from the **hypothalamus**. This system is shown below.

SUMMARY OF THE TEMPERATURE REGULATION OF A MAMMAL

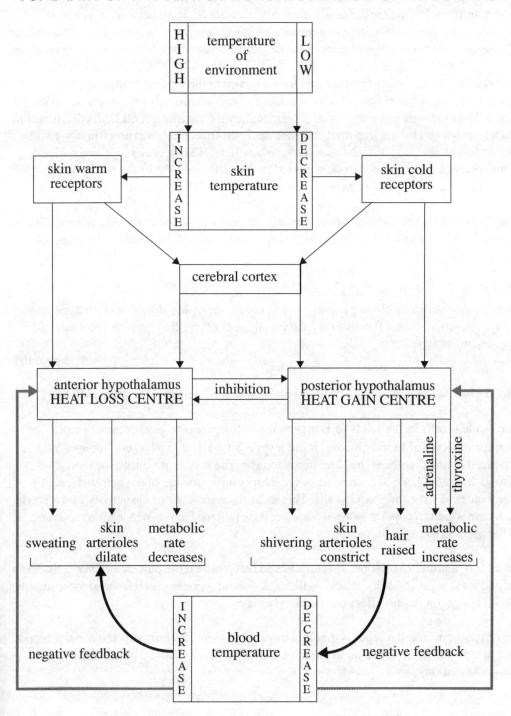

Option H: Further human physiology

H.1.4

Explain the control of water potential including the roles of the hypothalamus, the posterior pituitary gland and feelings of thirst (cross reference 15.2).

© IBO 1996

Water potential is a measure of the tendency of water to move between regions. In practice it is a force acting on water molecules in solution when separated from pure water by a membrane permeable to water only (i.e. partly permeable). See section 15 (excretion).

The kidneys are the organs responsible for maintaining the water potential in the mammal. Large volumes of water pass through the kidneys and are (mostly) reabsorbed into the blood. The exact amount reabsorbed is regulated by the hormone **Anti Diuretic Hormone** (ADH), produced by the hypothalamus and secreted from the posterior pituitary gland. The hypothalamus contains the osmoreceptors which will detect changes in the blood solute potential. Nerve impulses are passed to the posterior lobe of the pituitary gland, releasing or inhibiting the release of ADH.

You feel thirsty when the mucous membranes in your mouth are dryer than normal. This is caused by a lower amount of water in the body and your response will be to drink some fluid.

H.1.5

Explain the control of blood glucose concentration, including the roles of glucagon and insulin secretion, α and β cells in the pancreatic islets, hypothalamus and feelings of hunger and satiety.

© IBO 1996

Blood glucose: The regulation of blood glucose will be described in section H.4.3.

The α and β cells in the **Islets of Langerhans** in the pancreas have receptors which monitor the level of blood glucose. If the level is above 100 mg/100 cm^3, then insulin is produced by the β cells. If the level drops too low, the α cells produce glucagon. Insulin, a protein hormone, attaches to receptors on the outside of the cell surface membranes of liver and muscle cells. This leads to an increase of the activity of the carrier molecule which transports glucose across cell membranes and hence leads to an increase in the use and storage of glucose.

Glucagon, a protein hormone, also attaches to receptors on the outside of liver cells. Via a second messenger (cAMP), the cell will break down glycogen and release glucose into the blood. Glucagon has no effect on muscle glycogen.

One hypothesis about hunger is that it is caused by a low level of fatty acids in the blood. The 'norm' varies from person to person but a value lower than the norm will lead to hunger sensations.

H.2 DIGESTION

H.2.1
State that digestive juices are secreted into the alimentary, canal by glands including salivary, stomach wall, pancreas, and wall of small intestine (cross reference 5.1.4).

© IBO 1996

Digestive juices are secreted into the alimentary canal by glands which include :
- salivary gland.
- stomach wall .
- pancreas (NOT the islets of Langerhans).
- wall of the small intestine.

H.2.2
State the contents of saliva, gastric juice and pancreatic juice.

© IBO 1996

Salivary glands produce saliva which contains water, mucous, sodium bicarbonate ($NaHCO_3$) and salivary amylase.

Gastric pits produce gastric juice which contains water, mucous (from mucous cells), hydrochloric acid (HCl, from parietal cells) and pepsinogen (from chief cells).

The (exocrine part of the) **pancreas** produces pancreatic juice which contains water, sodium bicarbonate ($NaHCO_3$), trypsinogen, carboxypeptidase, deoxyribonuclease.

H.2.3
Outline the role of membrane-bound enzymes in the surface cells of the small intestine in completing digestion.

© IBO 1996

Some digestive enzymes are immobilised in the cell surface membrane of cells on the surface of intestinal villi. These enzymes continue working even if the cell is rubbed off the villus and mixed into the intestinal contents.

H.2.4
Draw the structural features of exocrine glands including secretory cells grouped into acini and ducts.

© IBO 1996

Exocrine glands have a duct into which they secrete their products. They are grouped around hollow spaces (**acini**) which join to form ducts. Examples of exocrine glands are sweat glands and glands producing digestive enzymes (e.g. chief cells).

Endocrine glands have 'internal secretion' which means that they do not posses a duct and

Option H: Further human physiology

secrete their product into the blood. Hormone producing cells form endocrine glands.

SCHEMATIC DIAGRAM OF AN EXOCRINE GLAND

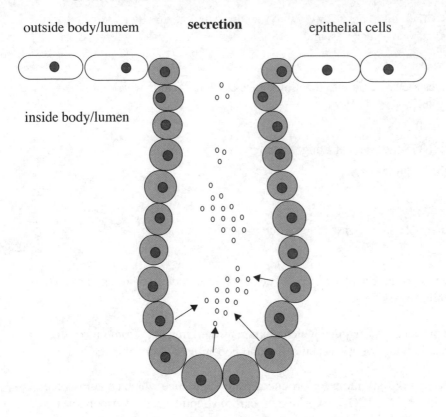

SCHEMATIC DIAGRAM OF AN ENDOCRINE GLAND

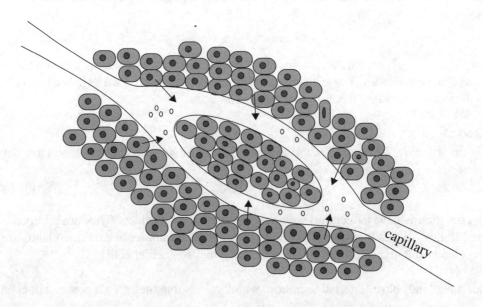

H.2.5
Explain the structural features of exocrine glands as seen in electronmicrographs.

© IBO 1996

The secretory cells of a gland produce the product and secrete it (into acinus or blood). The acini collect the product and pass it into the duct which leads into a lumen or out of the body.

H.2.6
State that secretion is under both nervous and hormonal control.

© IBO 1996

The secretion of saliva is triggered via the nervous system by the smell or taste of food.

The secretion of, for example, pepsinogen is stimulated by the presence of the hormone gastrin which is released from the pyloric section of the stomach when food enters the stomach.

H.2.7
Explain how soluble starch is completely digested as it passes along the alimentary canal.

© IBO 1996

Digestion of starch starts in the mouth. Salivary amylase breaks starch into maltose. The food does not stay in the mouth for very long and not all starch is digested.

The food is passed to the stomach where protein digestion takes place. The pH in the stomach is low and the salivary amylase does not work there. Starch is hardly changed.

In the duodenum, the pH is higher again and pancreatic amylase is active. The remainder of the starch is broken down to maltose, and maltase breaks maltose into glucose.

Glucose is absorbed by the epithelial cells of the villi and passed into the bloodstream which takes it to the liver.

H.2.8
Explain why cellulose remains undigested in the alimentary canal.

© IBO 1996

Like starch, cellulose is a polysaccharide. Cellulose does not dissolve in water which causes the first problem in its digestion.

While starch is made from long chains of a glucose (1,4 linkage), cellulose is made from β glucose (1,4 and 1,6 linkage). Cellulose cannot be digested by amylase. Mammals do not possess the enzyme to digest cellulose. Some bacteria do and they can live in a mutualistic relationship with herbivores.

Option H: Further human physiology

H.2.9
Explain why pepsin and trypsin are initially synthesized as inactive precursors and how they are subsequently activated.

© IBO 1996

Pepsin and **trypsin** are proteases. If they were produced in their active form, they would digest the cell which made them. So they are produced as inactive precursors (pepsinogen and trypsinogen) and are activated by the presence of hydrochloric acid (HCl) and enterokinase (an enzyme in the small intestine) respectively.

H.2.10
Outline the action of endo- and exopeptidases.

© IBO 1996

Two major kinds of protein digesting enzymes are recognised: endopeptidases and exopeptidases.

- **Endopeptidases:** they hydrolyse the peptide bond between amino acids located inside the chain but not the ends. They are responsible for the first stage of protein digestion. Examples are pepsin and trypsin.
- **Exopeptidases:** they hydrolyse the terminal peptide bonds. They are responsible for the final stages of protein digestion in the ileum. Examples are carboxypeptidase, aminopeptidase and dipeptidase.

H.2.11
Explain the problem of lipid digestion in a hydrophilic medium and the role of bile in overcoming this problem.

© IBO 1996

Lipids are difficult to digest since the food is in a hydrophilic medium. The lipid molecules will group together forming spheres of fat. An enzyme (**lipase**) is water soluble but has an active site for the hydrophobic lipid molecules to bind. It can only work on the surface of the lipid sphere. This makes digestion very slow.

Bile salts emulsify fats, meaning that they divide big drops of fats into smaller droplets. Bile salts have a hydrophilic end and a lipophilic (hydrophobic) end. The hydrophobic end will bind to the lipid, leaving the hydrophilic end to stick out and interact with water, thereby preventing other lipid molecules attaching there. This increases the surface area and therefore increases the speed with which the lipase can digest the lipid.

H.3 ABSORPTION OF DIGESTED FOOD

H.3.1
Draw a portion of the ileum (in transverse section), as seen under a light microscope (cross reference 5.1.5).

© IBO 1996

A diagram of a transverse section of a generalised portion of the small intestine can be found in section 5.1.

H.3.2
Explain the structural features of an epithelium cell of a villus as seen in electron micrographs including microvilli, mitochondria, pinocytotic vesicles and tight junctions.

© IBO 1996

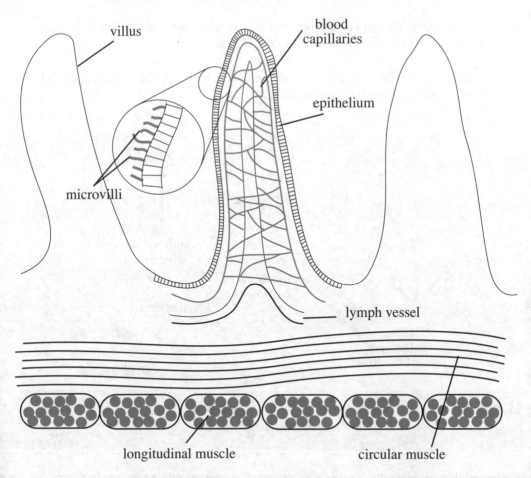

DIAGRAM OF THE LINING OF THE INTESTINE

You can see that the epithelial cell (positioned on the villus) has microvilli extending into the lumen. This is to further extend the surface area available to absorb the food. Mitochondria are present to provide the energy required for active transport by which

Option H: Further human physiology

some nutrients are absorbed. Vesicles are present since some food is taken up by endocytosis. **Pinocytosis** is common in young mammals feeding on milk. **Tight junctions** and desmosomes are connections between cells to maintain the integrity of the tissue. Tight junctions are fusions of adjacent cell surface membranes which form a continuous seal around each cell in the tissue. Desmosomes are spot welds involving a plaque of dense fibrous material between cells with filaments from cytoplasm of neighbouring cells looping.

H.3.3
Explain the mechanisms used by the ileum to absorb and transport food, including facilitated diffusion, active transport and endocytosis.

© IBO 1996

Glucose is absorbed into the epithelial cells by facilitated diffusion.

Amino acids and dipeptides are absorbed by active transport.

Fatty acids and glycerol diffuse into the epithelial cell and are combined to form fats which leave the cell via exocytosis to go to the lacteal.

Unicellular organisms take up their food via endocytosis and young mammals absorb some of the nutrients in milk via pinocytosis.

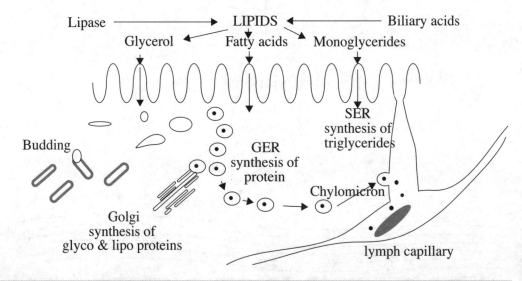

H.3.4
List the materials which are not absorbed and are egested.

© IBO 1996

All materials which are not absorbed in the small (or large) intestine must be egested.

These include: cellulose, lignin, bile pigments, bacteria and intestinal cells.

Biology

H.4 THE FUNCTIONS OF THE LIVER

H.4.1
Outline the circulation of blood through liver tissue including hepatic artery, hepatic portal vein, sinusoids and hepatic vein.

© IBO 1996

The liver, the largest organ in mammals, has an excellent blood supply. The hepatic artery and the hepatic portal vein deliver the blood and the hepatic vein takes is away from the liver back to the heart.

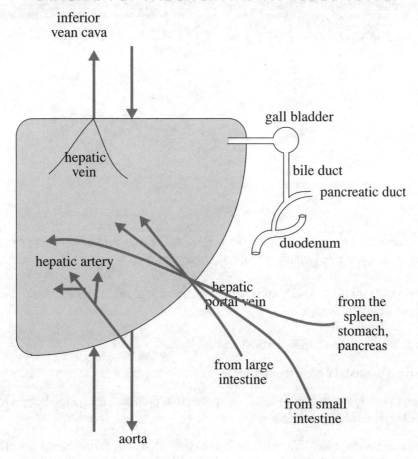

DIAGRAM OF THE LIVER AND ITS BLOOD SUPPLY

The liver is divided into many liver lobules. Between the lobules, branches of the hepatic artery and the hepatic portal vein are found, as well as bile tubules. A branch of the heptic vein is found in the centre of each lobule.

Option H: Further human physiology

AN IMPRESSION OF THE STRUCTURE OF THE LIVER

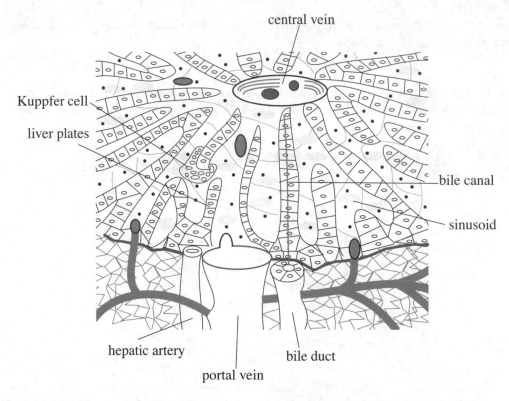

The functional unit of the liver is called the acinus. An acinus contains branches of the three bloodvessels, the bile ductule and many hepatocytes.

The blood flows from the hepatic arteriole and the hepatic portal venule through the sinusoids to the hepatic venule.

Sinusoids differ from capillaries in three ways:

1. they have a dilated, large, irregular lumen.

2. between the lining endothelial cells are spaces that facilitate exchange between the sinusoids and adjacent tissues.

3. the basement membrane-like material is not continuous but forms barrel hooplike rings around the endothelial walls.

The **Kuppfer cells** are lining the sinusoids. They ingest foreign particles and are involved in the breakdown of old erythrocytes. The **bile** is produced by the hepatocytes and moves in the opposite direction of the blood into the bile ductules.

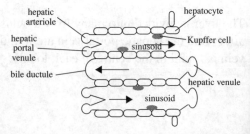

H.4.2
Explain the need for the liver to regulate levels of nutrients in the blood.

© IBO 1996

The liver regulates the levels of nutrients in the blood. This is important because the rate at which nutrients are used is not always constant and certainly the supply of nutrients depends on how recently a meal was ingested. Some time after eating, a lot of nutrients will enter the blood which would cause a sharp increase in, for example, the blood glucose level. Too high levels of glucose can do damage, for example, to the retina so a regulating mechanism which keeps levels of nutrients in the blood constant is required. The liver, receiving all the blood from the small intestine, will temporarily store excess glucose (see section H.1) and release it into the blood as blood glucose levels drop due to glucose being used (for cellular respiration).

Glucose is stored as glycogen under the influence of the hormone insulin (β cells of Islets of Langerhans in the pancreas). When blood glucose levels are low, glucagon (α cells of the Islets of Langerhans in the pancreas) stimulates the breakdown of glycogen and release of glucose to the blood.

Proteins are broken down by proteases and the individual amino acids can be used to build proteins. Some amino acids can be made from other amino acids in the process of trans amination and amino acids can also be used for energy after de-amination.

H.4.3
Outline the role of the liver in the storage of nutrients including carbohydrate, iron, the vitamins retinol and calciferol.

© IBO 1996

Carbohydrate storage:
The liver plays a vital role in carbohydrate metabolism. The role of **insulin** and glucagon in this will be further detailed in section 5.5.2 and H.1.5. Although the intake of carbohydrates during meals creates a very fluctuating blood glucose level in the hepatic portal vein (the blood vessel which takes the blood from the small intestine to the liver), the level of glucose in the blood at other places is remarkably constant (around 90 mg glucose/100 cm^3 blood). The liver converts all monosaccharides into glucose and stores any surplus as glycogen, an insoluble polysaccharide.

The following processes are involved in glucose metabolism:
Glycogenesis: for storing glucose

$$\text{glucose} \overset{\text{insulin}}{\Leftrightarrow} \text{glucose-6-phosphate} \Leftrightarrow \text{glucose-1-phosphate} \Leftrightarrow \text{glycogen}$$
$$\text{(phosphorylation)} \hspace{4cm} \text{(condensation)}$$

Remember that insulin is a hormone, NOT an enzyme.

The liver can store up to 100 gm of glycogen. The muscles also store glycogen.

Option H: Further human physiology

Glycogenolysis: for breaking down glycogen and mobilising glucose.

This process occurs in the liver when the blood glucose level falls below 60mg/100cm^3.

$$\text{glycogen} \underset{\text{(stored)}}{\overset{\text{phosphorylase}}{\Leftrightarrow}} \text{glucose-1-phosphate} \Leftrightarrow \text{glucose-6-phosphate} \Leftrightarrow \underset{\text{(free)}}{\text{glucose}}$$

The enzyme phosphorylase is activated by the presence of the hormones glucagon (from the pancreas), adrenalin (from the adrenal medulla) and nor-adrenalin (from the nerve endings of sympathetic neurons).

Muscles lack some of the required enzymes and convert glycogen into pyruvate which can be used for (an)aerobic respiration.

Gluconeogenesis: glucose can be produced from amino acids and glycerol in times of hypoglycaemia.

Below is a summary of the carbohydrate metabolism. What is the role of the liver in carbohydrate storage?

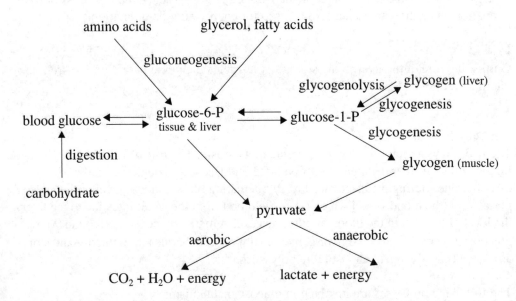

Storage of iron.
Iron is a component of haemoglobin. The process of **erythrocyte breakdown** is discussed in section H.4.5. After the haemoglobin from the erythrocytes has been broken down, the iron is carefully stored. Although several kinds of food contain iron, this element is difficult to absorb. The iron from the broken down haemoglobin will be used again to produce new haemoglobin. In the meantime, it will be stored in the liver in the form of ferritin, a complex of iron and β-globulin. The human liver contains approximately 1 mg iron per gm dry mass.

Biology

Storage of retinol and calciferol.
The liver is capable of storing water soluble vitamins but the main vitamins stored in the liver are fat soluble, such as retinol (vit. A) and calciferol (vit. D).

Retinol is found in dairy products and carrots. It is part of a visual pigment and a deficiency of retinol can lead to nightblindness.
Calciferol is found in cod liver oil and dairy products. It is also made by the skin under the influence of UV light. Calciferol helps in the uptake of calcium and a calciferol deficiency can lead to rickets in a child.

H.4.4
Describe the process of bile secretion.

© IBO 1996

Bile is a yellow/green fluid produced by the liver. It contains water, bile salts, bile pigments, inorganic salts and cholesterol. Bile is produced in the hepatocytes and travels via the bile canaliculi to the bile duct which empties into the gall bladder. Bile is stored and concentrated in the gall bladder. When acid chyme (partially digested food mass) comes into contact with the wall of the duodenum, it will secrete a hormone CCK. CCK promotes release of bile. Bile will travel through the bile duct which joins the pancreatic duct just before emptying into the duodenum.

Bile salts are derivatives of cholesterol. They have a hydrophobic side which will attach to a lipid, leaving the hydrophilic end to stick out and interact with water. This reduces the size of the lipid droplets and is called emulsification. Too little bile salts in the bile will raise the concentration of cholesterol and may cause them to precipitate, forming gall stones.

Bicarbonate (HCO_3^-) helps to neutralise the acid from the stomach as the food enters the duodenum.

Bile pigments are products of the breakdown of haemoglobin. The pigments have no function but in the small intestine, they are converted to a yellow pigment by the action of bacteria. This contributes to the colour of the faeces.

H.4.5
Describe the process of erythrocyte and haemoglobin breakdown in the liver including phagocytosis, digestion of globin, and bile pigment formation.

© IBO 1996

You have approximately 5×10^9 erythrocytes per cm³ blood.

This means a total of $5 \times 10^3 \times 5 \times 10^9 = 250 \times 10^{11}$ in your body.

Since erythrocytes on average last 120 days, every day $\frac{250 \times 10^{11}}{120} \approx 2 \times 10^{11}$ new erythrocytes are formed and the old ones are broken down.

Option H: Further human physiology

The breakdown of erythrocytes is one of the functions of the liver. Old erythrocytes are broken down by **phagocytosis** (macrophages) in the liver (Kupfer cells), spleen and bone marrow. The haemoglobin which was packed in the erythrocytes is released and broken down into haem and globin.

Haem is an iron containing prosthetic group. The iron is stored in the liver (see above) and the remainder of the group becomes biliverdin (a green bile pigment) which in turn becomes bilirubin (a yellow bile pigment). Bacteria in the gut change bilirubin into a yellow pigment which gives the characteristic colour to faeces.

Globin is a protein and is broken down to its constituent amino acids. These are then treated as any other amino acid and may be used to make a protein, trans-aminated or de-aminated for energy.

H.4.6
Outline the synthesis of plasma proteins by the liver.

© IBO 1996

Since the liver plays an important role in the amino acid metabolism, it stands to reason that it is involved in the production of plasma proteins.

Plasma proteins are an extremely important constituent of blood plasma. The most common plasma protein is albumin, which transports a variety of molecules such as calcium, some amino acids, some hormones. Other examples are fibrinogen and globulins such as gamma globulins which are antibodies. The concentration of plasma proteins determines the distribution of water between blood and interstitial (intercellular) fluid. Since this process takes place by hydrostatic pressure and osmosis, a small change in the number of dissolved particles can change the rate of movement of the water molecules.

Below is a summary of the carbohydrate and protein metabolism in the liver.

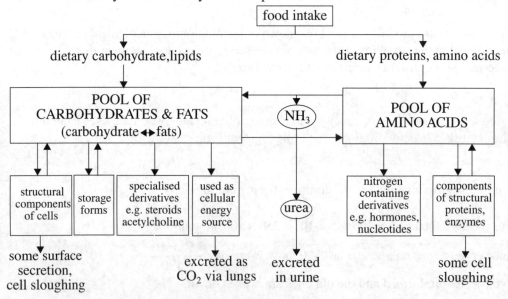

H.5 TRANSPORT

H.5.1
Explain the events of the cardiac cycle including atrial and ventricular systole and diastole, and heart sounds (cross reference 5.2.1 - 5.2.3).

© IBO 1996

One heartbeat is really one **cardiac cycle**. One cardiac cycle involves the following phases:
- blood enters the atria; the bicuspid and tricuspid valves open when atrial pressure exceeds ventricular pressure. This resting period is called **diastole**.
- then the two atria contract simultaneously (**atrial systole**), causing the blood to be pushed into the ventricles.
- almost immediately the ventricles contract (**ventricular systole**); this increases the pressure in the ventricle so closes the tricuspid and the bicuspid valves and opens the semilunar valves, pushing the blood into the aorta and pulmonary artery. The atria relax.
- when the ventricles then relax (**ventricular diastole**), some of the blood in the aorta and pulmonary artery will try to flow back and will close the semilunar valves.

The sound that the heart makes is usually considered to be a 'lub-dub' sound. The 'lub' is caused by the closing of the bicuspid and tricuspid and the 'dub' is caused by the closing of the semilunar valves in the arteries. So ventricular systole = 'lub'; ventricular diastole = 'dub'.

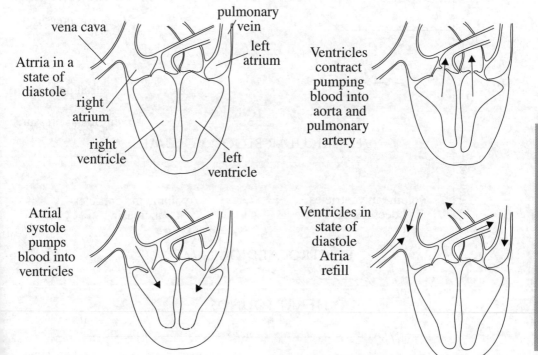

Option H: Further human physiology

H.5.2
Analyse data showing pressure and volume changes in the left atrium, left ventricle and aorta, during the cardiac cycle.

© IBO 1996

In the data below, you see that the volume and pressure inside the chambers of the heart changes as it goes through the cycle.

The atria fill up with blood during ventricular systole and diastole. As they fill up, both volume and pressure increase. When atrial systole starts, the volume of the atria decreases and the pressure goes up. After the blood flows into the ventricle, the atria relax and the volume increases while the pressure decreases. All these changes are small (from 0 kPa to 1 kPa) compared to the changes which occur in the ventricles and the changes in the left ventricle are much greater than in the right ventricle.

The left ventricle quickly fills with blood and then ventricular systole starts. The pressure in the ventricle increases from just above 0 kPa to almost 16 kPa in approximately 0.1 s. The pressure remains high for the next 0.2 s and the volume of the ventricle goes from 100% to almost 0%. During diastole, the ventricle also fills up with blood so its volume increases. The pressure remains low until atrial systole forces blood into the ventricle which also sharply increase the volume of the ventricles.

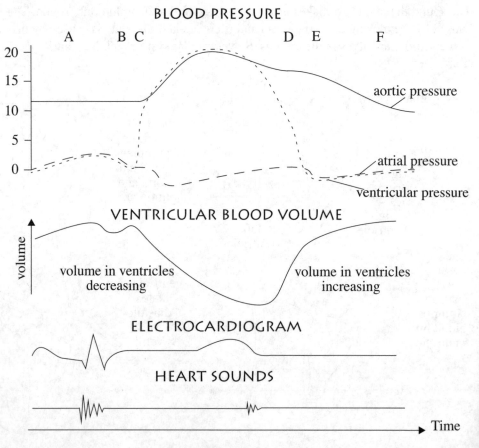

The diagram below illustrates the way in which the blood is distributed throughout the body (by volume):

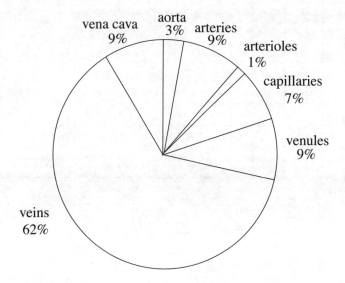

There is also a considerable variability in the speed of blood in various parts of the body. The velocity in a particular part of the body varies during the heartbeat. The following diagram compares the normal maximum velocities:

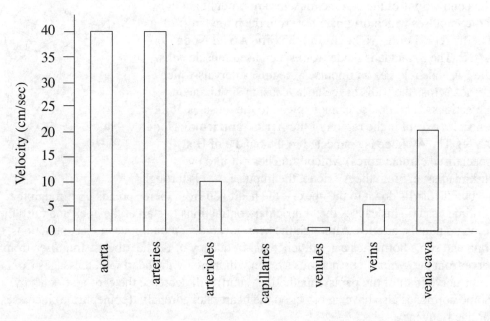

Finally, the pressure in the various blood vessels varies from type to type. Again the pressure depends upon the part of the heartbeat cycle we are considering. The diagram shows typical maximum pressures (kPa).

Option H: Further human physiology

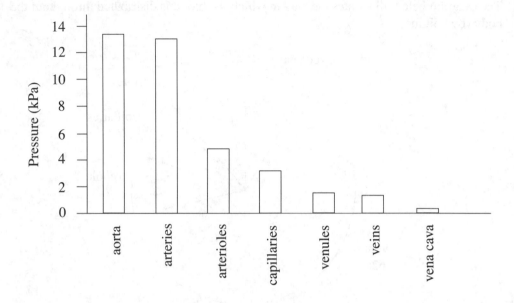

H.5.3
Outline the mechanisms that control the heart beat including SA node, AV node, and conducting fibres in the ventricular walls.

© IBO 1996

The contractions of the cardiac muscles are brought about by nerve impulses which originate not from the brain but from a specific region of the right atrium: the **Sino Atrial Node** (SAN). The SA node is made from specialised muscle cells. The SA node releases an impulse at regular intervals which spreads across the walls of the atria, causing simultaneous contractions. The impulse cannot spread to the muscles of the ventricles except in the region of the **Atrio Ventricular Node** (AVN). The AV node is connected to the bundle of His (specialised cardiac fibres) which branches out into the Purkinje tissue. From the AV node, the impulse travels through the bundle of His down to the apex of the heart and from there spreads up through the Purkinje tissue. This causes the ventricular contractions to start at the apex and push the blood up into the arteries. Although the heart is largely autonomous in its contractions, the brain and some hormones can influence the frequency of the heartbeats. Impulses from a nerve from the sympathetic nervous system will increase the heart rate, messages from the vagus nerve (part of the parasympathetic system) will decrease the cardiac frequency.
Some hormones also have an effect on the heart rate: adrenalin (epinephrine) increases cardiac frequency.

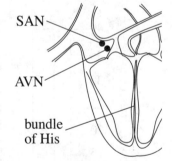

If the SA node does not function properly, it is quite easy to implant an artificial **pacemaker** to carry out this function. With a well adjusted pacemaker, a person with a malfunctioning SAN can live a long and active life.

H.5.4
Outline atherosclerosis and the causes of coronary thrombosis.

© IBO 1996

Atherosclerosis is the deposition of lipids on the inner surface of arteries. This hinders normal blood flow and may cause the formation of blood clots. All this can lead to a blockage in the coronary artery which causes a heart attack (myocardial infarction).

H.5.5
Discuss factors whioh affect rates of coronary heart disease.

© IBO 1996

Risk factors for **coronary heart disease** include:
- genetic factors.
- old age.
- gender (males are at greater risk).
- smoking.
- obesity.
- diet (saturated fat and cholesterol).
- lack of exercise.

H.5.6
Outline the way in which tissue fluid and lymph are formed in body tissues.

© IBO 1996

When the blood reaches the arteriole end of a capillary bed, the pressure of the blood is greater than the pressure of the fluid around the cells. Therefore a lot of fluid leaves the capillary to become **tissue fluid**. The composition of tissue fluid is very much like plasma except that the larger molecules (proteins) cannot pass through the capillary wall and that the number of leucocytes is variable. The fluid carries nutrients and oxygen which will be absorbed by the cells.

The blood remaining in the blood vessels has become much more concentrated (due to the blood cells and proteins staying behind) and soon this leads to a process of osmosis of the tissue fluid back into the blood vessels. The tissue fluid now carries carbon dioxide and other waste materials from the cells.

Almost all the tissue fluid will return to the blood vessels. The little bit that does not, drains into the lymph vessels.

H.5.7
Outline the transport functions of the lymphatic system.

© IBO 1996

Option H: Further human physiology

The lymphatic system is a system of blind ending tubes, collecting the surplus fluid (now called lymph). The lymph vesses eventually drain into the circulatory system in the subclavian vein. The composition of lymph is similar to that of tissue fluid.

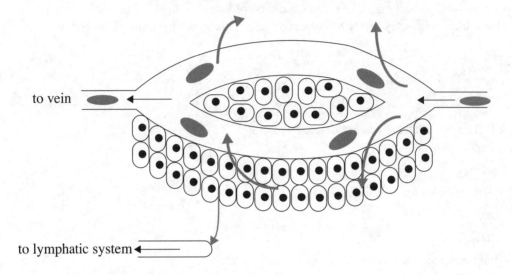

The lymphatic system has three major functions:
- it removes surplus fluid from the tissues (see above).
- it transports lipids absorbed in the small intestine (see section 5.1 and H.3).
- it plays a role in the body's defense.

H.6 GAS EXCHANGE

H.6.1
Define partial pressure.

© IBO 1996

Partial pressure: the pressure exerted by each component in a mixture. The pressure of a gas in a mixture is the same as it would exert if it occupied the same volume alone at the same temperature.

H.6.2
Explain the oxygen dissociation curves of adult and fetal haemoglobin, and myoglobin.

© IBO 1996

The oxygen that enters the blood in the lungs, will bind with the haemoglobin in the erythrocytes. The amount of saturation of the haemoglobin depends on the concentration of oxygen in the air. The first is expressed as a figure between 0 (no oxygen binding) and 1 (complete saturation, 4 oxygen bonded); the latter is noted as pO_2, the partial pressure of oxygen in the air.

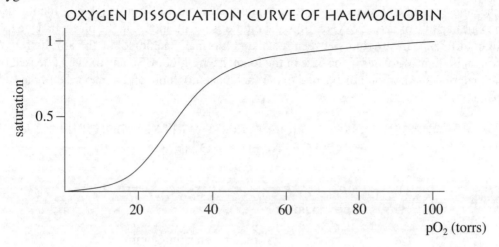

In the alveoli of the lungs, the pO_2 is 100 torrs (at sea level). In capillaries of active muscles, the pO_2 is approximately 20 torrs.

The pO_2 depends on the concentration of oxygen in the air and the air pressure. If the concentration of oxygen decreases, so does the pO_2. But if the air pressure decreases, as at high altitudes, there is also a drop in the pO_2. At sea level pO_2 is 100 torrs, at 6 000 m, pO_2 is 50 torrs. The saturation of haemoglobin is then lowered and hence the amount of oxygen transported. The body will, in time, compensate by increasing the number of erythrocytes.

Haemoglobin and Myoglobin

As haemoglobin carries oxygen around the bodies, when it reaches the muscles, the oxygen is taken over and stored by myoglobin. To be able to do this, myoglobin must have a higher affinity for oxygen than haemoglobin (at equal pO_2) but still be able to release the oxygen when the muscles need it.

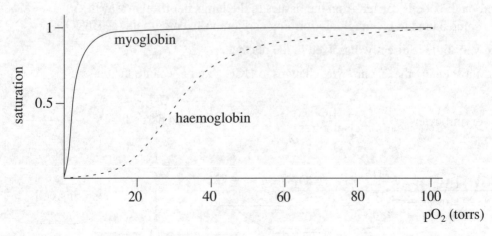

Option H: Further human physiology

Foetal Haemoglobin
Since mother and child have separate circulatory systems, the haemoglobin of the foetus must be capable of taking oxygen from the mother's haemoglobin in the placenta. Due to a slight difference in structure between foetal and 'normal' haemoglobin, the affinity for oxygen is slightly greater in the case of the foetal haemoglobin. So any oxygen released by the maternal haemoglobin is bonded to the foetal haemoglobin and transported to the foetus.

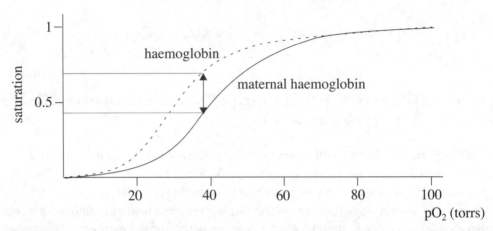

H.6.3
Describe the ways in which carbon dioxide is carried by the blood including the action of carbonic anhydrase, the chloride shift and buffering by plasma proteins.

© IBO 1996

Transport of oxygen and carbon dioxide:
Oxygen is transported from the lungs to the tissues by haemoglobin in the erythrocytes. It then becomes oxyhaemoglobin (HbO_2).

Carbon dioxide is carried from the tissues to the lungs in different ways:
- some bound to haemoglobin forming carbamino-haemoglobin (as $HbCO_2$).
- very little is directly dissolved in the plasma.
- most enters the erythrocyte, changes to HCO_3^- and goes into the plasma.

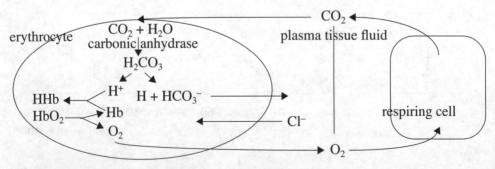

The CO_2 produced by the tissue cells diffuses into the plasma and directly into the erythrocyte. In the erythrocyte, the enzyme carbonic anhydrase turns it into H_2CO_3 which splits into H^+ and HCO_3^-. The HCO_3^- leaves the erythrocyte (and is exchanged for Cl^-, the so called chloride shift). The H^+ causes the HbO_2 to dissociate (releasing O_2 to the tissue cells) and binds with Hb to form HHb.

The HbO_2 is weakly acidic and associated with K^+. It can be referred to as $KHbO_2$. When oxygen is released and carbon dioxide taken up, the following buffer reaction takes place:

$$KHbO_2 \rightarrow KHb + O_2$$

$$H^+ + HCO_3^- + KHB \rightarrow HHb + KHCO_3$$

By acting accepting hydrogen ions, haemoglobin acts as a buffer molecule and so enables large quantities of carbonic acid to be carried to the lungs without any major change in blood pH.

H.6.4
Explain the role of the Bohr shift in the supply of oxygen to respiring tissues.

© IBO 1996

The Bohr effect
The pH of the blood (the amount of free H^+), is directly related to the carbon dioxide (CO_2) concentration. As the amount of CO_2 increases (and the pH lowers), the oxygen dissociation curve for haemoglobin shifts to the right. The effect, as you can see in the graph, is that the saturation of haemoglobin is reduced. In other words, oxygen is released from the haemoglobin. The CO_2 is produced in cellular respiration. The more energy the cells need, the more cellular respiration takes place and the more CO_2 is produced. The increased CO_2 concentration will reduce the saturation of haemoglobin and release oxygen to the cells.

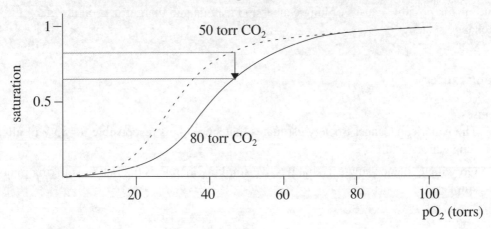

OXYGEN DISSOCIATION CURVES OF HAEMOGLOBIN IN DIFFERENT CO_2 CONCENTRATIONS

Option H: Further human physiology

H.6.5
Explain the mechanism of ventilation of the human lungs including the action of the internal and external intercostal muscles, the diaphragm and the abdominal muscles (cross reference 5.4.3).

© IBO 1996

Ventilation of the lungs involves the movement of the ribcage and the flattening of the diaphragm.

Between the ribs you find the external intercostal muscles, slanting forwards and downwards and the internal intercostal muscles, slanting backwards and downwards. Inhalation involves the increase of the lung volume (and hence decrease of pressure) by contracting the external intercostal muscles and relaxing the internal intercostal muscles. The outward and upward movement of the ribcage increases the volume of the thorax cavity.

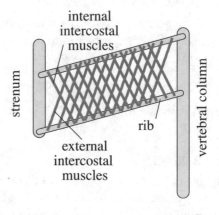

At the same time, the muscles in the diaphragm contract, flattening its usual dome shape. As the thorax expands, the lungs expand with it (pleural membranes, pleural fluid). The increase in the volume of the lungs leads to a decrease in the pressure. When the pressure inside the lungs is lower than that of the outside air, air flows in.

Exhalation is passive, brought about by the elasticity of the tissues. Exhalation can be forced by using the internal intercostal muscles to bring down the ribs and the abdominal muscles to push up the diaphragm.

H.6.6
Outline the possible causes of lung cancer and asthma and their effects on the gas exchange system.

© IBO 1996

Lung cancer

Causes:
- The American Cancer Society estimates that smoking is responsible for 83% of all lung cancer deaths.
- Growing evidence suggests that passive smoking increases the risk of smoking in non-smokers.

Effects:
Persistent coughing, coughing up mucous and blood
- Recurring pneumonia
- Smoke irritates the air passages causing them to produce more mucous.
- The cilia are (temporarily) paralysed and do not transport the mucous to the throat. This leads to coughing which further irritates air passages.
- The system of cleaning air passages fails and pathogens have a better chance of establishing themselves.
- Particles remaining in the lungs may contain carcinogens.

Asthma
Causes:
- Heriditary
- Attacks may be triggered by allergens e.g. pollen, house dust mites, certain foods etc.

Effect:
- Inflammation and constriction of bronchial tubes leading to wheezing, coughing and respiratory distress.

H.6.7
Describe the technique of mouth to mouth resuscitation.

© IBO 1996

Mouth to mouth resuscitation can be used to ventilate the lungs of a patient who is not breathing. It might be necessary to combine it with heart massage if the heart has ceased to function.

The idea behind mouth to mouth (or mouth to nose) resuscitation is that expired air still contains approximately 16% oxygen. This is enough for an inactive person to obtain the required amount of oxygen.

The first step is to find out if you are placing yourself in danger by assisting the casualty. If not, proceed as follows:

A. provide an open airway:
- remove obstructions over head, face, around neck, in mouth.
- place casualty on his or her back, lift the neck and tilt the forehead back, pushing the chin up.

B. ventilate the casualty's lungs
- take a deep breath.
- pinch the casualty's nose.

Option H: Further human physiology

- seal the lips around the casualty's mouth.
- blow into the casualty's lungs until you see the chest rise.
- remove mouth and watch the chest fall.

If no pulse can be found, it is necessary to also provide artificial compression of the heart to maintain circulation. You do this by placing the heel of your hand onto the lower half of the casualty's breastbone and press down sharply. Two lung ventilations and 15 compressions should be alternated by a single person. Two people working together can ventilate the lungs every fifth compression.

H.6.8
Explain the problem of gas exchange at high altitudes and the way the body acclimatises.
© IBO 1996

At high altitudes, the pressure is less and therefore the partial pressure of oxygen is also less. This makes it more difficult for the body to take up oxygen. As a result of lower levels of oxygen, the person might suffer from mountain sickness (fatigue, nausea, breathlessness and headaches). This may be caused by too rapid and deep breathing, losing too much carbon dioxde with a subsequent change in blood pH (**alkalaemia**).

This will be solved in a few days, by the kidneys excreting alkaline urine. Pulmonary ventilation increases and the bone marrow produces greater numbers of red blood cells to assist in the process of oxygen transport.

People who live at high altitudes have greater lung surface and larger vital capacity than those living at sea level.

Athletes competing in the Olympics in Mexico City had to come several weeks early to acclimatise to the altitude. Some athletes will reside at high altitudes prior to a competition, hoping that the improved capacity for oxygen transport will benefit them.

EXERCISE

1. The control of blood glucose concentration is an example of homeostasis.
   a. Which elements are required to maintain homeostasis?
   b. Identify these elements in the example of maintaining the blood glucose concentration.
   c. Describe what happens when blood glucose levels exceed the norm.

2. This question is about pepsin and trypsin.
   a. Where in the digestive system do you find them?
   b. What is/are their substrate(s)?
   c. What is/are the optimum pH?
   d. What is the name of the inactive precursor of each?

e. How is the inactive precursor activated?
  f. Why are they produced as inactive precursors?

3. a. Draw an annotated (=labelled) diagram of an epithelium cell of a villus as seen with the electron microscope.
  b. What is the role of the following in the above cell
   (i) microvilli.
   (ii) mitochondria.
   (iii) pinocytotic vesicles.
   (iv) tight junctions.

4. a. Draw a diagram of the liver and its blood supply.
  b. Compare and contrast the chemical composition of blood found in each of the major blood vessels associated with the liver.

5. Outline the sequence of events that leads to one complete contraction of all chambers of the heart. Name the structures involved.

6. a. Draw an oxygen dissociation curve of haemoglobin.
  b. Add to the same graph, the curve for foetal haemoglobin.
  c. If the foetus was extremely active, would the Bohr effect prevent the take up of oxygen from the mother? Explain your answer.

Option H: Further human physiology

Biology

THEORY OF KNOWLEDGE

25

Contributed by Barbara Free

Theory of knowledge

THE SCIENTIFIC METHOD

Most of the body of knowledge that comes under the name 'biology' has been acquired by the scientific method. There is no single correct model for this, but its main stages are summarised below:

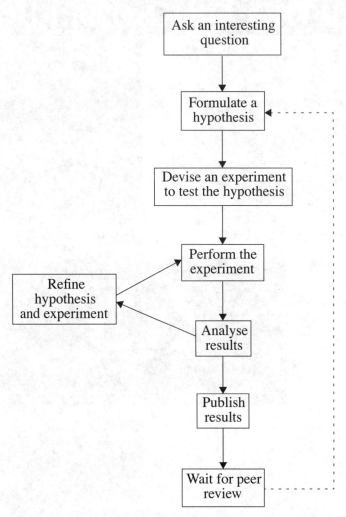

As an example of this process, consider the work of Mendel in discovering the principles of genetics.

The first stage was for Mendel (and many others) to notice that we inherit characteristics from our parents and to be interested in how this happens. Mendel then set up his experiment on his hypothesis that there were features of the peas that were transferred to their offspring.

The scientific method also allows for the refinement of the original hypothesis and the experiments designed to test it. This may result in several 'iterations' of the process.

Biology

Finally, when researchers feel they have the correct answer, they publish in scientific journals. This exposes the experiments to criticism. These days, other researchers will repeat the experiment to check the observations. This may result in the conclusions being challenged. Mendel's experiments and his results have been examined and, whilst it remains the case that his conclusions are still considered correct, there is a strong suggestion that his actual results are too good to be true and that, having realised his hypothesis was correct, Mendel massaged his results to fit the hypothesis.

This is the basis of the scientific method. Biologists, however, face rather larger problems than do, for example, physicists. Dealing as they do with living material, it is generally the case that their results are less reproducible. Thus when they, and others, repeat the experiment, the results are likely to be different from those originally obtained.

Suppose we take a 1 cm diameter rod of pure copper and clamp it 1 metre from its end. If a 1 kg weight is hung on the free end, the rod will bend.

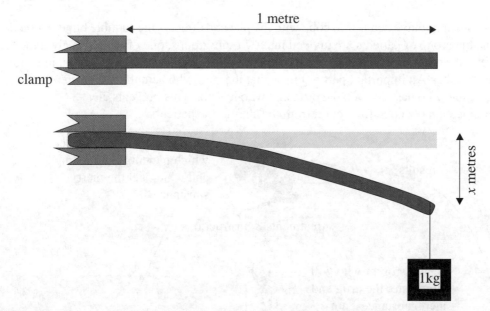

The displacement x has been exaggerated in the diagram. If this experiment is repeated with a second copper rod of the same size, the measured value of x is likely to be very similar to the first result.

If, instead, this experiment is performed using a series bamboo poles of the same diameter, the results are likely to be different because bamboo has a different strength to copper. Also, the values of x measured for the bamboo poles are likely to show a larger degree of variation than those measured for the copper rods. This is because biological specimens (such as bamboo poles) are generally more variable than inanimate objects such as copper rods. This has implications for biological researchers who are making measurements on living things whose inherent variaations will introduce variability into the results. This can make 'getting at the truth' particularly difficult for biologists.

This is particularly the case when we are trying to take measurements on human beings.

Theory of knowledge

An example of this is medical research. A typical problem is that of drug testing.

New drugs are generally examined for toxicity by testing them on animals. Whether or not they are effective at curing disease is much harder to determine. The basic principle of this testing is, of course to give the drug to people who are diseased and see what proportion of them recover. This needs to be compared with the proportion of people who would have recovered from the disease anyway. If the recovery proportion is significantly larger (significance testing was discussed in Chapter 10) in the group who are taking the drug than in the group who are not.

The major complicating factor here is the fairly well established fact that if people are given a 'cure' for a disease, this will improve their chances of recovery irrespective of the effectiveness of the 'cure'. This is known as the 'placebo effect' and is thought to result from the improved state of mind that results when people think that they have received treatment for their disease.

The most commonly used test to cope with the placebo effect is the 'double blind' method. In this, the group of patients are divided into two sub-groups, one of which receives the drug and the other of which receives an inert substance such as a sugar pill or an injection of sterile water. An important part of this is that the staff who actually administer the drug do not know who has received the drug and who has not. This prevents any communication of 'hope for a cure' from the staff to the patients.

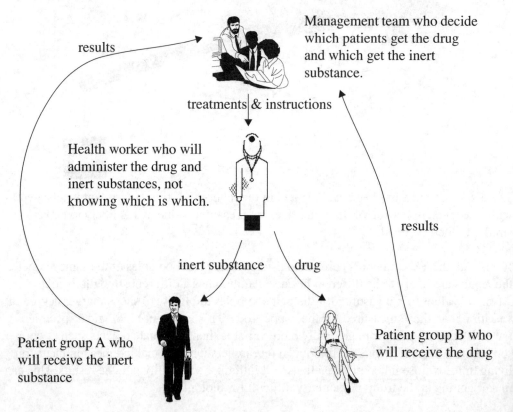

As the experiment proceeds, the health of both groups is followed and the results are fed back to the management team who are the only people who know which patients are receiving the drug and which are not. These results are then analysed using statistical techniques of the type discussed in the text. This is the basis for assessing the true effectiveness of the drug.

QUESTION

You are working as a member of the management team supervising the testing of a drug that, it is hoped, will prevent heart attacks. Two large groups of people who are thought to be at risk of heart attack are selected and divided into a group who will receive the drug and a group who will get sugar pills. During the experiment it becomes obvious that the drug is effective because the people in the group who are getting the sugar pills are suffering significantly more heart attacks than those who are receiving the drug. If you remain silent, you fear that a number of people in the 'sugar group' will die. You realise that, if the experiment is stopped and everyone gets the drug, lives will be saved. However, if the experiment is stopped early, the true effectiveness of the drug may never be established.

What should you do?

This is not a hypothetical question. A US group studying the effectiveness of aspirin in protection against heart attack stopped their experiment early when it became obvious that the treatment worked (to some extent). Similar ethical issues face groups who are currently working on cures for diseases such as AIDS and who may be forced to withold potentially effective treatments from dying patients.

BETRAYERS OF THE TRUTH?

The scientific method is a widely used and established method that seems to guarantee correct results. There are, however, suggestions that a very few scientists have 'massaged the evidence' to fit in with their hypothesis. The results of Mendel have already been mentioned and, if the suspicions are true they are minor, they do not affect the truth of Mendel's findings nor his place in history as one of the founders of modern biology.

This has not always been the case. It has been strongly suggested that the researches of the British psychologist Sir Cyril Burt were based on flawed data. Burt investigated the extent to which human intelligence is inherited and the extent to which it is determined by environmental factors such as quality of schooling. His experiment was based on studying twins who had been separated at birth and reared in different environments. Two identical twins share the same DNA, but if they are reared apart they have different environments. Burt performed a correlation study on the intelligences of these separated twins, calculating correlation coefficients. A correlation coefficient of 1 indicates perfect agreement between two sets of figures, 0 indicates no agreement and –1 indicates perfect disagreement.

Burt's initial figures were based on 21 pairs of twins reared apart and gave a correlation coefficient of 0.771. From this he concluded that about 75% of human intelligence is based on inherited factors. Several education systems based fundamental policy on this result. In the UK, children were divided into two groups at age 11 based on an intelligence test called 11+. The 'top group' then went on to receive an academic education in 'grammar schools'. The remainder were assumed to need a less academic program addressing basic skills and emphasising manual skills such as woodwork which they received in 'secondary modern schools'. Comparatively few students passed from the secondary modern system to universities.

Burt later claimed to have extended his study, first to 'over 30 pairs of twins' and, in 1966, to 53 pairs. In all three studies he claimed that the correlation coefficient between the intelligence measures of the twins was 0.771.

It has already been observed that the variability in measurements made on biological systems is inherently larger than that in physical systems. The chances that Burt's experiment would produce numerical results that were based on different groups of twins and yet which produced results that agreed to three decimal places is very small. It seems that there has to have been a degree of invention in Burt's results.

Unlike Mendel, Burt's reputation as a researcher is now discredited and the education system based on his conclusions has been dismantled in favour of one that assumes that education is capable of improving everyone's academic performance.

This and other cases of the failure of the scientific method are documented in: Broad & Wade, *Betrayers of the Truth*, Century Publishing, 1983.

QUESTIONS

1. You are a researcher working on a cure for malaria. Early results indicate that your team is working on a drug that is likely to work at some stage in the future, but the results are just not there at the moment. Your team leader is about to meet with the head of the government agency that is funding your research. It seems that the funding will be withdrawn unless the government sees evidence that the project is progressing well. It is put to you that unless some favourable results are sent to the government, the project will end and the cure will never eventuate. How do you respond?

2. You are an idealistic young researcher who believes very strongly in the christian ethic and the 'brotherhood of man'. You believe in the preservation of freedom. Your research is into the high incidence of diabetes in a group of islanders. During your research it becomes apparent that many of the islanders carry a gene that makes them very intolerant of alcohol. Since contact with westerners, alcohol has been freely available on the island, though it was not available before this. You become convinced that the high rates of diabetes are a direct result of alcohol abuse. You fear that release of your findings will result in the government imposing prohibition. You are conscious of the restriction of freedom and the crime that may result from this as well as the risk that the islanders may suffer a racial stigma. What do you do?

HOW IS DATA OBTAINED?

Biology, by its very nature deals with living things. As humans we are conscious of the rights of other living things to pursue their lives unmolested. This is, of course, not completely possible as all heterotrophs have to eat and they do so by eating other organisms. But to what extent are we justified in interfering with other living things in our search for knowledge? There are a range of cases to consider.

Firstly, it is unlikely that many would object to the preparation of a virus for study using an electron microscope on the grounds that it involved the death of the virus. How do you feel about the following cases?

1. A project to look at the effectiveness of a cure for cat 'flu that you intend to test on cats.

2. A project to look for a cure for AIDS that initially tests the drug for toxicity using mice and later uses a double blind test on humans.

3. An experiment to find out how the circulation of a horse works that involves dissecting a dead horse.

4. An experiment to find out how the circulation of a horse works that involves inserting probes into the veins of a live horse. This process may cause the horse some discomfort, but will not result in its death.

5. In the early days of scuba diving it was noticed that nitrogen dissolves in the blood and is released as a diver surfaces. If this release happens too rapidly, bubbles of nitrogen form in the blood causing pain, joint damage and sometimes death. Similar effects had been noticed in people working in pressurised environments such as the caissons used to sink bridge piles. This was known as caisson disease, 'the bends'
(joint pain caused victims to stoop), or decompression sickness. To determine safe times, depths and ascent rates, experiments were devised using animals in which these were put under pressure which was then released at various rates. Observations were made to see if the animals suffered pain. After the basic rules were established the experiments were repeated on humans. In one large study US navy divers were used. The results were used create 'dive tables' which are used today to protect both commercial and recreational divers.

6. During the period of atmospheric atomic testing, soldiers from several nations (as well as civilians) were exposed to quite high levels of radiation. It has always been denied by governments that this was done deliberately. However, a proportion of our knowledge of the harmful effects of radiation results from studies on the health of these people.

7. During the second world war, Nazi researcher performed experiments on the inmates. These were often extremely cruel and involved, for example, immersing people in iced water and observing their death from exposure.

Theory of knowledge

You probably responded to these as being of decreasing 'acceptability'. Most people would accept 1 but find 7 totally unacceptable. How did you feel?

QUESTIONS

1. Are any methods acceptable in obtaining biological knowledge? If not, how should we monitor the methods used by researchers?

2. If we feel very strongly that the Nazi experiments were immoral, should we use their results in, for example, designing exposure suits for people who may fall into very cold water? Should divers refuse to use dive safety tables?

ISSUES RELATING TO CLONING & GENETIC ENGINEERING

As biological knowledge advances, it has become more and more possible to 'interfere' with nature. In the early days, this was done by breeding. Both plants and animals have been altered by deliberately mating individuals with desirable characteristics. Many commercial crops are the products of centuries of such directed crossing of varieties over many generations. Racehorses and domestic dogs are also the products of centuries of deliberate breeding.

More modern techniques involve the grafting of plants. Many growers of tomatoes use plants that have roots from a variety known for the vigour of its root growth onto which they graft a stem taken from a variety selected for flavour and yield. The result is a plant which combines vigorous root growth with desirable fruiting properties.

Artificial insemination is now the standard method for reproducing farm animals. Very few of the lambs you see in fields are conceived completely naturally.

Cloning and genetic engineering promise to allow us enormously expanded control over the 'animal and vegetable kingdoms', up to and including the duplication of individual humans. There are immensely complex ethical issues here. Companies have already patented genetically modified seeds and have forbidden farmers from saving seed, requiring that they buy new supplies every year.

Assuming that it does become possible to clone and genetically alter humans, a huge range of unresolved issues are raised.

QUESTION

Which of the following futuristic operations on human foetuses would you find acceptable?
- the correction of an inherited disease such as haemophilia.
- the alteration of the genes of a foetus so that the baby will grow into an olympic shot-putter.
- the replacement of the nucleus so that the baby grows into a copy of you.

SPECIES EXTINCTION

As we become more environmentally conscious, issues of species extinction come more to the fore. As with the issues already considered, we are dealing with a range of 'moral acceptability'.

Here are some examples:

1. We have eliminated the scourge of smallpox by a program of immunisation. This has resulted in the effective extinction of the smallpox virus.

2. Would you like to see the elimination of all disease carrying insects if that were possible?

3. Would you like to see the extinction of the death adder snake?

4. Would you like to see the extinction of the great white shark?

5. Would you like to see the extinction of the giant panda?

We all respond in a variety of ways to these questions.

QUESTIONS

1. Given that we have a tendency to respond more to cuddly and furry animals and also to those that pose us no threat, should we afford greater 'rights of survival' to some species than to others? If so, what are reasonable criteria? You should also consider the degradation of the environment that might result from extensive species extinctions.

2. Man is thought to have originated in Africa and to have spread from there to other parts of the world. Large animals such as elephants and rhinoceros are very much more common in Africa than they are in other parts of the world. There is a lot of evidence that very large animals (mammoths, giant kangaroos etc.) were abundant in other parts of the world but have since become extinct. Do you think that there is any connection between these extinctions and the spread of man?

IMMUNISATION

Improvements in public sanitation, preventative measures such as immunisation and better treatments have resulted in a virtual doubling of human life expectancy in many parts of the world. One controversial aspect of these measures is immunisation. Immunisation only really works as a means of disease control if it is widespread in a population. If only 1% of babies are immunised against whooping cough, this will not prevent the spread of an epidemic. Levels of 90% immunisation almost certainly will protect both the immunised and the un-immunised by preventing its epidemic transmission. The difficulty is that there is a finite but very small

Theory of knowledge

chance that an individual child will be damaged as a result of the immunisation. In a very few cases children can develop a fever which has proved fatal. As this has become more widely known in some countries, more and more parents have not had their children immunised. They hope to avoid the risks of immunisation whilst benefiting from the fact that high rates of immunisation in the community have reduced the risks of epidemics.

This has become a problem in some developed countries in which falling rates of immunisation have resulted in childhood diseases such as diptheria making a comeback. Many more children are dying than were affected by immunisation.

In Holland some people refuse immunisation on the basis of religion. Some children (approximately 10 every 8 or so years) suffer from polio and are paralysed for life. Parents are legally allowed to make this choice for their children. Yet the government does force people to wear seatbelts (for their own good). Should the government 'protect' the children and insist on immunisation?

QUESTION

Given that some children will may as a result of immunisation, should it be compulsory? If you do not accept compulsion, what level of coercion is acceptable? Should this simply be through education or should it go further and include the withholding of government family support grants from families whose children have not been immunised?

# ANSWERS

26

ANSWERS
CHAPTER 1

1. B 2. B 3. C 4. D 5. A 6. C 7. B 8. D 9. B 10. C 11. D 12. A 13. B 14. C 15. A 16. A 17. D

18. a. 3 elements: cells are the units of life.
all organisms are made of cells.
all cells come from cells.

b. the cell theory was developed only after technology had made it possible to see the cells.
Hooke sees 'cells' in cork.
Van Leeuwenhoek describes cells using simple microscope.
Schleiden: all plants are made of cells.
Schwann : first concept of cell theory .
all plants and animals are made of cells.
all cells in an organism are identical.
Virchow showed that all cells arise from other cells.

19. eukaryotic cell, prokaryotic cell, cell organelle, membrane, DNA, atom.

20. a. child: ratio surface area/volume: $0.9 \text{ m}^2 / 0.024 \text{ m}^3 = 37.5 : 1$
dog: ratio surface area/volume: $0.13 \text{ m}^2 / 0.002 \text{ m}^3 = 65 : 1$

b. the dog

c. two possibilities:
- the dog because she has a greater surface area to volume ratio so she will e.g. loose more heat.
- he child because he is still growing/more active etc.

21. a. see section 1.2.1

b. see section 1.3.2

c. rough endoplasmic reticulum: system of membranes forming sheets continuous with the nuclear envelope.
- site for protein synthesis.

lysosome: small vesicle containing digestive enzymes
- intracellular digestion after fusion with food vacuole; "suicide bag"

Golgi apparatus: system of membranes
- intracellular transport - vesicle formation for exocytosis.

mitochondrion: double membrane around matrix
- site for Krebs cycle (matrix) and electron transport chain and oxidative phosphorylation (cristae) of cellular respiration.

nucleus: largest cell organelle, contains chromatin
- controls all activities of cell.

chloroplast: double membrane around stroma and thylakoids
- site for light dependent (thylakoids) and light independent (stroma) reactions of photosynthesis.

d. endosymbiont theory explains how eukaryotic cells arose from prokaryotic cells.
mitochondria and chloroplasts.
are thought to have been independent prokaryotes.
that were taken into another prokaryote via endocytosis.

and a symbiotic relationship developed.
where first the mitochondrion became the 'powerhouse'.
later the chloroplast became the 'food factory' of the new eukaryotic cell.

22. a. See section 1.4.1. for diagram.
Answer and diagram should include:
phospholipid bilayer.
cholesterol.
glycoproteins.
intrinsic and extrinsic proteins.
b. Membrane is phospholipid bilayer.
Phosphate part on outside of membrane.
Outside is hydrophilic/polar.
Lipid on inside of membrane.
Inside is hydrophobic/non-polar.
Interactions between non-polar amino acids and non polar lipids.
23. Interphase
biochemical reactions.
protein synthesis.
DNA replication.
Mitosis.
chromosomes moving to the equator.
separation of sister chromatids.
creation of two genetically identical nuclei.

CHAPTER 2
1. C 2. D 3. C 4. A 5. D 6. B 7. A 8. A 9. D 10. A 11. C 12. B 13. D 14. A 15. D
16. D 17. D
18. water has strong cohesion forces.
good solvent for polar molecules.
high specific heat.
high heat of melting/evaporation.
because:
water molecules are dipoles/polar.
negative and positive side.
therefore strong attraction forces between molecules.
takes a lot of energy to break bonds between molecules.
19. a. carbohydrate: carbon, oxygen, hydrogen.
lipid: carbon, oxygen, hydrogen.
protein: carbon, oxygen, hydrogen, nitrogen.
b. carbohydrate: monosaccharide.
diagram in section 2.2.2
lipid: glycerol and fatty acids.
diagram in section 2.2.3
protein: amino acid.
diagram in section 2.2.1

Answers

c. in forming disaccharides and lipids, two OH groups react.
they form water.
and link the two molecules via an O bridge.
in forming a dipeptide, an OH reacts with an NH_2
they form water
and link the two molecules via a peptide bond
20. a. increase in temperature means increase in speed of molecules
if molecules move faster, then they have more energy
enzyme lowers activation energy
higher temperature increases energy available
so more molecules react
until temperature becomes too high and enzyme is denatured
b. substrate must attach to active site for reaction to be catalysed
if a lot of substrate present then active site constantly "filled"
so reaction proceeds as rapidly as enzyme concentration allows
if substrate is not concentrated it may not "find" the active site
so active site is not always filled
so reaction is slow
c. high temperature denatures protein
drastic change of pH can denature protein
changes 3 dimensional shape
substrate must fit into active site
if shape of active site is changed, then substrate no longer fits
enzyme can not catalyse reaction
21. a. covalent bonds
b. covalent bonds
c. hydrogen bonds
d. covalent bonds
because they are based on the sharing of electrons
while hydrogen bonds are based on positive negative attraction forces.
e. because the hydrogen bonds need to "fit"
adenine and thymine can form 2 hydrogen bonds
cytosine and guanine can form 3 hydrogen bonds
because 3 rings are needed between the sugar-phosphate backbones
adenine and guanine have 2 rings each
cytosine and thymine have 1 ring each
22. Helicases separate ("unzip") the two complementary strands.
free DNA nucleotides then bind to the exposed bases via complementary base pairing
DNA polymerase connects one DNA nucleotide to the next
23. a. two (one for the a chain and one for the b chain)
b. twice (to make 2 a chains and 2 b chains)

CHAPTER 3.
1. B 2. B 3. D 4. D 5. B 6. A 7. C 8. B

9.
	possible phenotypes	possible genotypes
	tall	TT or Tt
	short	tt

P: tall   ×   short   Punnett square
        TT   tt

F1   tall (×   tall)
                    Tt   Tt

F2   tall + short
                       3 : 1
                      TT + Tt + tt
                       1 : 2 : 1

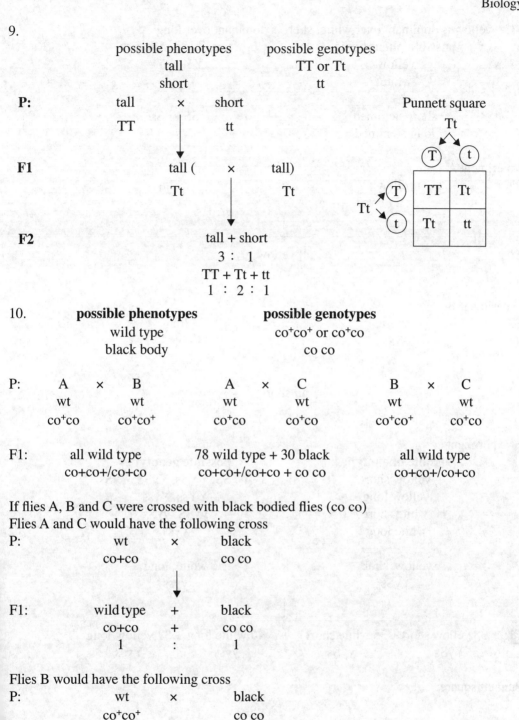

10.   **possible phenotypes**   **possible genotypes**
            wild type   co⁺co⁺ or co⁺co
            black body   co co

P: A   ×   B   A   ×   C   B   ×   C
       wt   wt   wt   wt   wt   wt
     co⁺co   co⁺co⁺   co⁺co   co⁺co   co⁺co⁺   co⁺co

F1:   all wild type   78 wild type + 30 black   all wild type
      co+co+/co+co   co+co+/co+co + co co   co+co+/co+co

If flies A, B and C were crossed with black bodied flies (co co)
Flies A and C would have the following cross
P: wt   ×   black
             co+co   co co

F1:   wild type   +   black
             co+co   +   co co
               1   :   1

Flies B would have the following cross
P: wt   ×   black
            co⁺co⁺   co co

F1:   all wild type
             co⁺co

Answers

11.a. yellow is dominant over white, short is dominant over long

b.
possible phenotypes	**possible genotypes**
yellow	YY or Yy
white	yy
short stemmed	SS or Ss
long stemmed	ss

experiment A
P: yellow   ×   white
    YY   yy
           ↓
F1:   all yellow
            Yy

experiment B
P: short   ×   long
     SS   ss
           ↓
F1:   all short
            Ss

c. experiment c

possible phenotypes	**possible genotypes**
yellow short	YYSS, YYSs, YySS, YySs
yellow long	YYss, Yyss
white short	yySS, yySs
white long	yyss

P: yellow, short   ×   white, long
        YySs   yyss
              ↓
F1: 38 yellow short + 35 white short + 40 white long + 37 yellow long
        YySs   yySs   yyss   Yyss

Punnett square:

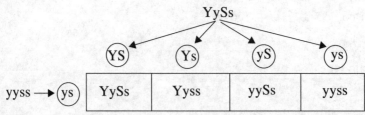

The only way to achieve equal amount of all 4 phenotypes is using the heterozygous parent YySs since the homozygous recessive parent is yyss.

12.
possible phenotypes	possible genotypes
coloured wildtype	CCAA, CCAa, CcAA, CcAa
coloured black	CCaa, Ccaa
albino	ccAA, ccAa, ccaa

('albino' can have the alleles for wildtype or black. These will not be expressed since the animal cannot make the pigment. If it could, it would be wildtype or black.)

P: black   ×   albino
F1: wildtype × wildtype

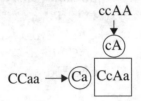

F2: wildtype + black + albino

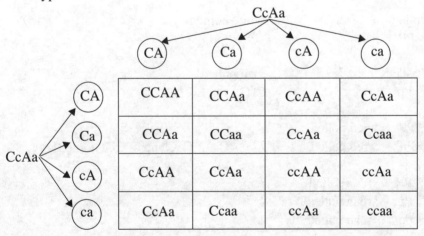

13.
possible phenotypes	possible genotypes
A	$I^A I^A$, $I^A i$
B	$I^B I^B$, $I^B i$
AB	$I^A I^B$
O	ii

P: grandfather   ×   grandmother
         A   B
        I^A.... I^B....

F1   father   ×   mother

Answers

F2:	I^B....	$\overset{A}{I^A i}$

baby
B
$I^B i$

a. Baby is type B. The genotype therefore is IBIB or IBi. Since Mother is type A, Baby must have genotype IBi, receiving IB from his father and i from his mother.

b. Mother therefore is IAi, Father has one IB allele, he could be type B or AB. No more details about Father or grandparents can be decided from this information.

14.
possible phenotypes	possible genotypes
red polled	$C^R C^R$ PP or $C^R C^R$ Pp
roan polled	$C^R C^W$ PP or $C^R C^W$ Pp
white polled	$C^W C^W$ PP or $C^W C^W$ Pp
red horned	$C^R C^R$ pp
roan horned	$C^R C^W$ pp
white horned	$C^W C^W$ pp

a. P :	red polled	X	white horned
	$C^R C^R$ PP		$C^W C^W$ pp

F1 :	roan polled	X	roan polled
	$C^R C^W$ Pp		$C^R C^W$ Pp

F2 :
$C^R C^R$ PP	red polled	1			
$C^R C^R$ Pp	red polled	2	$C^R C^R$ pp	red horned	1
$C^R C^W$ PP	roan polled	2			
$C^R C^W$ Pp	roan polled	4			
$C^R C^W$ pp	roan polled	2			
$C^W C^W$ PP	white polled	1			
$C^W C^W$ Pp	white polled	2			
$C^W C^W$ pp	white horned	1			

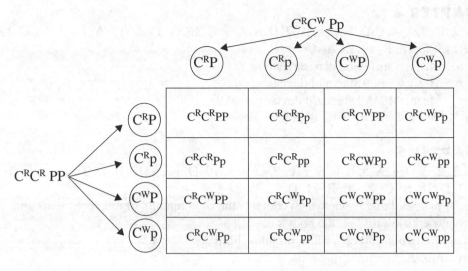

b. You would only use red polled cattle to breed but you woud first have to establish if they homozygous or heterozygous by doing a test cross (with a homozygous recessive individual).

15. **possible phenotypes**   **possible genotypes**
    normal male X^HY
    haemophiliac male X^hY
    normal female (including carriers) X^HX^H or X^HX^h

Haemophiliac females do not exist. This would require the recessive trait to be homozygous which in this case is lethal.

	P:	father	×	mother
		normal male		normal female
		X^HY		X^HX^h
	F1:	Mohammed	+ Latifa	+ brother
		normal male	normal female	haemophiliac male
		X^HY	X^HX^H or X^HX^h	X^hY

a. Mohammed does not carry the allele for haemophilia so will no chance of having a haemophiliac child (assuming his wife is normal)
b. Latifa might be a carrier. In that case and assuming a normal husband, a daughter would have 50% chance of being a carrier and a son would have 50% chance of being haemophiliac.

16. Out of 600 offspring, 72 are recombinants, i.e. they have a combination of traits not shown by their parents. In this case the recombinants are wild type eyes, vestigial wings and purple eyes, wild type wings.

The crossover frequency is (72/600) × 100% = 12%
The distance between the genes for eye colour and wing length is 12 cM.

Answers

CHAPTER 4
1. B 2. C 3. A 4. C/D 5. C 6. A 7. D 8. A 9. C 10. A 11. C 12. A 13. A 14. C 15. C 16. B 17. C 18. B 19. B 20. C
21. a. Mean: 56.1 Standard deviation: 2.8
    b. Mean: 134.9 Standard deviation: 4.6
       Mean: 1345.2 Standard deviation: 40.2
22. a. 15924 b. 1193 c. 1016 d. 22857

CHAPTER 5
1. C 2. B 3. D 4. A 5. B 6. B 7. C 8. C 9. A 10. D 11. A 12. B 13. D 14. B 15. B 16. A 17. D 18. B 19. A 20. B 21. D 22. B 23. C

24. a. A balanced diet is one which contains the right amount of macronutrients and micronutrients to satify all the body's needs.
b. Macronutrients: carbohydrates, proteins, lipids
Micronutrients: vitamins, minerals
c. Carbohydrates: can be used in cellular respiration to release energy.
Proteins: provide amino acids to make proteins for muscles, cell membranes, enzymes.
Lipids: can be used for energy or cell membranes
Vitamins: cannot be made by body, function as coenzymes
Minerals: can be structural part of the body or parts of molecules (e.g. Fe in haemoglobin)

25. a. The blood in arteries comes from the heart and is at high pressure. If the arteries walls were not thick, they might rupture.
b. Because the exchange of materials (food and oxygen to tissue, waste from the tissue into the blood) takes place between the capillary and the tissue cells. Blood flows slowly so that there is enough time for this exchange.
c. Because the pressure is low and the speed is moderate, blood could flow in the wrong direction (e.g. by gravitational pull) if not stopped by valves.

26. a. HIV virus will attack T-helper cells, this reduces the efficiency of the immune system so antibody production is reduced, which allows opportunistic diseases.
b. B cells are produced and mature in the bone marrow when mature they travel to lymph nodes antigen entering the body will be trapped by phagocyte phagocyte goes to lymp node to find B cell with correct binding site T helper cell will facilitate this and cause B cell to clone some cells become memory cells most become plasma cells and produce antibodies against antigen.
c. if T-cells are destroyed by HIV virus the process of antigen recognition and plasma cell production will not work/work less efficient so antibodies are not produced/produced less so opportunistic diseases can occur.

27. a. see section 5.4.3
b. large surface area, moist, thin walled (short diffusion distance), good blood supply (maintaining concentration gradient)
c. because the "old" air needs to be refreshed.

28. a. receptors: thermoreceptors in skin and brain (heat centre)
control centre: heat centre in the brain effectors: skin arterioles, sweat glands, muscles, cell metabolism, behaviour.
b. heat loss can be reduced by:
vasoconstriction: reducing the diameter of the bloodvessels to the skin so that less blood

flows to the skin, skin is less warm, less heat lost, 'fluffing' of hair or feathers will increase the thicknessof the insulating layer of air around the body, warm clothes do the same, a thick layer of subcutaneous fat

c. conduction: tranfer of thermal energy without movement of matter but in physical contact; e.g. increase in temperature of one end of a metal rod when the other end is heated.

convection: tranfer of thermal energy by movement of matter; e.g air currents (hot air rises)

radiation : tranfer of thermal energy without physical contact or movement of matter; e.g. Sun's radiation.

d. radiation

29.a. Gamete production via meiosis involved the pairing up and subsequent splitting up of homologous chromosomes. The arrangement of the chromosomes within each pair is random. Therefore, the gametes produced are different from eachother, contributing to variation.

b. oestrogen and progesterone inhibit FSH production, keeping them at high levels stops FSH from being released, follicles will not grow without FSH

no follicle - no eggcell- no ovulation - no pregnancy

CHAPTER 7

1. A 2. B 3. A 4. D 5. A 6. C 7. B 8. D 9. A

10. (digestive) enzymes are proteins
gene for polypeptide chain is found in the nucleus
transcription takes place in the nucleus
translation takes place in the cytoplasm
translation can take place on ribosomes of rER
protein goes into lumen of rER
vesicle is pinched off
vesicle travels to Golgi apparatus
vesicle fuses with Golgi apparatus moving content into Golgi
protein modified in Golgi (e.g. sugar added to make glycoprotein)
vesicle containing new product is pinched off Golgi and becomes lysosome.

11. a. Interphase because most of the nuclei are found in this stage.

b. Interphase: 25 minutes
Prophase: 16 minutes
Metaphase: 5 minutes
Anaphase: 6 minutes
Telophase: 7 minutes
(Add no. of nuclei; divide each stage by total and multiply by 60 to get the number of minutes.)

c. Rapidly dividing tissue was selected
therefore a large proportion of the nuclei was dividing
not representative for all tissue
underestimates the duration of Interphase

12. a. Digram measures 6 × 8 cm.
Cell is 20 mm

Answers

6×10^{-2} m / 20×10^{-6} m = 3000
magnification is 3000 times
b. $3*10^{-2}$ m / 3000 = 1×10^{-5} m
10 μm or 0.01 mm

CHAPTER 8
1. D 2. C 3. A 4. D 5. B 6. D 7. D 8. B 9. B 10. A 11. C 12. B 13.
13. Diagram : see 8.1.3
2 deoxyribose - phosphate backbones by C3 - P - C5 linkage
antiparallel (or 3'-5' direction indicated)
one base attached to each deoxyribose at C1
4 bases possible - combination purine -pyrimidine needed
A - T and C - G complementary pairs
A/G : purines (2 rings); C/T pyrimidines (1 ring)
A-T : 2 H bonds; C-G : 3 H bonds
twisted ladder - 10 nucleotides per turn
14. a. helicase separates strands ('unzips')
b. RNA primase catalases formation of RNA primer
c. DNA polymerase III : DNA synthesis
d. DNA polymerase I : removing RNA primers and filling the gaps thus created.
e. DNA ligase attaches Okazaki fragments
15. a. to code for repressor proteins
b. this is the binding site for RNA polymerase
c. to the operator
d. by preventing the RNA polymerase from moving from the promotor site to the structural gene and hence preventing transcription.
e. lactose will bind to the repressor protein. this stops the repressor protein from binding to the operator so transcription can take place. subsequent translation will result in the production of lactase which will digest the lactose.
16. a. GTP provides the energy for the two subunits of the ribosome to attach
b. ribosomes provide the environment where the tRNA anticodon can bind (complimentary base pairing) to the mRNA codon
c. polysomes are groups of ribosomes found on one mRNA. this means that several polypeptides chains are produced almost at the same time.
d. codons are mRNA triplets (3 nucleotides in a row) which code for a certain amino acid. The codon on the mRNA dictates which tRNA anticodon will bind and this in turn dictates which amino acid is brought in.
17. primary structure: covalent bonds/peptide bonds between amino acids.
secondary structure: hydrogen bonds
tertiary structure: hydrophilic and hydrophobic interactions, disulfide bridges
quaternary structure: hydrogen bonds, positive/negative attraction forces, hydrophilic and hydrophic forces, disulfide bridges.
18. a. to the active site
b. elsewhere
c. competitive inhibitor: prontosil
non-competitive inhibitor :cyanide

CHAPTER 9
1. A 2. C 3. D 4. A 5. A 6. C 7. B 8. A 9. A 10. B 11. A 12. C 13. D 14.
13. a. cytoplasm b. matrix of mitochondrion c. inner membrane of mitochondrion
d. see section 9.1.5 e. Krebs cycle is a cycle where one molecule is changed into another with the aid of enzymes. As long as the enzymes can be found in the matrix (in reasonable concentration) the Krebs cycle can proceed.
f. The sequence of electron carriers in the electron transport chain is very important. So they are situated in the inner membrane to keep their position to one another. The cristae are folds in the inner membrane to increase the surface area so that many electron carriers can be found there. g. no build up of protons in the intermembrane space, no ATP produced, eventually no respiration as the first steps of glycolysis require ATP
14. a. to provide the energy (and reducing power) needed for Calvin cycle/for fixing carbon dioxide. b. to produce ATP
c. non-cyclic photophosphorylation leads to the production of ATP and NADPH both of which are needed to fix carbondioxide and produce glucose. Glucose can be used for longer time storage of energy as well as for making other compounds e.g. cellulose.
d. to combine 3 carbondioxide into 1 triose phosphate fixing carbon dioxide
15. Similarities: both produce ATP, both involve movement of electrons through a membrane, both involve movement of protons across a membrane, both involve a built up of protons on one side of the membrane, both involve a series of electron carriers, both involve ATP synthetase
Differences: Photosynthesis, Respiration, cell organelle, chloroplast, mitochondria, ATP produced in stroma, matrix, source of electrons water, NADH, electron acceptor $NADP^+$ oxygen, energy source, light, bonds in glucose.

CHAPTER 10
1. C 2. B 3. C 4. A 5. A 6. D 7. B 8. D 9. A
10. a. body colour wild type and wing length wild type.
b. Possible phenotypes   Possible genotypes
    wild type - wild type $c^+c^+w^+w^+$ or $c^+cw^+w^+$ or $c^+c^+w^+w$ or c^+cw^+w
    wild type - vestigial c^+c^+ww or c^+cww
    black - wild type ccw^+w^+ or ccw^+w
    black - vestigial $ccww$

P :   black - vestigial   X   wild type - wild type
        $c^+c^+w^+w^+$ $c^+c^+w^+w^+$

F1 :   wild type - wild type   X   wild type - wild type
        c^+cw^+w c^+cw^+w

d.
F2 :   wild type - wild type
        wild type - vestigial
        black - wild type
        black - vestigial

Answers

c⁺c⁺w⁺w⁺	1	c⁺cw⁺w⁺	2	ccw⁺w⁺	1
c⁺c⁺w⁺w	2	c⁺cw⁺w	4	ccw⁺w	2
c⁺c⁺ww	1	c⁺cww	2	ccww	1

c.

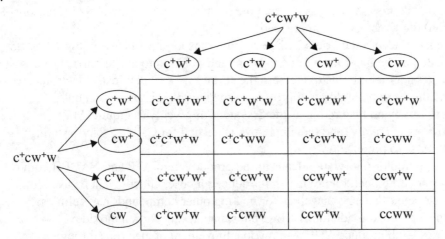

11. a. Mendel's second law = principle of independent assortment
Any one of a pair of characteristics may combine with either of another pair.
b. It does not. It only applies to genes on different chromosomes (not linked).

12. a.   grey straight   grey curly   ebony curly   ebony straight
b.   26   grey straight   26   grey curly
             26   ebony curly   26   ebony straight

c. use the equation to find chi squared is 34.9
degrees of freedom are 3
critical value is 7.81
difference between the observed and expected values is large and we reject the hypothesis on which the expected values were based. So the genes are likely to be linked.
d. no. of recombinants is 22. COV is 22/104 = 21% so the distance between body colour and wing shape is 21 cM.

13.   Possible phenotypes   Possible genotypes
         coloured grey   CCGG or CCGg or CcGG or CcGg
         coloured black   CCgg or Ccgg
         albino   ccGG or ccGg or ccgg

P :   coloured grey   X   albino (recessive)
       CCGG   ccgg

a. F1 :   coloured grey   X   coloured grey
           CcGg   CcGg

b. F2   1 CCGG   2 CcGG   1 ccGG
c.(genotypes) 2 CCGg   4 CcGg   2 ccGg
             1 CCgg   2 Ccgg   1 ccgg

d. F2
e. (phenotypes)   3   coloured black
                                   9   coloured grey
                                   4   albino

'Albino' overrules the other gene. No matter if the other gene has alleles for black or grey, they will not be expressed if the individual cannot make the pigment. 'Albino' contains the genotypes for albino-grey and albino-black. Note the effect is has on the phenotype ratio

14. Any of the examples described in 10.6.7 or other relevant example.

CHAPTER 11

1. A   2. C   3. C   4. D   5. C   6. D   7. B   8. A   9.
10. a. the development of egg cells (ova)   b. to produce the primordial germ cell layer.
c. to produce the gametes by reducing the amount of genetic material in the nucleus.
d. spermatozoa contain little other than nucleus. Ovum need to keep maximum amount of cell material to start growth with after fertilisation. Two or three polar bodies are produced to remove excess genetic material.
11. a. the trophoblastic cells of the blastocyst   b. it sustains the corpus luteus which will continue to produce progesterone which will maintain the endometrium.
c. because later on the placenta will produce sufficient amounts of progesterone so the corpus luteus is no longer needed which means that HCG is no longer needed.
d. Commercial pregnancy tests are sticks that are dipped in urine. The urine will travel up the stick encountering HCG antibodies which have a dye attached to them. If HCG is present, the antibodies with the dye will dissolve and travel up the stick with the urine. They will meet a second set of antibodies and precipitate into a coloured line. If HCG is not present, the first set of antibodies will not dissolve so no coloured line will be formed. These kits are quite reliable when they indicate that the woman is pregnant. When they do not show this, the woman may not be pregnant or may not have enought HCG in her urine yet. So a 'negative' test should be repeated after several days.

CHAPTER 12

1. A   2. D   3. C   4. B   5. A   6. D   7. B   8. A
9. a. the HIV virus wich destroys T-helper cells and therefore makes the immune system less efficient.
b. via blood and sexual contact
c. increased use of condoms, decreased sexual promiscuity, homophobia, other relevant changes.
10. a. B cells will form plasma cells which produce antibodies
b. MHC proteins are needed for the T helper cell to respond to the antigen on the macrophage
c. T helper cells activate B cells
d. Cytotoxic T cells destroy cells that have been invaded by pathogens
e. Immunoglobulins are antibodies; they will make the pathogen harmless by puncturing the membrane or sticking them together.

CHAPTER 13

1. B   2. A   3. D   4. A   5. C   6. D   7. B

Answers

8.a. Several keys are possible. One possibility is given below.

1.	Insects have wings	go to 2
	Insect does not have wings	go to 6
2.	Insect has short wings	go to 3
	Insect has long wings	go to 4
3.	Insect striped body	insect E
	Insect has grey body	insect F
4.	Insect has long antennae	go to 5
	Insect has short antennae	insect G
5.	Insect has 'feathered' antennae	insect H
	Insect has single antenna	insect I
6.	Insect has "tails"	go to 7
	Insect has no "tails"	go to 8
7.	Insect has antennae	insect C
	Insect has no antennae	insect D
8.	Insect has 2 tails	insect A
	Insect has 3 tails	insect B

9. a. Bryophyta and Filicinophyta
b. Filicinophyta, Coniferophyta and Angiospermatophyta
c. Angiospermatophyta
d Angiospermatophyta
e. Filicinophyta, Coniferophyta and Angiospermatophyta

CHAPTER 14

1. D 2. B 3. B 4. D 5. C 6. C 7. A 8. B 9. C
10. a. Na^+ on the outside, Cl^- and negatively charged organic ions on the inside of the axon.
b. The Na channels (sodium channels) open.
c. It makes the membrane potential become less negative, appraching zero and will then reverse the membrane potential, becoming positive.
d. As long as the membrane potential is below zero the forces are diffusion and electromagnetic forces. When the potential is above zero, it is only diffusion forces.
e. By opening the K^+ channels (potassium channels) and allowing K^+ to move out of the axon.
f. By the sodium potassium pump (active transport, requires energy)
11. a. a nerve impulse arriving at the muscle will cause a release of Ca^{2+} into the cytoplasm of the muscle cell. Ca^{2+} will attach to troponin which will change the shape of tropomyosin and uncover the binding sites for the myosin hooks.

b. troponin and tropomyosin are proteins attached to actin filaments. They cover/uncover the binding sites for the myosin hooks.
c. actin an myosin filament can slide in between eachother and make the muscle shorter. Myosin has little hooks and actin has binding sites for these hooks so they can slide using a ratchet mechanism, forming and breaking bridges between them as they slide.
d. the above mechanism requires energy provided by ATP

CHAPTER 15

1. D 2. C 3. A 4. C 5. A 6. D 7. B 8. B
9. a. ammonia, urea, trimethylamine oxide, uric acid.
b. ammonia: freshwater fish
urea: mammals
trimethylamine oxide: marine organisms
uric acid: birds
c. ammonia is toxic and requires a lot of water to flush it out of the system; therefore it is found only in fresh water organisms.
urea is less toxic and requires less water but still some. Found in organisms which can allow moderate amounts of water loss.
trimethylamine oxide: see urea
uric acid is insoluble and requires very little water to be excreted; this makes it suitable for birds which cannot carry extra water just for excretion.
10.

	proximal convoluted tubule	descending branch of Henle's loop	ascending branch of Henle's loop	distal convoluted tubule	collecting duct
glucose	out - all * active transport	-------------	-------------	-------------	-------------
amino acids	out - all active transport	-------------	-------------	-------------	-------------
Na$^+$	out - almost all active transport	in diffusion	out diffusion and active transport	-------------	-------------
Cl$^-$	out - almost all passive	in diffusion	out diffusion and active transport	-------------	-------------

Answers

	proximal convoluted tubule	descending branch of Henle's loop	ascending branch of Henle's loop	distal convoluted tubule	collecting duct
water	out - almost all osmosis	out osmosis	almost impermeable	some out ADH dependent osmosis	some out ADH dependent osmosis
urea	out - half diffustion	in diffusion	in diffusion	-------------	some out ** ADH dependent diffusion
foreign substance	out - all active transport	-------------	-------------	-------------	-------------

CHAPTER 16

1. C 2. A 3. C 4. D 5. A 6. B 7. D 8. A 9. B
10. a. see section 16.1.2
b. cuticle: reduce evaporation of water
epidermis: barrier against infection
palisade mesophyll: photosynthesis
spongy mesophyll: allow rapid diffusion of gases and photosynthesis
stomata: open and close pore to allow gaseous exchange
c. cuticle: waxy - impermeable to water
epidermis: cells in contact, not allowing anything to pass in between
palisade mesophyll: closely packed cells near top surface of the leaf with many chloroplasts per cell
spongy mesophyll: loosely organised cells with many air spaces in between; cells have chloroplasts.
stomata : open and close according to turgor in the guard cells; they will close when too much water is lost.
11. a. Roots will take up water by osmosis. The cells in the root will have a higher concentration of dissolved particles than the water in the soil. As a result, water moves into the roots. The water will then be transported to the xylem and to other areas of the plant so that the concentration gradient is maintained and the roots can take up more water.
b. Water is moved through xylem vessels. They are made of columns of cells which died, breaking down the walls between one cell and the next. This makes transport through xylem vessels fast.
The water in the xylem is moved against gravity by a combination of forces :
- evaporation, causing a
- transpiration pull which only works because of the
- strong cohesion forces between water molecules.
Water evaporates from the large surface area of the leaves

Water will evaporate from the spongy mesophyll cell into the intercellular spaces to replace the water lost by diffusion through the stomata.
A water molecule from the xylem will move into the mesophyll cells to replace the water lost by evaporation.
Cohesion forces will make the next water molecule want to follow it and it will move up; ditto for the molecule after it and the one after that.
Eventually this will make the root take up water.
12. a. water and oxygen
b. water will activate the enzymes which will start metabolic processes
oxygen is required for cellular respiration which will release energy
c. see 16.3.3
13. a. selective breeding - hybridisation - genetic engineering
b. selective breeding: choosing which seeds to use for next year's crop or which animal to breed.
hybridisation : crossing two strains of a species or two different species with the aim to combine desirable characters of both.
genetic engineering : to implant a useful gene from one species into another.

CHAPTER 17

1. a. carbohydrates, proteins, lipids, minerals, vitamins, water, fibre.
b. carbohydrates: potatoes, apple
proteins : beef, beans
lipids: butter, peanuts
minerals : e.g. calcium in milk and iodine in sea food
vitamins : e.g. ascorbic acid in citrus fruit and retinol in liver.
c. carbohydrate: provide energy
protein: growth and repair, to make enzymes
lipids: energy, component of cell membranes, insulation
minerals: e.g. calcium to harden bones/teeth and iodine as component of thyroxine
vitamins: act as co-enzymes, allowing reactions to proceed
2. a. The baby is much smaller than the mother so needs less food. At first, it requires very little, being very small. Later it needs a bit more but most women by then are less active which reduces their enery requirements. Near the end of pregnancy, a woman should eat approximately 25% extra.
b. The 15 year old boy.
c. If the boy was small and not active and the woman was large and very active.
3. a. vegans do not eat or use any animal product
lacto-vegetarian will eat plant products and milk
pesco-vegetarians will eat plant products and fish
ovolacto-vegetarians will eat plant products, milk and eggs
b. bones becoming brittle and breaking easily, caused by insufficient calcium in the bones.
c. older women because of the change in their hormones, vegans and pesco-vegetarians because they do not eat dairy products which are a good source of calcium.

CHAPTER 18

1. a. Thin actin filaments and thick myosin filaments

Answers

b. The filaments slide relative to each other so that they overlap more. This shortens the muscle. ATP is required as an energy source.

2. a. receptors in muscles, tendons and joints; they provide information on how much the muscle is stretched.

b. no

c. motor area of the cerebral cortex: "decision" to move a muscle

motor neuron: transmits impulse from cortex to muscle

synapses: connections between neurons or neuron-muscle; transmit impulse in chemical form

muscle fibres: sliding actin and myosin filaments shorten muscle

proprioceptors and sensory neuron: provide feedback.

3. a. adrenalin improves the oxygen and glucose supply to the muscles

b. by increasing the rate and depth of ventilation

by increasing cardiac frequency which increases the flow of blood

by increasing the level of glucose in the blood

4. heartrate: as stroke volume increases, resting cardiac frequency should decrease; frequency during exercise should also be lower than before the training programme.

muscles become bigger and stronger.

5. a. Rest - Ice - Compression - Elevation will reduce pain and the severity of the injury.

b. Rest : to avoid further damage and give the body a chance to heal

Ice : to deaden pain, reduce bleeding and swelling

Compression : to reduce swelling

Elevation : to reduce swelling, to assist drainage from the area and it often reduces pain.

CHAPTER 19

1. (digestive) enzymes are proteins

gene for polypeptide chain is found in the nucleus

transcription takes place in the nucleus

translation takes place in the cytoplasm

translation can take place on ribosomes of rER

protein goes into lumen of rER

vesicle is pinched off

vesicle travels to Golgi apparatus

vesicle fuses with Golgi apparatus moving content into Golgi

protein modified in Golgi (e.g. sugar added to make glycoprotein)

vesicle containing new product is pinched off Golgi and becomes lysosome.

2. a. GTP provides the energy for the two subunits of the ribosome to attach

b. ribosomes provide the environment where the tRNA anticodon can bind (complimentary base pairing) to the mRNA codon

c. polysomes are groups of ribosomes found on one mRNA. this means that several polypeptides chains are produced almost at the same time.

d. codons are mRNA triplets (3 nucleotides in a row) which code for a certain amino acid. The codon on the mRNA dictates which tRNA anticodon will bind and this in turn dictates which amino acid is brought in.

3. primary structure: covalent bonds/peptide bonds between amino acids.

secondary structure: hydrogen bonds

tertiary structure: hydrophilic and hydrophobic interactions, disulfide bridges
quaternary structure: hydrogen bonds, positive/negative attraction forces, hydrophilic and hydrophic forces, disulfide bridges.
4. a. to provide the energy (and reducing power) needed for Calvin cycle/for fixing carbon dioxide. b. to produce ATP
c. non-cyclic photophosphorylation leads to the production of ATP and NADPH both of which are needed to fix carbondioxide and produce glucose. Glucose can be used for longer time storage of energy as well as for making other compounds e.g. cellulose.
d. to combine 3 carbondioxide into 1 triose phosphate fixing carbon dioxide
5. a. cytoplasm b. matrix of mitochondrion c. inner membrane of mitochondrion
d. see section 9.1.5 e. Krebs cycle is a cycle where one molecule is changed into another with the aid of enzymes. As long as the enzymes can be found in the matrix (in reasonable concentration) the Krebs cycle can proceed.
f. The sequence of electron carriers in the electron transport chain is very important. So they are situated in the inner membrane to keep their position to one another. The cristae are folds in the inner membrane to increase the surface area so that many electron carriers can be found there. g. no build up of protons in the intermembrane space, no ATP produced, eventually no respiration as the first steps of glycolysis require ATP
6. Similarities: both produce ATP, both involve movement of electrons through a membrane, both involve movement of protons across a membrane, both involve a built up of protons on one side of the membrane, both involve a series of electron carriers, both involve ATP synthetase
Differences: Photosynthesis, Respiration, cell organelle, chloroplast, mitochondria, ATP produced in stroma, matrix, source of electrons water, NADH, electron acceptor $NADP^+$ oxygen, energy source, light, bonds in glucose.

CHAPTER 20
1.a. Lynn Margulis in the early 1970s
b. Chloroplasts and mitochondria originally were independent prokaryotic cells that formed a symbiotic relationship with another prokaryotic cell which eventually turned into a eukaryotic cell.
c. *Cyanophora* (a unicellular eukaryote) has this kind of symbiotic relationship with a cyanobacterium which produces food by photosynthesis.
Similarities between prokaryotes and mitochondria and chloroplasts :
circular DNA rather than linear
DNA not in chromosomes
DNA not contained in nucleus
70S ribosomes (rather than 80S)
similar in size and smaller than eukaryotic cells
2. a. Jean Baptiste de Lamarck
b. That the behaviour of parents determines the phenotype of the offspring.
c. Docking of ears and tails in some breeds of dogs
Circumcision in human males
Neither of these had led to the production of offspring born with short ears, no tail or no foreskin.
Other examples showing the same

Answers

d. Most of the changes in phenotype are not caused by changes in the genetic material. E.g. a bodybuilder increases the size of his muscles by exercise, no change in the genetic material is involved. Even in those cases where the change in phenotype is caused by a change in genotype (e.g. a tumor), this change is in somatic cells. Only changes in gametes or gamete producing cells can be inherited by the next generation.

3. a.

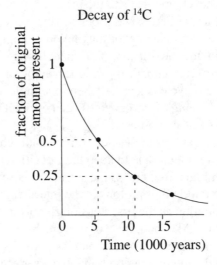

b. At least 40 000 years but with the accuracy of the method between 20 000 and 50 000 years accurate, it would really be impossible to tell.

4. a. Because the fossils of *A. afarensis* had not been found yet when these models were proposed.

b. The size of the brain. *Australopithecus* had a much smaller brain than *Homo*.

c. Because there is more variation in the genetic material of the mitochondria in Africa than anywhere else. If *H. erectus* has evolved into *H. sapiens* at various places (only requiring one wave of migration from Africa: that of *H. erectus*), the differences between the various groups of *H. sapiens* would have been bigger and we would not have found a marked difference in the amount of variation in the gnetic material of the mitochondria.

5. a. 8 (2^3)

b. 4 (2^2)

c. 7.04×10^{13} (2^{46})

6. Dominant allele is normal wings and its frequency is p
Recessive allele is vestigial wings and its frequency is q.
Given: $q = 0.3$; therefore $p = 0.7$
Normal wings is vg^+vg^+ or vg^+vg
frequency of $vg^+vg^+ = 0.7 \times 0.7 = 0.49$
frequency of $vg^+vg = 0.7 \times 0.3 = 0.31$
total frequency for normal wing is $0.49 + 0.31 = 0.80$
out of a population of 125 flies, $0.8 \times 125 = 100$ flies are expected to have normal wings.

CHAPTER 21

1. a. Innate behaviour : behaviour which normally occurs in all members of the species despite natural variation in environmental influences.

b. Innate behaviour is genetically inherited.
c. The hedgehog rolling up when it perceives danger.
d. Innate behaviour will be affected by natural selection as any other inherited factor that affects the survival of the individual. If the behaviour is to the advantage of the individual, the individual is likely to produce (and have survive) more offspring which may carry the advantageous gene.
2. a. the retina
b. rods and cones
c. the rods because they are linked several to a neuron which means that the impulses are 'added up'.
d. Rods and cones contain photosensitive pigments that are broken down in light. This causes an action potential to be sent to the brain.
e. It interprets the action potentials coming from different light sensitive cells to establish a 'picture'.
3. When the animal has not had a chance to learn the proper behaviour, e.g. the kittiwake chicks not moving about so that they do not fall of the ledge on which the nest is built
When the price of learning is too high, e.g. a Thompson gazelle should not wait around to learn if the stalking hyenas are dangerous because most of them would not be around to benefit from the learning.
4. a. Pavlov's dogs beginning to salivate at the sound of a bell which they had learned to associate with food.
b. Rats in Skinner boxes which learn to press a lever to obtain food.
c. young geese following Lorenz when he was the first moving thing they saw after hatching.
d. a chimpazee stacking up boxes in a cage to reach the food at the ceiling.
e. Operant conditioning is trial and error learning. Insight learning aims to avoid most of the errors and as such is usually less 'costly' to the animal as an error could mean death. So insight learning enhances chances of survival.
5. a. Behaviour that benefits others and involves risk or cost the performer.
b. The lioness putting herself at risk by going hunting for her young.
The honey bee stinging an intruder to the hive which will cause the bee to die.
c. The 'selfish gene'. Altruism is directed at those who share genes with the 'giver'. This will increase the chances of the gene in the population. Altruism is mostly directed to related individuals.
6. a. It prepares for action.
b. heart: cardiac frequency increases
salivary glands: little saliva is produced
iris : radial muscles contract, enlarging the pupil
c. increasing cardiac frequency increases the supply of glucose and oxygen to the muscles which will therefore be better able to contract, e.g. leg muscles when trying to escape.
the production of saliva is not vital when attempting to save ones live, so only little is produced when the radial muscles in the iris contract, the size of the pupil increases so that visual perception is improved which may be useful in escape or fight response
7. a. An excitatory presynaptic neuron will depolarise the postsynaptic membrane by opening the sodium channels on the postsynaptic membrane : the difference in potential of the postsyntaptic membrane will become closer to zero.
an inhibitory presynaptic neuron will hyperpolarise the postsynaptic membrane by

Answers

opening its potassium channels : the difference in potential will become bigger, away from zero.

b. If the postsynaptic membrane has been depolarised, another, even small, excitatory neuron may be able to depolarise the postsynaptic membrane enough to pass the threshold potential and set off an action potential in the postsynaptic neuron.

If the postsynaptic membrane has been hyperpolarised, an excitatory neuron many not be able to depolarise the postsynaptic membrane enough to pass the treshold potential and no action potential in the postsynaptic neuron is set off so the impulse fades out.

CHAPTER 22

1. a. 22kg per person b. 20 cattle per 100 persons.
2. Population (not linear)

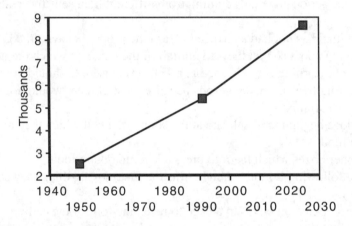

There is some evidence of exponential growth.

3. Holland is cool, short of cultivation areas and needs to use land as effectively as possible. Australia has large areas of cultivated land and is comparatively warm.
Food (almost linear?)

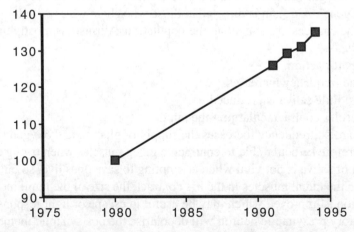

Interpreting these data is difficult. It seems that, as a world community, we should take note of the Mathusian warning and limit our rate of population growth.

7. This is almost exclusively propagated by using cuttings.

CHAPTER 23

1. a. H_0: there is no significant difference between the leaves of the wooddaisy from site 1 and site 2.
H_1: there is a significant difference between the leaves of the wooddaisy from site 1 and site 2.
therefore we do a two-tailed test.
df = 20 + 20 -2 = 38

$$t = \frac{|\bar{x}_1 - \bar{x}_2|}{\sqrt{\frac{s_1^2}{n_1} + \frac{s_2^2}{n_2}}}$$

= 7.15

the critical value for t (p=0.05) at over 30 degrees of freedom is 1.960
value of t exceeds the critical value so we reject H_0
so we can say that there is a significant difference between the wooddaisies of site 1 and 2.
b. In an open area, the leaves receive a lot of sunlight and can have 2 layers of palisade mesophyll to allow for enough photosynthesis.
In a area with bushes, the wood daisy leaves receive less light and will need to grow bigger to have enough photosynthesis.

2. Similarities:
both do not kill the plant/host
both do damage to plant/host
Differences:
herbivory is not a long term relationship/parasitism is a long term relationship herbivores will eat part of the plant/organ of the plant, parasites take nutrients/suck blood other good arguments

3. Primary succession is a progression of communities over time in an area that has never sustained life before.
A number of steps are involved :
weathering of rock
lichens use root-like rhizoids and acid to further break up rock
their remains add organic nutrients
process continues
with mosses, ferns, grasses, shrubs, trees
each breaking rock further and adding organic material

4. a. $D = \dfrac{N(N-1)}{\sum n(n-1)}$

site 1: D = 1.59
site 2: D = 2.37
site 2 has a higher diversity index.
b. site 1 is likely to be more polluted since it has a lower species richness
5.a. because chemoautotrophs are always bacteria which are small and do not provide a lot

Answers

of food for further trophic levels.
b. adding manure/fertiliser
expensive
can leach out
eutrophication
crop rotation with leguminous plants
takes time away from growing cash crop
deep ploughing
may upset soil structure and cause drainage problems
6. a. from reactive free oxygen atoms
b. it stops 99% of the UV radiation from reaching the Earth's surface
UV radiation is harmful
it can cause skin cancer/mutations/damage plankton/lower crop yield
c. mainly CFC's
d. use alternatives in fridges/AC's
recycle CFC's
e. dissolved CO_2, H_2SO_4 and HNO_3
f. damages conifers makes lakes/streams acidic and no longer support life
g. fuel desulfurisation
reducing flue gas
use of alternative energy

CHAPTER 24

1. a. a receptor to check the actual value a control centre which has information about the desired (range of) value(s) and which can initiate action if the actual value is different from the norm
an effector which can change the actual value to come closer to the norm
b. receptor: a and b cells in Islets of Langerhans of pancreas
control centre : same cells
effector: glucagon and insulin
c. when glucose level exceeds the acceptable range
(e.g. after a carbohydrate rich meal)
the b cells in Islets of Langerhans of pancreas
are stimulated to release insulin.
Insulin promotes the take up of glucose from the blood, e.g. in the liver and muscles where it is turned into glycogen and stored and in the other body cells where glucose is used for metabolism.
this lowers the glucose level in the blood and insulin is no longer released.
2. a. pepsin in the stomach
trypsin in the small intestine (secreted by pancreas)
b. polypeptides
c. pepsin : pH 2-3 trypsin : pH 8
d. pepsinogen trypsinogen
e. pepsinogen: by HCl and by other pepsins
trypsinogen: by enterokinase (enzyme from intestinal glands)
f. because they are protein digesting enzymes

if they were produced in their active form
they would digest the cell that made them.
3. a. diagram to include structures listed under b
b. microvilli: increase surface area to facilitate absorption
mitochondria: to provide energy for active transport for absorption of amino acids and dipeptides.
pinocytotic vesicles: result of pinocytosis of e.g. milk in young mammals.
tight junctions: seal the spaces between epithelial cells to ensure that all matter passes through the cells and not in between.
4.a. see section H.4.1
b.

	hepatic artery	hepatic portal vein	hepatic vein
oxygen	high	low	low
carbon dioxide	low	high	high
glucose	ok	variable	ok

5. Sino Atrial Node sends impulse
which spreads across the walls of the atria
atria contract and push blood into ventricles
impulse goes via Atrio Ventricular Node
and bundle of His and Purkinje tissue
to ventricles
which contract starting at apex
pushing blood into aorta and pulmonary artery
6. a. see section H.6.2
b. foetal Hb is to the left (above) maternal Hb
c. Bohr effect shifts curve to the right
active baby would cause Bohr effect but in its tissues not in the placenta
so in placenta the curve of foetal vs maternal Hb would still be the same
but in baby's tissues the Bohr effect would cause the (foetal) Hb to release extra oxygen it would shift the curve to the right.

Answers

INDEX

27

Index

A

Absorption 114
Acetylcholine 256
Acid Rain 94, 98, 459
Acini 469
Actin 259, 317
Action Potential 253
Action Spectrum 193, 343
Activation Energy 333
Active Immunity 234
Active Site 35
Active Transport 14
ADH 270
Adrenalin 117, 323
Aerobic Respiration 185
Afferent Vessel 268
AIDS 121, 231
Algae 245
Algal Bloom 457
Aliphatic 32
Alkalaemia 492
Allele 59
Allostery 174, 335
Altruistic Behaviour 402
Alveoli 122, 124
Amniocentesis 58, 136
Amniotic Sac 134
Amoebic Dysentery 119
Amphetamine 408
Amphipatic 12
Anaerobic Respiration 186
Anaphase 149, 150, 200
Anemophily 423
Aneuploidy 64
Angiospermatophyta 246
Animalia 243, 445
Anorexia Nervosa 115, 307
Antagonist 315
Anti Diuretic Hormone 468
Antibody 120, 171
Antibody Recognition Site 149, 331
Antigen 120
Apical Dominance 424
Apoplast Pathway 279
Appendicar Bone 314
Aqueous Humour 391

Aromatic 32
Artificial Fertiliser 456
Artificial Immunity 234
Asthma 126, 491
Atherosclerosis 485
Atoms 26
Atrial Systole 481
Atrio Ventricular Node 117, 484
Autonomic Nervous System 404
Autosome 204
Autotroph 80
Auxin 426
Axial Bone 314

B

Bacterial Disease 119
Bacteriophage 45
Bacterium 230
Balanced Diet 114, 296
Basement Membrane 220, 269
B-cell 234
Benzodiazepine 409
Bernal 356
Big Bang 354
Bile 479
Biomass 86, 461
Biome 447
Blastocyst 134
Blastomere 134
Blind Spot 391
Blood Glucose 468
B-lymphocyte 120
Body Mass Index 304
Bohr Effect 489
Bowman's Capsule 269
Breathing 125
Bryophyta 246
Bundle of His 117

C

C_3 Plants 193
C_4 Pathway 191
C_4 Plants 193
Caffeine 408
Cairns-Smith 356
Calciferol 479

Calendar Method 136
Calvin 189
Calvin Cycle 189
CAM Plant 193
Cannabis 409
Cap 135
Carbohydrate 29, 114, 294, 295
Carbon Dating 363
Carbon-Oxygen Cycle 444
Carcinogen 18
Cardiorespiratory Endurance 326
Carrier 65
Carrier Proteins 14
Carrying Capacity 92
Casparian Strips 280
Catalyst 331
Cell Cycle 16
Cell Membrane 332
Cell Surface Membrane 7
Cell Wall 7
Cell-mediated Response 121
Cellular Respiration 85
Cellulose 31
Centimorgan 206
Central Nervous System 129
Centriole 17
Centromere 17
Centrosome 17
CFC 97, 459
Chemical Synapse 256, 320
Chemoautotroph 453
Chemoautotrophy 454
Chemoreceptor 389
Chiasma 201
Chief Cell 113
Chi-squared Test 207
Chlorofluorocarbon 97
Chlorophyll 81
Chlorophyll A 188
Chloroplast 10, 336
Cholera 231
Cholesterol 305
Chorionic Villus Sampling 136
Choroid 391
Chromatin 158
Chromatography 342
Ciliary Body 391

Ciliary Muscle 391
Cites 452
Class 368
Classification 242
Cleavage Division 134
Cline 379
Clonal Expansion 233
Clonal Selection 233
Clone 71
Cloning 426, 502
Coacervate 357
Cocaine 408
Codominance 69
Codominant Allele 65
Cohesion 282
Communication 387
Community 78
Companion Cell 283
Comparative Anatomy 366
Competition 440
Competitive Inhibitor 335
Condensation 30
Conditioning 400
Condom 135
Conduction 128
Cones 393
Coniferophyta 246, 247
Conjunctiva 391
Consumer 80
Convection 128
Copulation 134
Cornea 391
Corona Radiate 224
Coronary Heart Disease 119, 305, 485
Corpus Luteum 132
Cortical Granule 225
Cortical Reaction 225
Courtship 387
Crassulaean Acid Metabolism 192
Cristae 185
Cro Magnon 374
Crop Rotation 456
Cross Over Value 206
Crossing-over 201
Curare 408
Cutting 424
Cyclic Photophoshorylation 188

Index

Cytokinesis 16
Cytoplasm 7
Cytotoxic T-cell 235

D

Darwin 359
Dawkins 403
Deep Ploughing 457
Degenerate 43
Denitrification 456
Deoxyribonucleic Acid 37
Detritivore 80, 90
Developing Oocyte 221
Diaphragm 135
Diastole 481
Diet 294, 304
Differential Survival 93
Diffusion 13
Dihybrid Cross 203
Disacharide 30, 300
Disease 119
Disjunction 64
Dislocation 327
Distal Convoluted Tubule 270
Disulfide Bridge 331
Diversity 245
DNA 37, 158
    Ligase 160
    Polymerase 41, 160
    Polymerase I 160
    Replication 160
Dominant Allele 65
Dopamine 256
Double Blind Method 498
Down Syndrome 64, 375
Drosophila 205
Dry Biomass 416
Ductless Gland 130

E

Ecology 78
Ecosystem 78
Elbow Joint 257
Eldredge 378
Electrical Synapse 255
Electroreceptor 389

Elements 26
Elongation 169
Emigration 90
Endocrine System 130
Endocytosis 14
Endopeptidase 472
Endorphin 407
Endosymbiont Theory 8
Energy 297, 322
    Hydroelectric 460
    Renewable 460
    Wind 461
Enkephalin 407
Enzyme 35, 171, 332
Essential Amino Acids 302
Euchromatin 159
Eukaryotic Chromosome 58
Eukaryotic Kingdom 357
Eutrophication 449
Evolution 354
Excretion 130, 266
Exocrine Gland 469
Exon 165
Exopeptidase 472
Expiration 125
Expiratory Reserve Volume 125
Extinction 503
Eye 120, 390

F

Facilitated Diffusion 14
Family 368
Famine 412
Fat 114
Fatigue 324
Fertilisation 134, 224
Fibre 294, 295
Filicinophyta 246, 247
Fitness 324
Flagellum 7
Flatworm 230
Flexibility 326
Foetal Haemoglobin 488
Foetus 134
Food Chain 442
Fossil 362

Fovea 391
FSH 221
Fulcrum 316
Fungus 230, 243, 445

G

Galapagos Islands 94
Gastric Juice 113
Gastric Pit 469
Gender Determination 67
Gene 58, 158
Gene Mutation 60
Gene Transfer 45
Genetic Engineering 45, 502
Genetic Screening 70
Genome 59, 70
Genotype 65
Genus 368
Geometric Curve 91
Gibberellin 427
Globin 480
Globular Protein 332
Glomerulus 268
Gluconeogenesis 478
Glycogen 31
Glycogenesis 477
Glycolysis 181, 345
Glycoproteins 13
Golgi Apparatus 9, 330
Gonadotropin 225
Gould 378
Graafian Follicle 221
Grause 439
Greenhouse Effect 85, 89, 94, 444
Greenpeace 452
Grooming 387
Gross Production 441

H

Habitat 78
Haem 480
Haemoglobin 332, 487
Haldane 5
Half Life 363
Hardy 380
Hardy - Weinberg Equation 380

Hardy- Weinberg Principle 381
Health 304
Helicase 41
Henle's Loop
    Ascending Limb 270
    Descending Limb 270
Herbivory 440
Herd Immunity 237
Hering-Breuer Reflex 397
Heterochromatin 159
Heterotroph 80
Heterozygosity 382
Heterozygous 65
Histone 158
HIV 121, 231
Homeostasis 466
Homologous Chromosome 200
Homozygous 65
Hooke 2
Hormone 171, 332
Hormone Binding Site 149, 331
Host Cell 46
Hybrid Vigour 212, 382
Hydrogen Bond 27
Hydrophobic Group 331
Hydrophyte 278
Hyperventilation 126

I

Immigration 90
Immunisation 236, 503
Immunity
    Artificial Active 234
    Artificial Passive 234
    Natural Active 234
    Natural Passive 234
Imprinting 400
In Vitro Fertilisation 137
Inbreeding 212
Indicator Species 449
Induced Fit Model 173, 334
Inhibitor 173, 335
Initiation 169
Innate Behaviour 386, 395
Inner Membrane 185
Insight Learning 401

Index

Inspiration 125
Inspiratory Reserve Volume 125
Insulin 477
Intermembrane Space 185
Interphase 16, 200
Interspecific Hybridisation 212
Intron 165
Iodopsin 393
Ion 26
Iris 391
Islets Of Langerhans 468
IUCN 452
IUD 136
IVF 137

J

Jansen 2
Joint 314

K

Karyotyping 58
Katchalsky 356
Key 244
Kidney 267
Kidney Dialysis 271
Kinesis 398
Kinetin 426
Kingdom 368
Klinefelter Syndrome 376
Krebs Cycle 182, 347
Kupffer Cell 476

L

Lac Operon 163
Lacenta 134
Lacto-vegetarian 306
Lagging Strand 161
Lamark 358
Law Of Independent Assortment 202
Law Of Segregation 69
Layering 424
Leading Strand 161
Leaf Area Index 415
Leeuwenhoek 2
Lens 391
Leucocyte 120, 166

LH 221
Ligament 257
Light Dependent Stage 187
Light Independent Stage 187
Limiting Factor 85, 101, 344, 417, 441
Line Transect 101
Link Reaction 182
Linkage Group 204
Linnaeus 242
Lipase 472
Lipid 32, 294, 295
Lock And Key Model 35
Locus 65
Long Sight 392
Lorenz 401
Lucy 372, 373
Lung 120, 124
Lung Cancer 490
Lysis of Water 188
Lysosome 9

M

Macronutrient 115, 294
Magnification 152
Major Histocompatability Complex 234
Malaria 119
Malnutrition 114, 306
Malthus 413
Malthusian Theory 414
Manure 456
Margulis 357
Mate Selection 387
Matrix 185
Mechanoreceptor 389
Median 102
Meiosis 62, 202
Membrane 148, 330
Memory Cell 121
Mendel 64, 202, 496
Mendel's First Law 69
Mesophyte 282
Mesosome 6
Messenger RNA 42
Metabolic Pathway 171, 333
Metaphase 150, 200
MHC 234

Micronutrient 115, 294
Micropropagation 426
Microsphere 357
Microtubule 17, 332
Migration 387
Miller 355
Mineral 114, 294, 295
Mitchell 183
Mitochondrion 9
Mode 102
Monoclonal Antibody 236
Monoculture 418
Monosaccharide 29, 300
Monotreme Mammal 361
Mortality 90
Morula 134
Motor Neuron 252, 319
Mucoprotein 224
Multiple Allele 67
Muscle 314, 322
Mutualism 440
Myoglobin 487
Myosin 259, 317

N

Natality 90
Natural Immunity 234
Neanderthal 372
Negative Feedback 129
Neo-Darwinism 375
Nephron 268
Nerve 252
Net Assimilation Rate 416
Net Production 441
Neuron 129
    Excitatory 321
    Inhibitory 321
Niche 438
Nicotine 408
Nitrification 456
Nitrogenous Base 37
Nomenclature 242
Non-competitive Inhibitor 335
Non-cyclic Photophosphorylation 188
Noradrenaline 256
Nucleosome 158

Nucleus 9, 148, 330
Nutrient 294
Nutrition 114

O

Okazaki Fragment 161
Omeostasis 127
Oogenesis 222
Operant Conditioning 400
Operon Model 163
Optic Nerve 391
Order 368
Organ 151
Organ System 151
Organelle 3
Organic Base 37
Organic Molecule 26
Osmoregulation 130
Osmosis 13
Osteoporosis 307
Outbreeding 212
Outer Membrane 185
Overload 324
Ovolacto-vegetarian 306
Ovulation 221
Oxygen Debt 323
Ozone Layer 94, 96, 458

P

Pacemaker 484
Pancreas 469
Panspermia 360
Parasitism 440
Parasympathetic System 405
Parental Care 387
Parkinson's Disease 407
Partial Pressure 486
Passive Immunity 234
Passive Transport 14
Pathogen 230
Pavlov 400
Pedigree Chart 66
Peppered Moth 360, 379
Pepsin 472
Pesco-vegetarian 306
pH 36

Phagocytosis 14, 120, 480
Phenlyketonuria 375
Phenotype 65
Phosphorylation 181
Photoreceptor 390
Photorespiration 192
Photosynthesis 81, 336
    Light Dependent Stage 82
    Light Independent Stage 83
Photosystem I 188
Photosystem II 188
Phylogeny 366
Phylum 246, 368
Pill 136
Pinocytosis 14, 474
Placental Mammal 362
Plant 243
Plant
    Dicotyledonous 276
    Monocotyledonous 285
Plantae 445
Plasma Protein 480
Plasmid 7
Plasmodium 119
Podocyte 269
Polar Body 221
Pollination 423
Polygenic Inheritance 210
Polymerase Chain Reaction 48
Polyploidy 64, 212, 213
Polysaccharide 31, 300
Polysome 168
Population 78
Predation 440
Primary Structure 170, 331
Producer 80
Progressive Overload 325
Prokaryota 445
Prokaryote 243
Promoter Region 164
Prophase 149
Prophase I 200
Prostaglandin 135
Protein 114, 294, 295
Proteins
    Extrinsic 13
    Intrinsic 13

Protoctista 243, 445
Proton Gradient 183
Protozoa 119, 230, 245
Proximal Convoluted Tubule 269
Psychoactive Drug 408
Punctuated Equilibria 378
Punnett Grid 65
Pupil 391
Pupil Reflex 398
Pupil Size 391
Purkinje 2
Purkinje Tissue 117

Q

Quadrat 101
Quarternary Structure 170, 332

R

Random Sample 99
Reabsorption 268
Recessive Allele 65
Recombinant 201
Recombinant DNA 45
Recombination 202
Red Data Book 452
Redox Reaction 180
Repetitive Sequence 159
Resting Potential 253
Restriction Enzymes 46
Retina 391, 392
Retinol 479
Retrovirus 166
Reverse Transcriptase 166
Rhodopsin 393
Rhythm Method 136
Ribosomal RNA 42
Ribosome 6, 9
Rice 419
Rickettsiae 119
Ring Species 379
Rio Convention 452
RNA 37
RNA Polymerase 164
Rods 393
Rough Endoplasmic Reticulum 9, 330
Roundworm 230

S

Saliva 113
Salivary Gland 469
Saprobe 90
Saprotroph 80, 90, 246
Sarcolemma 259
Sarcomere 259
Sarcoplasm 259
Sarcoplasmic Reticulum 259
Schleiden 2
Schwan 2
Scientific Method 496
Sclera 391
Secondary Oocyte 221
Secondary Structure 170, 331
Secretion 268
Selfish Gene 403
Semi-conservative Mechanism 160
Seminiferous Tubule 220
Sensory Neuron 319
Serotin 256
Sertoli Cell 220
Sex Linkage 68
Shivering 128
Short Sight 392
Sieve Plate 283
Sieve Tube Member Cell 283
Singer 12
Sino Atrial Node 117, 484
Sinusoid 476
Skeleton 257, 314
Skinner 400
Skinner Box 400
Sleeping Sickness 119
Slime Capsule 6
Small Intestine 113
Special Creation 360
Speciation 378
Species 78, 242 368, 378
Specificity 324
Spermatogenesis 220
Spermicide 136
Spinal Reflex 395
Sprain 327
Standard Deviation 102
Standing Crop 86

Starch 31
Sterilisation 135
Stomach 120
Strength 326
Structural Protein 171, 332
Students' t-test 434
Subclass 368
Subcutaneous Fat 304
Suborder 368
Subphylum 368
Subspecies 368
Succession 444
Suspensory Ligament 391
Sweating 128
Sympathetic System 405
Symplast Pathway 280
Synapsis 64
Synovial Fluid 258

T

Target Cell 130
Taxis 398
T-cell 121
Telophase 150, 200
Termination 169
Tertiary Structure 170, 331
Test Cross 65
Testosterone 221
Tetrad 64
t-helper Cell 235
Thermoreceptor 390
Thermoregulation 127
Thickness of the Lens 391
Threshold Potential 255
Tight Junction 474
Tissue 151
Tissue Culture 425
Tissue Fluid 485
Torn Ligament 327
Torn Muscle 327
Training 324
Trans-amination 302
Transcription 42, 163
Transfer RNA 42
Translation 42, 167
Translocation 283

Index

Transpiration 281
Trophic Level 80
Trophoblast 134
Tropomysosin 260
Troponin 260
Trypsin 472
Tumour 18
Turgor 11, 280

U

Ultra-filtration 268
Urey 355
Urine 266
Urinogenital System 131
UV Radiation 458

V

Vaccination 237
Vacuolar Pathway 280
Vagina 120
Vasa Recta 270
Vasectomy 135
Vasoconstriction 128
Vasodilation 127
Vegan 306
Ventricular Diastole 481
Ventricular Systole 481
Vesicle 330
Villus 113
Viral Disease 119
Virchow 2
Virus 230, 445
Vital Capacity 125
Vitamin 114, 294, 296
Vitreous Humour 391

W

Wall of the Glomerulus 269
Wallace 359
Waste 266
Water Cycle 443
Water Potential 279, 468
Weinberg 380
Wheat 419
World Wildlife Fund 452

X

Xerophyte 277
Xylem Vessel 281

Z

Zona Pellucida 221, 224